**IF FOUND, please notify and arrange return to owner.**

Pilot's Name _____

Address _____

_____

City                                                    State      Zip Code

Telephone  ( ____ ) _____

Additional copies of *Aviation Weather and Weather Services* are available from

Gleim Publications, Inc.
P.O. Box 12848
University Station
Gainesville, Florida 32604
(352) 375-0772
(800) 87-GLEIM or (800) 874-5346
FAX: (352) 375-6940
Internet: www.gleim.com
E-mail: sales@gleim.com

The price is $22.95 (subject to change without notice). Orders must be prepaid. Use the order form on page 433. Shipping and handling charges will be added to telephone orders. Add applicable sales tax to shipments within Florida.

Gleim Publications, Inc. guarantees the immediate refund of all resalable materials returned within 30 days. Shipping and handling charges are nonrefundable.

## ALSO AVAILABLE FROM GLEIM PUBLICATIONS, INC.

*ORDER FORM ON PAGE 433*

*Private Pilot and Recreational Pilot FAA Written Exam*
*Private Pilot Practical Test Prep and Flight Maneuvers*
*Pilot Handbook*
*FAR/AIM*
*Private Pilot Syllabus and Logbook*
*Flight Computer*
*Navigational Plotter*
*Private Pilot Kit*
*Flight Bag*

### Advanced Pilot Training Books

*Instrument Pilot FAA Written Exam*
*Instrument Pilot Practical Test Prep and Flight Maneuvers*

*Commercial Pilot Practical Test Prep and Flight Maneuvers*
*Commercial Pilot FAA Written Exam*

*Flight/Ground Instructor FAA Written Exam*
*Fundamentals of Instructing FAA Written Exam*
*Flight Instructor Practical Test Prep and Flight Maneuvers*

*Airline Transport Pilot FAA Written Exam*
*Flight Engineer FAA Written Exam*

# REVIEWERS AND CONTRIBUTORS

Ray Carr, CFII, MEI, B.S. in Mass Communications, University of Washington, is a charter pilot and flight instructor at Gulf Atlantic Airways in Gainesville, FL. Mr. Carr assisted with the previous edition.

Karen A. Hom, B.A., University of Florida, is our book production coordinator. Ms. Hom coordinated the production staff and reviewed the final manuscript.

Barry A. Jones, ATP, CFII, MEI, B.S. in Air Commerce/Flight Technology, Florida Institute of Technology, is our aviation project manager and also a charter pilot and flight instructor with Gulf Atlantic Airways in Gainesville, FL. Mr. Jones updated the text, incorporated numerous revisions, assisted in assembling the manuscript, and provided technical assistance throughout the project.

Travis A. Moore, M.B.A., University of Florida, provided production and editorial assistance throughout the project.

Nancy Y. Raughley, B.A., Tift College, is our editor. Ms. Raughley reviewed the entire manuscript, revised it for readability, and assisted in all phases of production.

John F. Rebstock, B.S., School of Accounting, University of Florida, reviewed portions of the edition and composed the page layout.

The many FAA and NWS employees who helped, in person or by telephone, primarily in Oklahoma City, OK; Kansas City, MO; Gainesville, FL; and Washington, DC. In particular, we would like to thank Kathleen Schlachter who had primary responsibility for revisions to *Aviation Weather Services* (AC 00-45E).

The CFIs who have worked with us throughout the years to develop and improve over pilot training materials.

# A PERSONAL THANKS

This manual would not have been possible without the extraordinary efforts and dedication of Terry Hall and Deborah West, who typed the entire manuscript and all revisions, as well as prepared the camera-ready pages.

The author also appreciates the proofreading and production assistance of Shannon Bunker, Jerry Coutant, Tiffany Dunbar, Chad Houghton, Mark Moore, Jan Morris, and Tricia Slaton.

Finally, I appreciate the encouragement, support, and tolerance of my family throughout this project.

# THIRD EDITION

# AVIATION WEATHER AND WEATHER SERVICES

### by Irvin N. Gleim, Ph.D., CFII

**with the assistance of**
**Barry A. Jones, ATP, CFII, MEI**

## ABOUT THE AUTHOR

Irvin N. Gleim earned his private pilot certificate in 1965 from the Institute of Aviation at the University of Illinois, where he subsequently received his Ph.D. He is a commercial pilot and flight instructor (instrument) with multiengine and seaplane ratings, and is a member of the Aircraft Owners and Pilots Association, American Bonanza Society, Civil Air Patrol, Experimental Aircraft Association, and Seaplane Pilots Association. He is also author of flight maneuvers and practical test prep books for the private, instrument, commercial, and flight instructor certificates/ratings, and study guides for the private/recreational, instrument, commercial, flight/ground instructor, fundamentals of instructing, airline transport pilot, and flight engineer FAA knowledge tests. Another pilot training book, *Pilot Handbook*, is also available.

Dr. Gleim has also written articles for professional accounting and business law journals, and is the author of widely used review manuals for the CIA (Certified Internal Auditor) exam, the CMA (Certified Management Accountant) exam, the CPA (Certified Public Accountant) exam, and the EA (IRS Enrolled Agent) exam. He is Professor Emeritus, Fisher School of Accounting, University of Florida, and is a CIA, CMA, CFM, and CPA.

**Gleim Publications, Inc.**
P.O. Box 12848
University Station
Gainesville, Florida 32604

(352) 375-0772
(800) 87-GLEIM
FAX (352) 375-6940

Internet: www.gleim.com
E-mail: admin@gleim.com

ISSN 1098-450X

ISBN 1-58194-117-X

First Printing: July 2000

This is the first printing of the third edition of *Aviation Weather and Weather Services*. Please e-mail update@gleim.com with AWWS 3-1 in the subject or text. You will receive our current update as a reply.

EXAMPLE:

| | |
|---|---|
| To: | update@gleim.com |
| From: | your e-mail address |
| Subject: | AWWS 3-1 |

# HELP !!

Please send any corrections and suggestions for subsequent editions to me, Irvin N. Gleim, c/o Gleim Publications, Inc. • P.O. Box 12848 • University Station • Gainesville, Florida • 32604. The last page in this book has been reserved for you to make your comments and suggestions. It can be torn out and mailed to me.

Also, please bring this book to the attention of flight instructors, fixed-base operators, and others interested in flying. Wide distribution of our books and increased interest in flying depend on your assistance and good word. Thank you.

## NOTE: UPDATES

If necessary, we will develop an update for *Aviation Weather and Weather Services*. Send e-mail to "update@gleim.com" as described at the top right of this page and visit our Internet site for the latest updates and information on all of our products. To continue providing our customers with first-rate service, we request that questions about our books and software be sent to us via mail, e-mail, or fax. The appropriate staff member will give it thorough consideration and a prompt response. Questions concerning orders, prices, shipments, or payments will be handled via telephone by our competent and courteous customer service staff.

# TABLE OF CONTENTS

# AUTHOR'S MESSAGE

The purpose of this book is to provide an up-to-date compilation of all the FAA's weather publications in one easy-to-understand and easy-to-use book. This book is largely a reformation of *Aviation Weather* (AC 00-6A-1975) and *Aviation Weather Services* (AC 00-45E -- December, 1999), but other ACs are covered as well.

Parts I and II contain 16 chapters which are a restatement and amplification of each of the 16 chapters in AC 00-6A: *Aviation Weather*. We have broken *Aviation Weather* into Parts I and II to differentiate the advanced topics in Part II from the more basic topics in Part I.

The FAA's AC 00-6A: *Aviation Weather* is 219 pages in length. After editing the material in *Aviation Weather* and presenting it in larger type and in outline format, our outline is 210 pages in length. Our outline is much easier to read and study. We explain how to study each chapter, including the material to be scanned vs. the important material to be studied.

More importantly, you learn more from our outline. Your mind is not a word processor. It does not deal in terms of sentences and paragraphs. Your mind works with concepts, and our outline presents concepts and their relationship to other concepts.

Similarly, the FAA's AC 00-45E: *Aviation Weather Services* (AWS) is 210 pages in length. Our Part III of this book, which outlines AWS, is 166 pages in length. Our outline includes all tables, diagrams, weather maps, charts, etc., which appear in AWS (AC 00-45E).

NOTE: We provide guidance on topics that are important and need to be studied vs. topics that are not necessary for most pilots to study. We also have 100% coverage of all material in the FAA's AC 00-6A and AC 00-45E. Our index is very thorough (beginning on page 435) so you can research any topic, concept, etc., you desire.

Enjoy Flying -- Safely!
*Irvin N. Gleim*
July 2000

---

### THIRD EDITION (07/00) CHANGES

1. Part III, Aviation Weather Services, has been extensively revised due to changes by the National Weather Service.

2. Please visit our web site at **http://www.gleim.com/Aviation/Updates/books/aws/** for an overview of the changes from the Second to the Third edition. This address is case sensitive.

# PART I
# *AVIATION WEATHER*

*Aviation Weather* was published jointly by the FAA and National Weather Service (NWS) in 1975. The book had previously included weather services (e.g., then available reports and forecasts); however, these prior editions quickly went out of date. In 1975, *Aviation Weather* was accompanied by *Aviation Weather Services* (AC 00-45A). Subsequent editions of *Aviation Weather Services* were published in 1979, 1985, and 1995. AC 00-45E was published in December 1999.

Note: Gleim Publications, Inc. does not have the FAA's problem in staying up-to-date because we revise and reprint our books frequently. Thus, we can and will provide pilots with all current relevant weather information in one book.

*Aviation Weather* (AC 00-6A) contains 16 chapters which are outlined on pages 3 through 210. Most figures in *Aviation Weather* are reproduced. Only those that do not add to or facilitate understanding of the topics explained in the outline are omitted. Conversely, additional information and explanations have been added where they facilitate learning and understanding of weather by pilots. You, as a pilot, should learn and understand aviation weather as presented in Chapters 1 through 12 in Part I of this book.

Part I: Aviation Weather
1. The Earth's Atmosphere
2. Temperature
3. Atmospheric Pressure and Altimetry
4. Wind
5. Moisture, Cloud Formation, and Precipitation
6. Stable and Unstable Air
7. Clouds
8. Air Masses and Fronts
9. Turbulence
10. Icing
11. Thunderstorms
12. Common IFR Producers

Chapters 13 through 16, which are presented as Part II of this book, cover specialized topics that may be beyond your present interest.

Part II: Aviation Weather -- Over and Beyond
13. High Altitude Weather
14. Arctic Weather
15. Tropical Weather
16. Soaring Weather

You do **not** have to read the FAA's *Aviation Weather* (AC 00-6A) in conjunction with this outline. We have presented all of the information that is in *Aviation Weather*. Our contribution is to present the information in a format that will facilitate **knowledge transfer** to you. Our outline is easy to study because you seek a working knowledge of weather as it relates to flying.

Our outline is effective. Remember that human minds are not word processing computers that deal with sentences and paragraphs. Human minds deal with concepts and how they relate to other concepts -- which is exactly what our outline presentation facilitates for you.

Of the 16 chapters in *Aviation Weather*, the FAA has "In Closing" sections at the back of eight chapters. In addition to a summary, the FAA's "In Closing" section provides additional comments on and insight into chapter discussion. We have provided an "In Closing" section at the end of each chapter which is a summary of the FAA's "In Closing" section if one exists and our summary of the chapter.

## HOW TO STUDY EACH CHAPTER

1.   Each chapter begins with a listing of the chapter's topics.

2.   Before beginning to study (i.e., learn from) the outline, think about the subject of the chapter and how the chapter's topics relate to the overall subject of the outline. Anticipate what will be said in the outline about these topics (even if you have never studied the topics). This step should require only 3 to 5 minutes.

3.   Study the outline. Studying differs from reading in that studying, by definition, results in understanding. In other words, by studying you are internalizing weather information so you can use it as a pilot in the future.

4.   In each chapter, ask yourself

     a.   What each topic is about and how it affects your present and future flying.

     b.   What was covered on each page as you complete each page.

     c.   How well the "In Closing" section of the outline summarized and gave you further insight into the chapter.

     d.   How you would have improved the chapter if you had to explain the topics in the chapter to your flight instructor or another pilot.

5.   Relate each chapter's coverage to the expected coverage in other chapters by looking back to the Table of Contents on page v.

6.   Make notes as appropriate on pages 441 and 442 to be sent to your author upon completion of your study of *Aviation Weather and Weather Services*.

# CHAPTER ONE
# THE EARTH'S ATMOSPHERE

> Please take a few minutes to study each of the concepts listed above and anticipate/imagine what they are and how they relate to the other listed concepts.

A. **Introduction** -- Our restless **atmosphere** is almost constantly in motion as it strives to reach equilibrium.

　1. Because the sun heats the atmosphere unequally, differences in pressure result, which cause a series of never-ending air movements.

　2. These air movements set up chain reactions which culminate in a continuing variety of weather.

　3. Virtually all of our activities are affected by weather, but aviation is affected most of all.

B. **Composition.** Air is a mixture of several gases.

　1. It is about 78% nitrogen and 21% oxygen.

　2. The remaining 1% is other gases such as argon, carbon dioxide, neon, helium, and others.

　3. However, air is never completely dry. It always contains some water vapor, in amounts varying from *almost* zero to about 5% by volume.

C.  **Vertical Structure.**  The atmosphere is classified into layers, or spheres, by the characteristics exhibited in these layers.  See Figure 2 below.

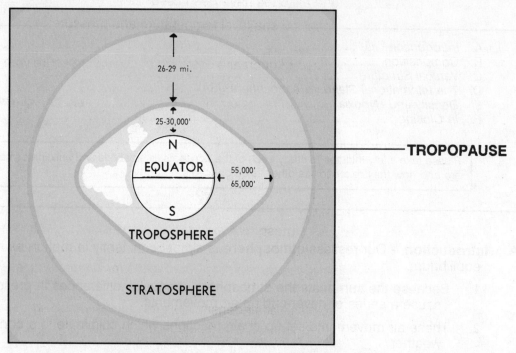

Figure 2.  Earth's Atmosphere to 29 Miles High

1.   The **troposphere** is the layer from the surface to an average altitude of about 7 mi.

   a.   It is characterized by an overall decrease of temperature with increasing altitude.

   b.   The height of the troposphere varies with latitude and seasons.

   c.   It slopes from about 20,000 ft. over the poles to about 65,000 ft. over the Equator (4-12 mi.);

   d.   It is higher in summer than in winter.

2.   At the top of the troposphere is the **tropopause**, a transition zone marking the boundary between the troposphere and the layer above, the stratosphere.

   a.   The thickness of the tropopause varies from several discrete layers in some regions to entirely broken up in others.

   b.   The tropopause acts to trap most of the water vapor in the troposphere.

   c.   It is characterized by decreasing winds and constant temperatures with increases in altitude.

3.   Above the tropopause is the **stratosphere**, i.e., beginning at 4-12 mi.

   a.   This layer is characterized by relatively small changes in temperature with height except for a warming trend near the top (26-29 mi.) and a near absence of water vapor.

D. **The International Standard Atmosphere (ISA)** is a fixed standard of reference for engineers and meteorologists. To arrive at a standard, the average conditions throughout the atmosphere for all latitudes, seasons, and altitudes have been used.

    1. ISA consists of a specified sea-level temperature and pressure: 15°C (59°F) and 29.92 in. of mercury (Hg).

        a. And specific rates of decrease of temperature and pressure with height: 2°C per 1,000 ft. and 1 in. Hg per 1,000 ft.

    2. These standards are used for calibrating pressure altimeters and developing aircraft performance data.

E. **Density and Hypoxia**

    1. Air is matter and has weight.

        a. Since it is a gas, it is compressible.
        b. Thus, as its pressure increases, its **density** also increases.

    2. The pressure that the atmosphere exerts on a given surface is the result of the weight of the air above that surface.

        a. Thus, air near the surface of the Earth exerts more pressure, and is more dense than air at high altitudes.

        b. There is a decrease of air density and pressure with height.

    3. The rate at which the lungs absorb oxygen depends on the pressure exerted by oxygen in the air. Oxygen makes up about 20% of the atmosphere.

        a. Since air pressure decreases as altitude increases, the oxygen pressure also decreases.

        b. A pilot continuously gaining altitude or making a prolonged flight at high altitude without supplemental oxygen will likely suffer from a deficiency of oxygen.

        c. This deficiency of oxygen results in **hypoxia**, which may cause

            1) A feeling of exhaustion,
            2) Impairment of vision and judgment, and
            3) Unconsciousness.

    4. FAR 91.211 requires you to use aviation breathing oxygen for any flight in excess of 30 min. at cabin pressure altitudes above 12,500 ft. MSL up to and including 14,000 ft. MSL.

        a. You are required to use oxygen continuously when piloting an airplane at cabin pressure altitudes above 14,000 ft. MSL.

        b. Your passengers are required to have continuous oxygen provided at cabin pressure altitudes above 15,000 ft. MSL.

        c. FARs concerning commercial flight require that supplemental oxygen be provided to pilots and used at even lower cabin pressure altitudes.

        d. Hypoxia may affect night vision adversely at altitudes as low as 5,000 ft. MSL.

        e. If you feel fatigued, drowsy, etc., descend and see if you begin to feel better.

F. **In Closing**

1. Troposphere extends upward to an average altitude of about 7 mi.

    a. Tropopause is a transition zone between the troposphere and stratosphere.
    b. Stratosphere is next with small changes in temperature.

2. International standard atmosphere (ISA)

    a. Sea level 59°F (15°C) and 29.92 in. Hg

    b. Average lapse rate (i.e., decrease of temperature and pressure with altitude) is 2°C per 1,000 ft. and 1 in. Hg per 1,000 ft.

3. Air pressure decreases with altitude and insufficient oxygen is available to pilots beginning about 10,000 to 12,000 ft. MSL.

    a. Oxygen deficiency is called hypoxia.

# END OF CHAPTER

# CHAPTER TWO
# TEMPERATURE

> Please take a few minutes to study each of the concepts listed above and anticipate/imagine what they are and how they relate to the other listed concepts.

A. **Introduction** -- Temperature affects every aspect of weather and aviation.

  1. Temperature variation is the cause of most weather phenomena.
  2. It is a major factor in the determination of aircraft performance.

B. **Temperature and Heat**

  1. The temperature of a particular substance, such as air or water, is a measurement of the average kinetic energy of the molecules in that substance, relative to some set standard.

      a. Heat is energy which is being transferred from one object to another because of the temperature difference between them.

      b. After being transferred, this heat is stored as internal energy.

  2. A specific amount of heat absorbed by or removed from a substance raises or lowers its temperature a definite amount.

      a. Each substance has its unique temperature change for the specific change in heat.

      b. EXAMPLE: If a land surface and a water surface have the same temperature and an equal amount of heat is added, the land surface becomes hotter than the water surface. Conversely, with equal heat loss, the land becomes colder than the water.

  3. The Earth receives energy from the sun in the form of solar radiation.

      a. The Earth, in turn, radiates energy, and this outgoing radiation is "terrestrial radiation."

      b. The average heat gained from incoming solar radiation must equal heat lost through terrestrial radiation in order to keep the Earth from getting progressively hotter or colder.

          1) This balance is world-wide; we must consider regional and local imbalances which create temperature variations.

C. **Temperature Scales.** Two commonly used temperature scales are Celsius (Centigrade) and Fahrenheit.

  1. The Celsius scale is used exclusively for upper air temperatures and is rapidly becoming the world standard for surface temperatures also.

      a. Celsius is used to report the surface and dew point temperatures in the aviation routine weather report (METAR).

2.   Two common temperature references are the melting point of pure ice and the boiling point of pure water at sea level.

    a.   The boiling point of water is 100°C or 212°F.

    b.   The melting point of ice is 0°C or 32°F.

       1)   NOTE:  Although 0°C (32°F) is generally the freezing point of water as well as the melting point of ice, this is not always so.

         a)   Under certain conditions, supercooled (not yet frozen) droplets of water can often be found in clouds at temperatures down to −15°C (5°F). This will be further explained in Part I, Chapter 5, Moisture, Cloud Formation, and Precipitation, on page 45.

         b)   Supercooled water freezes on impact with an airplane.

    c.   Figure 3 below compares the two scales.

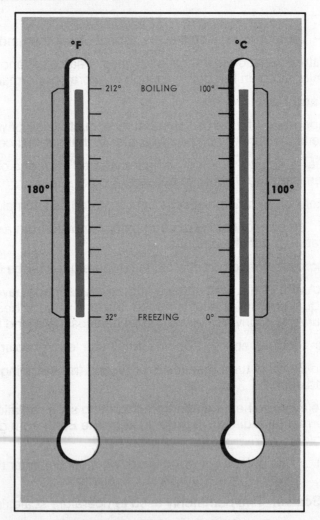

Figure 3.  Fahrenheit vs. Celsius

3.   The difference between the boiling point of water and the melting point of ice (the temperature range between which water is a liquid) is 100°C or 180°F.

    a.   The ratio between degrees Celsius and Fahrenheit is 100/180 or 5/9.

    b.   Since 0°F is 32 Fahrenheit degrees colder than 0°C, you must apply this difference when comparing temperatures on the two scales.

c.  Conversion formulae:

$$°C = \frac{5}{9}(F - 32) \quad or \quad °F = \frac{9}{5}C + 32$$

d.  EXAMPLES:

| °Celsius | °Fahrenheit | °Fahrenheit | °Celsius |
|---|---|---|---|
| 0 | + 32 | 0 | −18 |
| 10 | 50 | 10 | −12 |
| 20 | 68 | 20 | − 7 |
| 30 | 86 | 30 | − 1 |
| 40 | 104 | 40 | + 4 |
| 50 | 122 | 50 | 10 |
| 60 | 140 | 60 | 16 |
| 70 | 158 | 70 | 21 |
| 80 | 176 | 80 | 27 |
| 90 | 194 | 90 | 32 |
| 100 | 212 | 100 | 38 |

e.  Most flight computers provide for direct conversion of temperature from one scale to the other.

D.  **Temperature Variations.**  The amount of solar radiation received by any region varies with time of day, with seasons, with latitude, with differences in topographical surface, and with altitude. These differences in solar radiation create temperature variations.

1.  **Diurnal variation** is the change in temperature from day to night brought about by the daily rotation of the Earth.

a.  The Earth receives heat during the day by solar radiation and continuously dissipates heat by terrestrial radiation.

b.  During the day, solar radiation exceeds terrestrial radiation and the surface becomes warmer.

c.  At night, solar radiation ceases, but terrestrial radiation continues and cools the surface.

1)  Cooling continues after sunrise until solar radiation again exceeds terrestrial radiation.

a)  Minimum temperature usually occurs after sunrise, sometimes by as much as 1 hr.

2)  The continued cooling after sunrise is one reason that fog sometimes forms shortly after the sun is above the horizon.

2.  **Seasonal Variation.**  Since the axis of the Earth tilts to the plane of orbit, the sun is more nearly overhead in one hemisphere than in the other, depending upon the season.

   a.  The Northern Hemisphere is warmer in June, July, and August because it receives more solar energy than does the Southern Hemisphere.

   b.  The Southern Hemisphere receives more solar radiation and is warmer during December, January, and February.

   c.  Figures 4 and 5 on page 11 show these seasonal surface temperature variations.

3.  **Variation with Latitude.**  Since the Earth is essentially spherical, the sun is more nearly overhead in equatorial regions than at higher latitudes.

   a.  Equatorial regions receive the most solar radiation and are warmest.

   b.  Slanting rays of the sun at higher latitudes deliver less energy over a given area with the least being received at the poles.

   c.  Thus, temperature varies with latitude from the warm Equator to the cold poles.

      1)  You can see this average temperature variation in Figures 4 and 5 on page 11.

4.  **Variations with Topography.**  Not related to the movement or shape of the Earth are temperature variations induced by water and terrain.

   a.  Large, deep water bodies tend to minimize temperature changes, whereas continents favor large changes.

      1)  Wet soil such as in swamps and marshes is almost as effective as water in suppressing temperature changes.

      2)  Thick vegetation tends to control temperature changes since it contains some water and also insulates against heat transfer between the ground and the atmosphere.

      3)  Arid, barren surfaces permit the greatest temperature changes.

   b.  These topographical influences are both diurnal and seasonal.

      1)  EXAMPLE:  The difference between a daily maximum and minimum may be 10° or less over water, near a shore line, or over a swamp or marsh, while a difference of 50° or more is common over rocky or sandy deserts.

      2)  Figures 4 and 5 on page 11 also show the seasonal topographical variation.

         a)  Note that in the Northern Hemisphere in July, temperatures are warmer over continents than over oceans; in January they are colder over continents than over oceans.

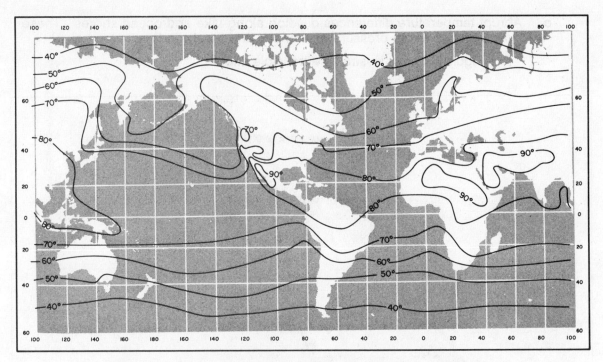

Figure 4.  World-Wide Average Surface Temperatures (°F) in July

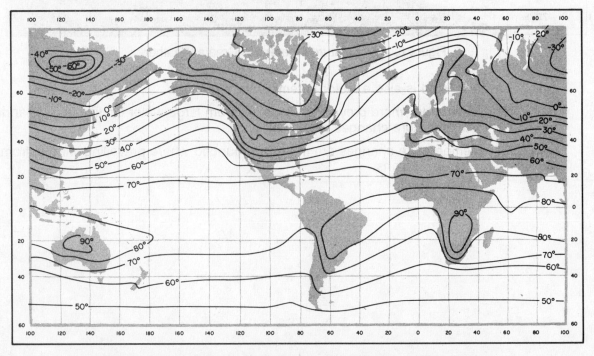

Figure 5.  World-Wide Average Surface Temperatures (°F) in January

c.    Abrupt temperature differences develop along lake and ocean shores.  These
variations generate pressure differences and local winds.

1)    Figure 6 below illustrates a possible effect.

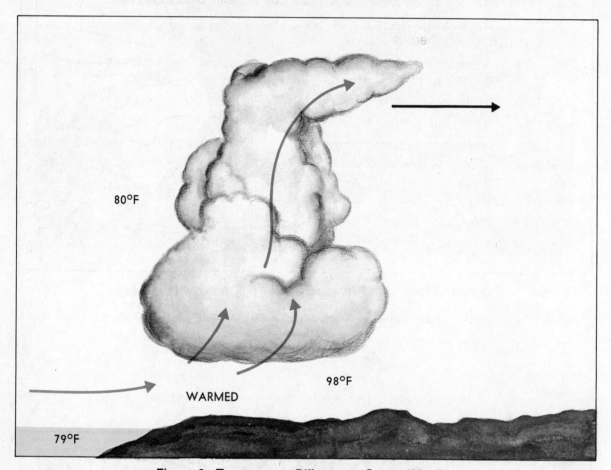

80°F

WARMED

98°F

79°F

Figure 6.  Temperature Differences Create Weather

d.    Prevailing wind is also a factor in temperature controls.

1)    In an area where prevailing winds are from large water bodies, e.g., islands,
temperature changes are rather small.

2)    Temperature changes are more pronounced where prevailing wind is from dry,
barren regions.

5.  **Variation with Altitude.**  Temperature normally decreases with increasing altitude throughout the troposphere.  This is due primarily to solar radiation heating the surface, and the surface, in turn, warming the air above it by terrestrial radiation.

a.  This decrease of temperature with altitude is defined as lapse rate.

1)  The average decrease of temperature -- average lapse rate -- in the troposphere is 2°C per 1,000 ft.

b.  An increase in temperature with altitude is defined as an inversion, i.e., the lapse rate is inverted.

1)  An inversion often develops near the ground on clear, cool nights when wind is light.

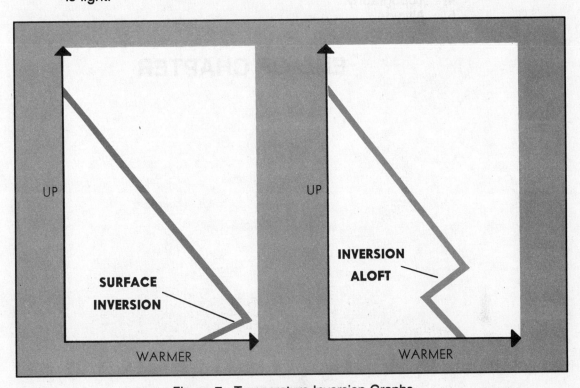

Figure 7.  Temperature Inversion Graphs

a)  The ground radiates and cools much faster than the overlying air.

b)  Air in contact with the ground becomes cold while the temperature a few hundred feet above changes very little.

c)  Thus, temperature increases with height.

2)  Inversions may also occur at any altitude when conditions are favorable.

a)  EXAMPLE:  A current of warm air aloft overrunning cold air near the surface produces an inversion aloft.

3)  Inversions are common in the stratosphere.

E.  **In Closing**

1.  The two commonly used scales for measuring temperature are Celsius and Fahrenheit.

2.  Land absorbs and radiates heat much faster than water.

3.  Variations in solar radiation create temperature variations which are the primary cause of weather.

    a.  The amount of solar radiation received varies with

        1)  Day-night.
        2)  Season.
        3)  Latitude.
        4)  Topography.
        5)  Altitude.

# END OF CHAPTER

# CHAPTER THREE
# ATMOSPHERIC PRESSURE AND ALTIMETRY

> Please take a few minutes to study each of the concepts listed above and anticipate/imagine what they are and how they relate to the other listed concepts.

A. **Atmospheric Pressure** -- the force per unit area exerted by the weight of the atmosphere.

    1. The instrument designed for measuring pressure is the **barometer**.

        a. Weather services and the aviation community use two types of barometers in measuring pressure -- mercurial and aneroid.

    2. The **mercurial barometer** diagrammed in Figure 8 on page 16 consists of an open dish of mercury into which we place the open end of an evacuated glass tube.

        a. Atmospheric pressure forces mercury to rise in the tube.
        b. Near sea level, the column of mercury rises on the average to a height of 29.92 in.
        c. The height of the mercury column is a measure of atmospheric pressure.

    3. An **aneroid barometer** (illustrated in Figure 9 on page 16) uses a flexible metal cell and a registering mechanism.

        a. The cell is partially evacuated of air and contracts or expands as the pressure outside it changes.

            1) One end of the cell is fixed, while the other end moves the registering mechanism.

            2) This mechanism drives an indicator hand along a scale graduated in pressure units.

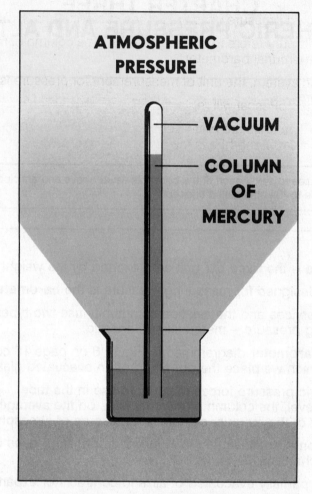

Figure 8.  The Mercurial Barometer

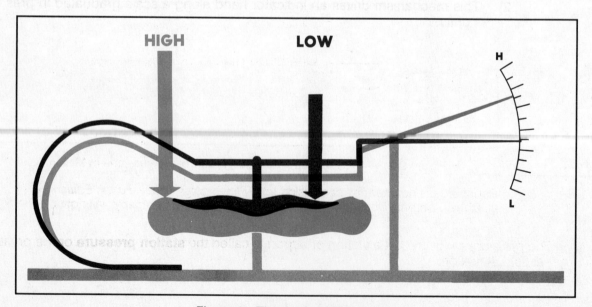

Figure 9.  The Aneroid Barometer

4.  Two commonly used pressure units are inches of mercury (in. Hg) or hectoPascal (hPa).

    a.  Inches of mercury refers to the height to which a column of mercury would be raised in the basic mercurial barometer.

    b.  In the metric system, the unit of measurement for pressure is the hectoPascal (hPa).

        1)  The hectoPascal will replace millibars (mb), which is currently in use.

            a)  The units are equivalent; i.e., 1013.2 mb is 1013.2 hPa.
            b)  1013.2 mb/hPa is the ISA sea-level pressure.

        2)  In this book, both millibars and hectoPascals will be used.

    c.  Note the relationship between inches and millibars in Figure 9(a) below.

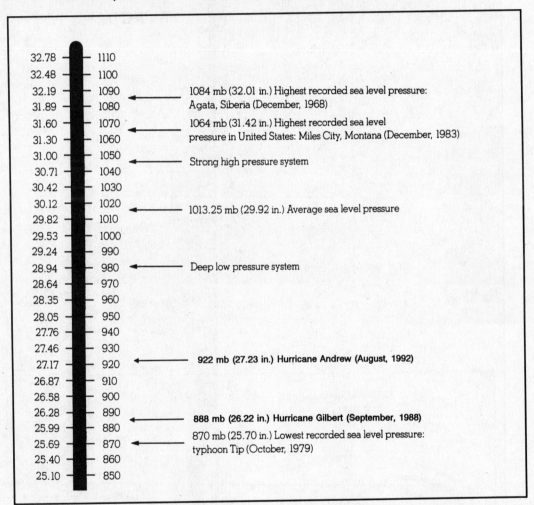

Figure 9(a).  Atmospheric Pressure in Inches of Mercury and in Millibars

Reprinted (as adapted) by permission from *Meteorology Today, Fourth Edition* by
C. Donald Ahrens, Copyright © 1991 by West Publishing Company.  All rights reserved.

5.  The pressure measured at a station or airport is called the **station pressure** or the pressure at field elevation.

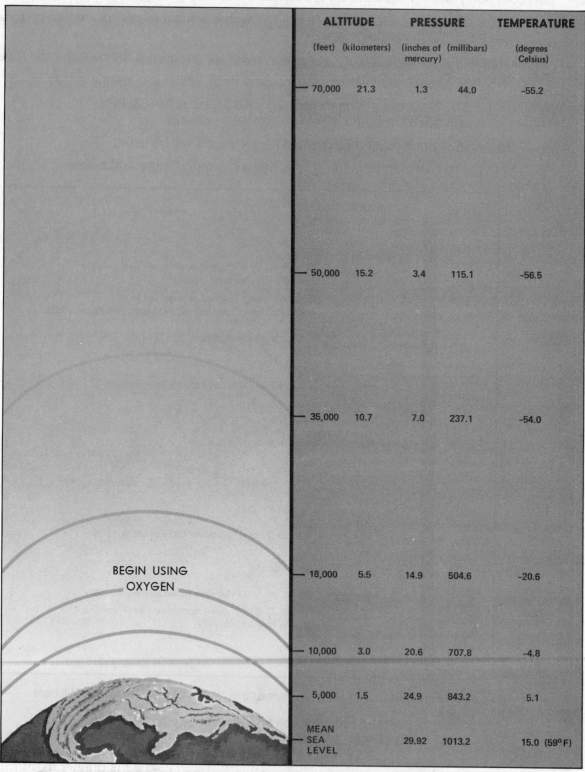

| ALTITUDE | | PRESSURE | | TEMPERATURE |
|---|---|---|---|---|
| (feet) | (kilometers) | (inches of mercury) | (millibars) | (degrees Celsius) |
| 70,000 | 21.3 | 1.3 | 44.0 | -55.2 |
| 50,000 | 15.2 | 3.4 | 115.1 | -56.5 |
| 35,000 | 10.7 | 7.0 | 237.1 | -54.0 |
| 18,000 | 5.5 | 14.9 | 504.6 | -20.6 |
| 10,000 | 3.0 | 20.6 | 707.8 | -4.8 |
| 5,000 | 1.5 | 24.9 | 843.2 | 5.1 |
| MEAN SEA LEVEL | | 29.92 | 1013.2 | 15.0 (59°F) |

BEGIN USING OXYGEN

Figure 10.  The Standard Atmosphere -- Decrease in Pressure with Increasing Heights

B.  **Pressure Variation.**  Pressure varies primarily with altitude and temperature of the air.

1.  **Altitude.**  As we move upward through the atmosphere, the amount, and thus the weight, of the air above us becomes less and less.

    a.  If we carry a barometer with us, we can measure a decrease in pressure as weight of the air above decreases.

      1)  Within the lower few thousand feet of the troposphere, pressure decreases roughly 1 in. for each 1,000-ft. increase in altitude.

      2)  The higher we go, the slower is the rate of decrease with height.

    b.  Figure 10 on page 18 shows the pressure decrease with height in the standard atmosphere.

2.  **Temperature.**  Like most substances, air expands as it becomes warmer and contracts as it cools.

    a.  Figure 11 below shows three equal columns of air -- one colder than standard, one at standard temperature, and one warmer than standard.

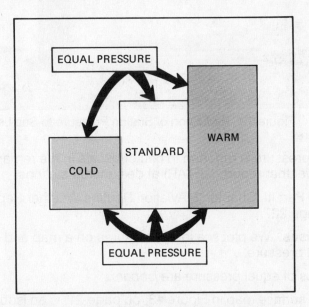

Figure 11.  Effect of Temperature on Pressure

      1)  Because each column contains the same amount of air, pressure is equal at the bottom of each column and equal at the top of each column.

      2)  Thus, pressure decrease upward through each column is the same.

    b.  However, vertical expansion of the warm column has made it higher than the column at standard temperature.

      1)  Shrinkage of the cold column has made it shorter.

    c.  Thus, the rate of decrease of pressure with height in warm air is less than standard.

      1)  The rate of decrease of pressure with height in cold air is greater than standard.

3.   **Sea Level Pressure.**  To readily compare pressure between stations at different altitudes, we must adjust them to some common level, i.e., mean sea level (MSL).

   a.   In Figure 12 below, pressure measured at a 5,000-ft. station is 25 in. Hg; pressure changes about 1 in. Hg for each 1,000 ft. or a total of 5 in. Hg.  Sea level pressure at this station is thus approximately 25 + 5, or 30 in. Hg.

      1)   The weather observer takes temperature and other effects into account, but this simplified example explains the basic principle of sea level pressure.

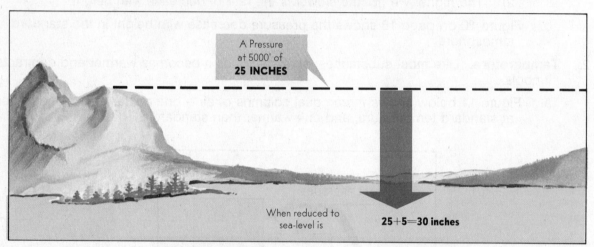

A Pressure
at 5000' of
**25 INCHES**

When reduced to
sea-level is

**25+5=30 inches**

Figure 12.  Reduction of Station Pressure to Sea Level

   b.   Sea level pressure is reported in hectoPascals in the remarks section of the aviation routine weather report (METAR) at designated stations.

      1)   See Part III, Chapter 2, Aviation Routine Weather Report (METAR), beginning on page 227.

4.   **Pressure Analyses.**  We plot sea level pressures on a map and draw lines connecting points of equal pressure.

   a.   These lines of equal pressure are *isobars*.

      1)   The surface map in Figure 13, on page 21, is an isobaric analysis showing identifiable, organized pressure patterns.

         a)   See Part III, Chapter 16, Surface Analysis Chart, beginning on page 296.

      2)   The five pressure systems shown are defined as follows:

         a)   **Low** -- an area of pressure surrounded on all sides by higher pressure, also called a cyclone.  In the Northern Hemisphere, a cyclone is a mass of air which rotates counter-clockwise (i.e., a low-pressure system), viewed from above.

         b)   **High** -- a center of pressure surrounded on all sides by lower pressure, also called an anticyclone.  In the Northern Hemisphere, an anticyclone is a mass of air which rotates clockwise (i.e., a high-pressure system), viewed from above.

         c)   **Trough** -- an elongated area of low pressure with the lowest pressure along a line marking maximum cyclonic curvature

         d)   **Ridge** -- an elongated area of high pressure with the highest pressure along a line marking maximum anticyclonic curvature

e)   **Col** -- the neutral area between two highs or two lows.  It also is the intersection of a trough and a ridge.  The col on a pressure surface is analogous to a mountain pass on a topographic surface.

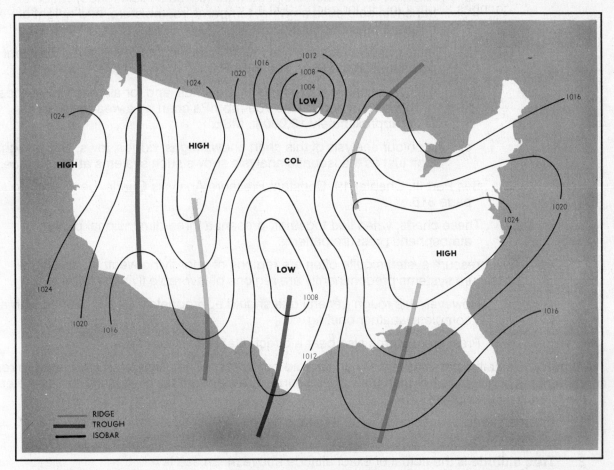

Figure 13.  Pressure Systems

b.   Upper air weather maps reveal these same types of pressure patterns aloft for several levels.

1)   They also show temperature, moisture, and wind at each level.

c.   An upper air map is a *constant pressure analysis*.  Constant pressure simply refers to a specific pressure (e.g., 700 mb/hPa).

1)   Everywhere above the Earth's surface, pressure decreases with height.  At some height, it decreases to this constant pressure of 700 mb/hPa (about 10,000 ft. MSL, depending upon temperature).

a)   Thus, there is a "surface" throughout the atmosphere at which pressure is 700 mb/hPa.

b)   This is called the 700-mb/hPa constant pressure surface.

c)   However, the *height* of this surface is *not* constant.

i)    Rising pressure pushes the surface upward into highs and ridges.

ii)   Falling pressure lowers the height of the surface into lows and troughs.

iii)  These systems migrate continuously as "waves" on the pressure surface.

2)   The National Weather Service (NWS) and military weather services take routinely scheduled upper-air observations, sometimes called soundings.

   a)   A balloon carries aloft an instrument (called a radiosonde) which transmits data such as wind, temperature, moisture, and height at selected pressure surfaces.

   b)   These observations are used to plot the heights of a desired constant pressure surface (e.g., 700 mb/hPa).

      i)   Variations in these heights are small, and for all practical purposes, you may regard the 700-mb/hPa chart as a weather map at approximately 10,000 ft. MSL.

   c)   A contour analysis of this chart shows highs, ridges, lows, and troughs aloft just as the isobaric analysis shows such systems at the surface.

3)   See Part III, Chapter 19, Constant Pressure Analysis Charts, beginning on page 316.

4)   These charts, when tied together, present a three-dimensional picture of atmospheric pressure patterns.

d.   Low-pressure systems quite often are regions of poor flying weather, and high-pressure systems predominantly are regions of favorable flying weather.

   1)   However, this rough general rule should appropriately be supplanted with a complete weather briefing.

   2)   Pressure patterns also bear a direct relationship to wind.

C.  **Altimetry.** The altimeter is essentially an aneroid barometer. The altimeter is graduated to read increments of height rather than units of pressure. The standard for graduating the altimeter is the standard atmosphere.

1.   Altitude seems like a simple term; it means height. But in aviation, it can have many meanings.

2.   **True altitude** is the actual or exact altitude above mean sea level.

   a.   Since existing conditions in a real atmosphere are seldom standard, altitude indications on the altimeter are seldom actual or true altitudes.

3.   **Indicated altitude** is the altitude above mean sea level indicated on the altimeter when set at the local altimeter setting.

   a.   The height indicated on the altimeter changes with changes in surface pressure.

   1)   Figure 15 on page 23 shows this effect.

   2)   A movable scale on the altimeter permits you to adjust for variations in surface pressure.

   3)   **Altimeter setting** is the value to which the scale of the pressure altimeter is set so as to indicate true altitude at field elevation.

      a)   This value is the atmospheric pressure adjusted to sea level in the region in which you are flying.

   4)   You must keep your altimeter setting current by adjusting it frequently in flight to the setting reported by the nearest tower or weather reporting station.

   5)   If an altimeter setting is not available before takeoff, you can set the altimeter to read field elevation, i.e., true altitude.

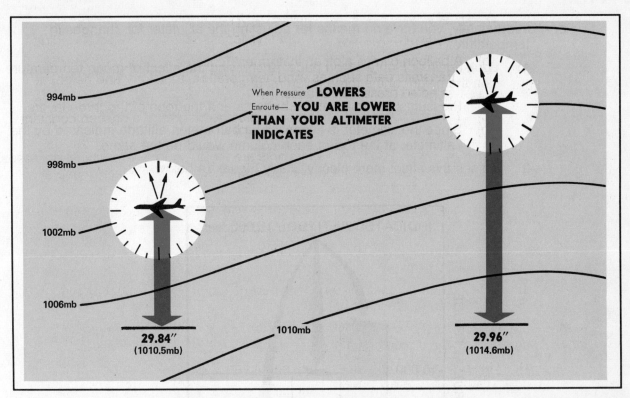

Figure 15.  Effect of Pressure on Indicated Altitude

b.   Unfortunately, you have no means for adjusting the altimeter for changes in temperature.

   1)   Look again at Figure 11 on page 19 showing the effect of mean temperature on the height of the three columns of air.

      a)   Pressures are equal at the bottoms and the tops of the three layers.

      b)   Since the altimeter is essentially a barometer, altitude indicated by the altimeter at the top of each column would be the same.

   2)   To see this effect more clearly, study Figure 14 below.

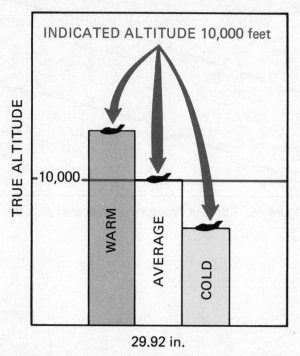

Figure 14.  Effect of Temperature on Indicated Altitude

      a)   Note that in the warm air, true altitude is higher than indicated.
      b)   In the cold air, true altitude is lower than indicated.

4.  **Corrected (approximately true) altitude** is indicated altitude corrected for the temperature of the air column below the aircraft.

    a.  If it were possible for a pilot to determine mean temperature of the column of air between the aircraft and the surface, flight computers would be designed to use this mean temperature in computing true altitude.

        1)  However, the only guide a pilot has to temperature below him/her is the outside air temperature at his/her altitude.

        2)  Thus, the flight computer uses outside air temperature to correct indicated altitude to approximate true altitude.

            a)  It is close enough to true altitude to be used for terrain clearance provided you have your altimeter set to the value reported from a nearby reporting station.

            b)  Pilots have met with disaster because they failed to allow for the difference between indicated and true altitude.  In cold weather when you must clear high terrain, take time to compute true altitude.

    b.  EXAMPLE:  If you are flying at an indicated altitude of 5,000 ft. MSL with an altimeter setting of 29.92 (standard pressure), your pressure altitude is also 5,000 ft.  Once you determine pressure altitude and know the outside air temperature (OAT), use your flight computer to determine the corrected altitude.

        1)  If your OAT is +5°C (standard temperature at 5,000 ft. MSL), your corrected altitude is 5,000 ft. MSL.

        2)  However, if your OAT is −15°C (20° below standard), your corrected altitude is only 4,640 ft. MSL.

5.  **Pressure altitude** is the altitude in the standard atmosphere where pressure is the same as where you are currently flying.

    a.  In the standard atmosphere, pressure at sea level is 29.92 in. Hg and decreases approximately 1 in. per 1,000 ft. of altitude.

    b.  Thus, for example, if the pressure at your altitude of flight is 27.92 in. Hg, you are flying at a pressure altitude of approximately 2,000 ft. higher than your indicated altitude.

    c.  You can determine pressure altitude by setting your altimeter to the standard altimeter setting of 29.92 in.

    d.  All flights at and above 18,000 ft. MSL, i.e., in Class A airspace, are flown at pressure altitudes.

6. **Density altitude** is the altitude in the standard atmosphere where air density is the same as where you are.

    a.   Pressure, temperature, and to a lesser extent, humidity determine air density.

        1)   On a hot day, the air becomes "thinner" or lighter, and its density where you are is equivalent to a higher altitude in the standard atmosphere -- thus the term "high density altitude."

        2)   On a cold day, the air becomes heavy; its density is the same as that at an altitude in the standard atmosphere lower than your altitude -- "low density altitude."

    b.   Density altitude is not a height reference; rather, it is an index to aircraft performance.

        1)   Low density altitude increases performance.

        2)   High density altitude is a real hazard since it reduces aircraft performance. It affects performance in three ways.

            a)   It reduces power because the engine takes in less air to support combustion.

            b)   It reduces thrust because the propeller gets less grip on the light air or a jet has less mass of gases to spit out the exhaust.

            c)   It reduces lift because the light air exerts less force on the airfoils.

    c.   You cannot detect the effect of high density altitude on your airspeed indicator. Your aircraft lifts off, climbs, cruises, glides, and lands at the prescribed indicated airspeeds.

        1)   But at a specified indicated airspeed, your true airspeed and your groundspeed increase proportionally as density altitude becomes higher.

    d.   The net results are that high density altitude lengthens your takeoff and landing rolls and reduces your rate of climb.

        1)   Before liftoff, you must attain a faster groundspeed, and therefore, you need more runway; your reduced power and thrust add a need for still more runway.

        2)   You land at a faster groundspeed and, therefore, need more room to stop.

        3)   At a prescribed indicated airspeed, you are flying at a faster true airspeed, and therefore, you cover more distance in a given time, which means climbing at a more shallow angle.

            a)   Add to this the problems of reduced power and rate of climb, and you are in double jeopardy in your climb.

            b)   Figure 17 on page 27 shows the effect of density altitude on takeoff distance and rate of climb.

               i)   It shows the takeoff roll increasing from 1,300 ft. to 1,800 ft. and the rate of climb decreasing from 1,300 fpm to less than 1,000 fpm.

e.   High density altitude also can be a problem at cruising altitudes. When air is abnormally warm, the high density altitude lowers your service ceiling.

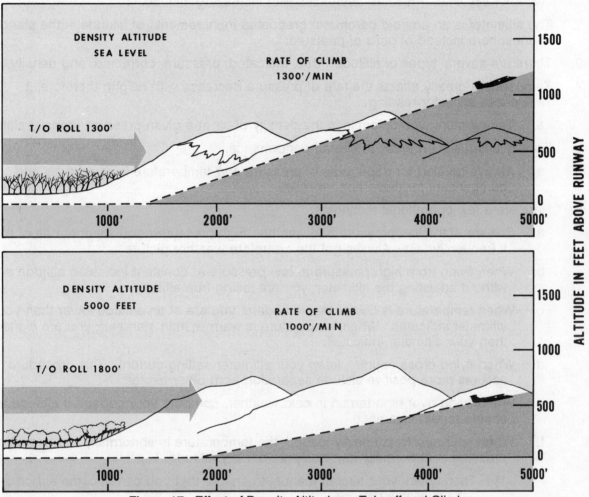

Figure 17.  Effect of Density Altitude on Takeoff and Climb

f.   To compute density altitude:

1)   Set your altimeter at 29.92 in. and read pressure altitude from your altimeter.

   a)   Read outside air temperature, and then use your flight computer or a graph to get density altitude.

   b)   EXAMPLE:  If temperature at 10,000 ft. pressure altitude is 20°C, density altitude is 12,700 ft.  (Check this on your flight computer.)  Your aircraft will perform as though it were at 12,700 ft. MSL indicated with a standard temperature of −8°C.

2)   At an airport served by a weather observing station, you usually can get density altitude for the airport from the weather observer.

3)   A graph for computing density altitude is in Part III, Chapter 26, Other Weather-Related Information, on page 367.

D.  **In Closing**

1.  Pressure patterns (highs, lows, etc.) can be a clue to weather causes and movement of weather systems.  Pressure decreases with increasing altitude.

2.  The altimeter is an aneroid barometer graduated in increments of altitude in the standard atmosphere instead of units of pressure.

3.  There are several types of altitude:  true, indicated, pressure, corrected, and density.

4.  Temperature greatly affects the rate of pressure decrease with height; therefore, it influences altimeter readings.

    a.  Temperature also determines the density of air at a given pressure (density altitude).

5.  Density altitude is an index to aircraft performance.

    a.  Always be alert for departures of pressure and temperature from normals and compensate for these abnormalities.

6.  Here are a few operational reminders:

    a.  Beware of the low pressure-bad weather, high pressure-good weather rule of thumb.  It frequently fails.  Always get the **complete** weather picture.

    b.  When flying from high pressure to low pressure at constant indicated altitude and without adjusting the altimeter, you are losing true altitude.

    c.  When temperature is colder than standard, you are at an altitude *lower* than your altimeter indicates.  When temperature is warmer than standard, you are *higher* than your altimeter indicates.

    d.  When flying cross country, keep your altimeter setting current.  This procedure assures more positive altitude separation from other aircraft.

    e.  When flying over high terrain in cold weather, compute your corrected altitude to ensure terrain clearance.

    f.  When your aircraft is heavily loaded, the temperature is abnormally warm, and/or the pressure is abnormally low, compute density altitude.

        1)  Then check your aircraft manual to ensure that you can become airborne from the available runway.

        2)  Check further to determine that your rate of climb permits clearance of obstacles beyond the end of the runway.

        3)  This procedure is advisable for any airport regardless of altitude.

    g.  When planning takeoff or landing at a high altitude airport regardless of load, determine density altitude.

        1)  The procedure is especially critical when temperature is abnormally warm or pressure abnormally low.

        2)  Make certain you have sufficient runway for takeoff or landing roll.

        3)  Make sure you can clear obstacles beyond the end of the runway after takeoff or in event of a go-around.

# END OF CHAPTER

# CHAPTER FOUR
# WIND

> Please take a few minutes to study each of the concepts listed above and anticipate/imagine what they are and how they relate to the other listed concepts.

A. **Introduction** -- Differences in temperature create differences in pressure. These pressure differences drive a complex system of winds in a never-ending attempt to reach equilibrium.

B. **Convection**

    1. When two surfaces are heated unequally, they heat the overlying air unevenly, resulting in a circulatory motion called convection. Figure 18 on page 30 shows the convective process.

        a. The warmer air expands and becomes lighter or less dense than the cooler air.

        b. The dense, cooler air sinks to the ground by its greater mass, forcing the warmer air upward.

        c. The rising air spreads and cools, eventually descending to complete the convective process.

        d. As long as the uneven heating persists, a continuous convective current is maintained.

    2. The horizontal air flow in a convective current is wind. Convection of both large and small scales accounts for systems ranging from hemispheric circulations down to local eddies.

        a. This horizontal flow, wind, is sometimes called **advection**.

        b. However, the term advection more commonly applies to the transport of atmospheric properties by the wind, i.e., warm advection, cold advection, advection of water vapor, etc.

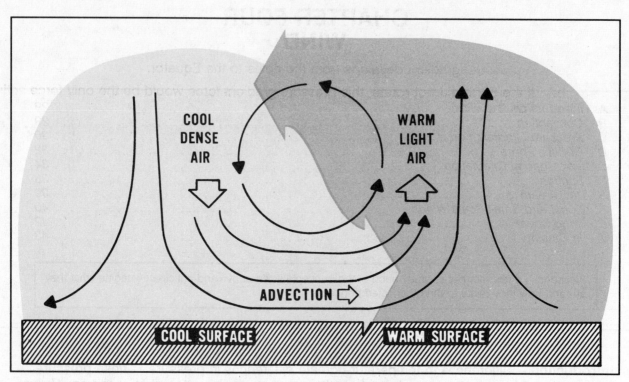

Figure 18.  The Convective Process

C.  **Pressure Gradient Force**

1.   Whenever a pressure difference (or gradient) develops over an area, the air begins to move directly from the higher pressure to the lower pressure.

   a.   The pressure gradient is the rate of decrease of pressure per unit of distance at a fixed time.

   b.   The force that causes the air to move is called the pressure gradient force.

2.   Recall that areas of high and low pressure are encircled by isobars, or lines of equal pressure.  (See Figure 13 on page 21.)

   a.   The pressure gradient force moves the air directly across the isobars.

   b.   Closely spaced isobars indicate a strong pressure gradient, or a sharp change in pressure over a short distance.

      1)   Thus, the pressure gradient force is also strong.

   c.   The stronger the pressure gradient force, the stronger the wind

   d.   From a pressure analysis, you can get a general idea of wind speed from the isobar spacing.

      1)   Closely spaced isobars mean strong winds.
      2)   Widely spaced isobars mean lighter winds.

3.  Because of uneven heating of the Earth, surface pressure is low in warm equatorial regions and high in cold polar regions.

    a.  A pressure gradient develops from the poles to the Equator.

    b.  If the Earth did not rotate, this pressure gradient force would be the only force acting on the wind.

        1)  Circulation would be two giant hemispheric convective currents as shown in Figure 19 below.

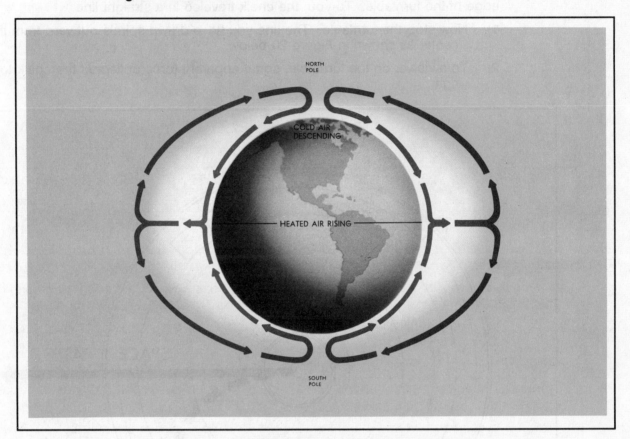

Figure 19.  Circulation Pattern of a Nonrotating Earth

        2)  Cold air would sink at the poles, wind would blow straight from the poles to the Equator, warm air at the Equator would be forced upward, and high level winds would blow directly toward the poles.

    c.  However, the Earth does rotate, and because of its rotation, this simple circulation is greatly distorted.

## D.  **Coriolis Force**

1.    A moving mass (e.g., air) travels in a straight line until acted upon by some outside force (Newton's First Law of Motion).

   a.    However, if you viewed the moving mass from a rotating platform, the path of the moving mass relative to your platform would appear to be deflected or curved.

   b.    EXAMPLE:  Imagine the turntable of a record player rotating counterclockwise.  Use a piece of chalk and a ruler to draw a "straight" line from the center to the outer edge of the turntable.  To you, the chalk traveled in a straight line.

      1)    Now stop the turntable.  The line you have drawn spirals outward from the center as shown in Figure 20 below.

      2)    To a viewer on the turntable, some apparent force deflected the chalk to the right.

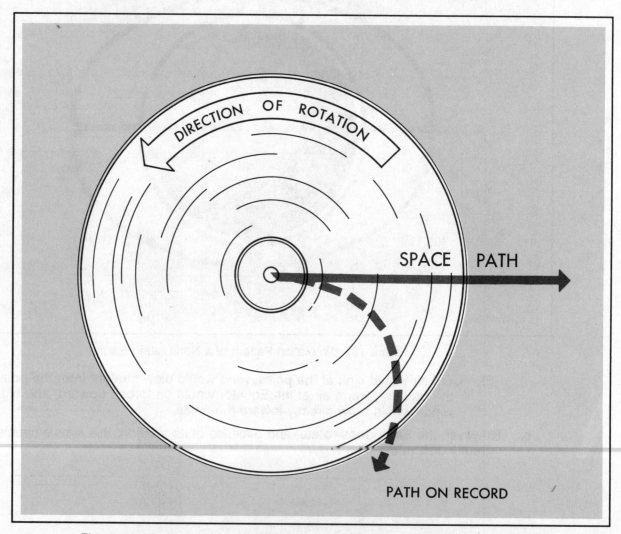

Figure 20.  Apparent Deflective Force Due to Rotation of a Horizontal Platform

2.    Similarly, air moving across the surface of the Earth seems to us to be deflected by some force.

   a.    This principle was first explained by a Frenchman, Gaspard Coriolis, and carries his name -- the Coriolis force.

   b.    Because the Earth is spherical, the Coriolis force is much more complex than the simple turntable example.

3.    The Coriolis force affects the paths of aircraft, missiles, flying birds, ocean currents, and most important to the study of weather, air currents.

    a.    It deflects air to the right in the Northern Hemisphere and to the left in the Southern Hemisphere.

    b.    This book concentrates on deflection to the right in the Northern Hemisphere.

4.    The Coriolis force varies directly with wind speed and latitude.

    a.    The stronger the wind speed, the greater the deflection.
    b.    Coriolis force varies with latitude from zero at the Equator to maximum at the poles.

5.    When a pressure gradient force is first established, wind begins to blow from higher to lower pressure directly across the isobars.  In Figure 21 below it would be from bottom to top.

    a.    However, the instant the air begins moving from high pressure to low pressure, the Coriolis force deflects it to the right, curving its path.

    b.    As the speed of this moving air increases, the Coriolis force increases, deflecting the wind more and more to the right.

    c.    Eventually, the wind speed increases to a point where the Coriolis force balances the pressure gradient force, as shown in Figure 21 below.

        1)    At this point, the wind has been deflected 90° and is now parallel to the isobars (depicted as "Resultant Wind" in Figure 21).

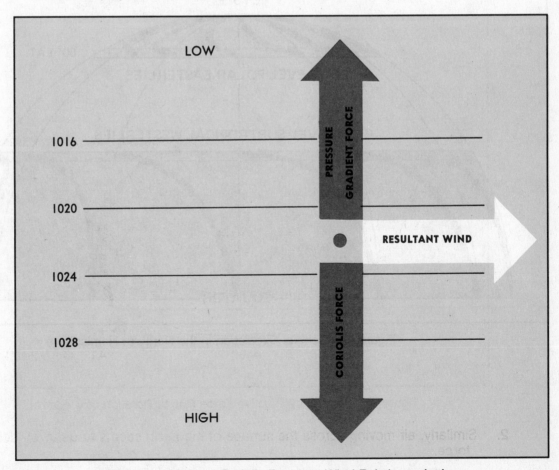

Figure 21.  Effect of Coriolis Force on Wind Relative to Isobars

## E.  The General Circulation

1.  As air is forced aloft at the Equator and begins its high-level trek northward, the Coriolis force turns it to the right or to the east as shown in Figure 22 below.

   a.  Wind becomes westerly (i.e., moves from west to east) at about 30° latitude temporarily blocking further northward movement, as shown in Figure 22 below.

   b.  Similarly, as air over the poles begins its low-level journey southward toward the Equator, it likewise is deflected to the right and becomes an east wind, halting for a while its southerly progress at about 60° latitude, also shown in Figure 22.

2.  As a result, air piles up between 30° and 60° latitude in both hemispheres.

   a.  The added weight of the air increases the pressure into semipermanent high pressure belts.

      1)  Figures 23 and 24 on page 35 are maps of mean surface pressure for the months of July and January.

         a)  The maps show clearly the subtropical high pressure belts near 30° latitude in both the Northern and Southern Hemispheres.

   b.  These high pressure belts create a temporary impasse disrupting the simple convective transfer between the Equator and the poles.

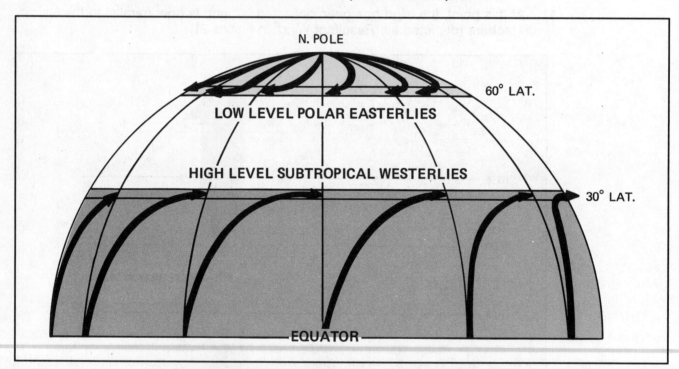

Figure 22.  General Circulation in the Northern Hemisphere

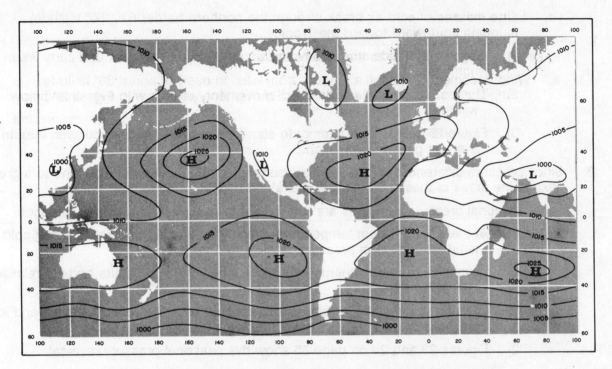

Figure 23.  Mean World-Wide Surface Pressure Distribution in July

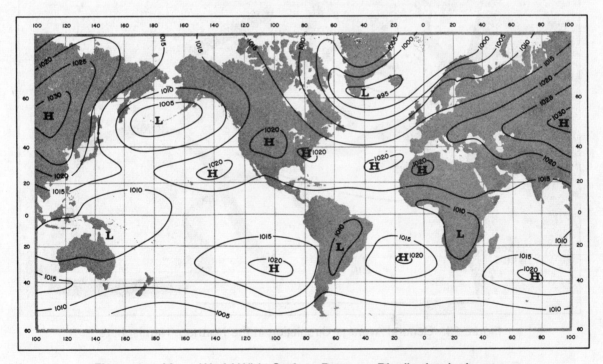

Figure 24.  Mean World-Wide Surface Pressure Distribution in January

    c.    Large masses of cold air break through the northern barrier (i.e., 60° latitude) plunging southward toward the Tropics.

        1)    Large mid-latitude storms develop between cold outbreaks and carry warm air northward.

        2)    The result is a mid-latitude band of migratory storms with ever-changing weather.

        3)    Figure 25 below is an attempt to standardize this chaotic circulation into an average general circulation.

3.    Since pressure differences cause wind, seasonal pressure variations determine to a great extent the areas of these cold air outbreaks and mid-latitude storms.

    a.    Seasonal pressure variations are largely due to seasonal temperature changes.

        1)    At the surface, warm temperatures largely determine low pressure and cold temperatures, high pressure.

    b.    During summer, warm continents tend to be areas of low pressure and the relatively cool oceans, high pressure.

        1)    In winter, the reverse is true -- high pressure over the cold continents and low pressure over the relatively warm oceans.

        2)    Figures 23 and 24 on page 35 show this seasonal pressure reversal.

Figure 25.  General Average Circulation in the Northern Hemisphere

    c.    Cold outbreaks are strongest in the winter and are predominantly from the colder continental areas.

        1)    Outbreaks are weaker in the summer, and more likely to originate from the cooler water surfaces.

        2)    Since these outbreaks are masses of cool, dense air, they characteristically are high-pressure areas.

4.    As the air moves outward from high pressure, it is deflected to the right by the Coriolis force.

    a.    Thus, the wind around a high moves clockwise.

        1)    The high pressure with its associated wind system is called an *anticyclone*.

    b.    The storms that develop between high pressure systems are characterized by low pressure.

        1)    As wind moves inward toward low pressure, it is also deflected to the right.
        2)    Thus, the wind around a low is counterclockwise.
        3)    The low pressure and its wind system are called a *cyclone*.

    c.    Figure 26 below shows winds moving parallel to isobars (called contours on upper level charts).  The winds are clockwise around highs and counterclockwise around lows.

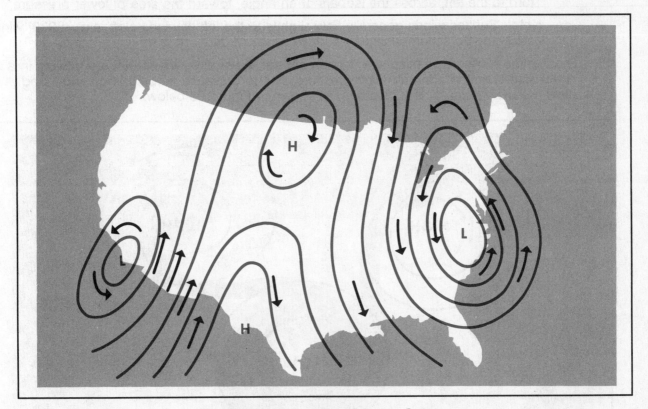

Figure 26.  Airflow around Pressure Systems

5.    The high pressure belt between 30° and 60° latitude forces air outward to the north and to the south, resulting in three major wind belts, as shown in Figure 25 on page 36.

    a.    The southward moving air is deflected by the Coriolis force, becoming the well-known subtropical northeast trade winds (see Part II, Chapter 15, Tropical Weather, beginning on page 163).

        1)    These winds carry tropical storms from east to west, e.g., hurricanes from the ocean off North Africa west into the Caribbean.

    b.    The northbound air becomes entrained into the mid-latitude storms.

        1)    High-level winds are deflected to the right and are known as the prevailing westerlies.

        2)    These westerlies drive mid-latitude storms generally from west to east.

    c.    Polar easterlies dominate low-level circulation north of about 60° latitude where few major storm systems develop.

## F.    **Friction**

1.    Friction between the wind and the terrain surface slows the wind.

      a.    The rougher the terrain, the greater the frictional force
      b.    The stronger the wind speed, the greater the friction
      c.    Frictional force always acts in opposition to the wind direction.

2.    When the wind is within approximately 3,000 ft. of the surface, friction reduces the wind speed, which in turn reduces the Coriolis force.

      a.    Recall that in the absence of surface friction, the Coriolis force turns the wind 90° to the right to parallel the isobars.

      b.    Surface friction slows the wind, which reduces the Coriolis force, causing the wind to turn to the left, across the isobars at an angle, toward the area of lower pressure.

      c.    Note:  Surface winds generally flow slightly to the left of winds aloft, e.g., a 270° wind aloft may be accompanied by surface winds of 240°.

3.    Thus, in the Northern Hemisphere, friction causes the surface winds that are moving in a clockwise direction around a high pressure area to cross the isobars at an angle and flow toward an area of low pressure, as shown in Figure 28 below.

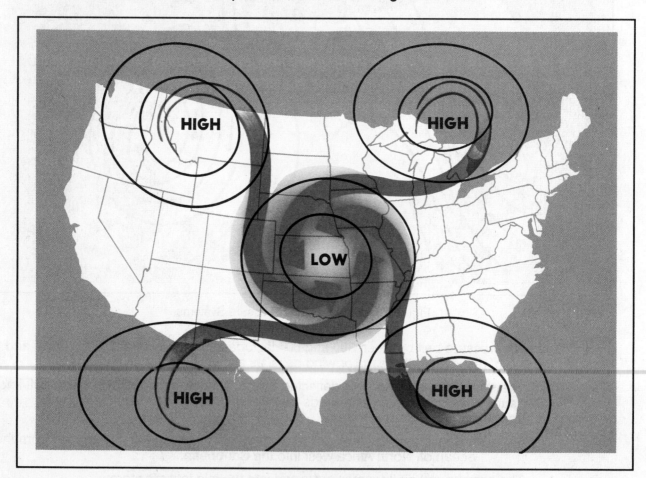

Figure 28.  Circulation around Pressure Systems at the Surface

4.   The angle of surface wind to isobars is about 10° over water, increasing with roughness of terrain.

   a.   In mountainous regions, one often has difficulty relating surface wind to pressure gradient because of immense friction and also because local terrain effects on pressure could cause the angle to be 40° or more.

   b.   The average angle across the isobars is about 30°, for all surfaces.

   c.   The speed of the wind will also affect this angle.

      1)   Normally, the angle is smaller for high winds (i.e., due to a stronger Coriolis force) and large for slower winds (i.e., weaker Coriolis force).

G.   **Jet Stream**

   1.   A discussion of the general circulation is incomplete when it does not mention the **jet stream**.

      a.   Winds on the average increase with height throughout the troposphere, culminating in a maximum near the level of the tropopause.

      b.   These maximum winds tend to be further concentrated in narrow bands.

   2.   A jet stream, then, is a narrow band of strong winds meandering through the atmosphere at a level near the tropopause.

      a.   Since it is of interest primarily to high-level flight, further discussion of the jet stream is reserved for Part II, Chapter 13, High Altitude Weather, beginning on page 144.

## H.  Local and Small-Scale Winds

1.  The previous discussion has dealt only with the general circulation and major wind systems.  Local terrain features such as mountains and shore lines influence local winds and weather.

2.  **Mountain and Valley Winds**

    a.  In the daytime, air next to a mountain slope is heated by contact with the ground as the ground receives radiation from the sun.

        1)  This air usually becomes warmer than air at the same altitude but farther from the slope.

        2)  Colder, denser air in the surroundings settles downward and forces the warmer air near the ground up the mountain slope.

            a)  This wind is a "valley wind," so called because the air is flowing up out of the valley.

    b.  At night, the air in contact with the mountain slope is cooled by terrestrial radiation and becomes heavier than the surrounding air.

        1)  It sinks along the slope, producing the "mountain wind" which flows like water down the mountain slope.

    c.  Mountain winds are usually stronger than valley winds, especially in winter.

        1)  The mountain wind often continues down the more gentle slopes of canyons and valleys, and in such cases takes the name "drainage wind."

        2)  It can become quite strong over some terrain conditions and in extreme cases can become hazardous when flowing through canyon restrictions as discussed in Part I, Chapter 9, Turbulence, beginning on page 93.

3.  **Katabatic Wind**

    a.  A **katabatic wind** is any wind blowing down an incline when the incline is influential in causing the wind.

        1)  Thus, the mountain wind is a katabatic wind.

    b.  Any katabatic wind originates because cold, heavy air spills down sloping terrain displacing warmer, less dense air ahead of it.

        1)  Air is heated and dried as it flows down slope.
        2)  Sometimes the descending air becomes warmer than the air it replaces.

    c.  Many katabatic winds recurring in local areas have been given colorful names to highlight their dramatic local effect.

        1)  Some of these include:

            a)  The Bora, a cold northerly wind blowing from the Alps to the Mediterranean coast;

            b)  The Chinook, a warm wind down the east slope of the Rocky Mountains often reaching hundreds of miles into the high plains;

            c)  The Taku, a cold wind in Alaska blowing off the Taku glacier; and

            d)  The Santa Ana, a warm wind descending from the Sierras into the Santa Ana Valley of California.

4.   **Land and Sea Breezes**

   a.   As frequently stated earlier, land surfaces warm and cool more rapidly than do water surfaces.

   b.   Land is warmer than the sea during the day.

      1)   Wind blows from the cool water to warm land -- the "sea breeze," so called because it blows from the sea.

      2)   At night, the wind reverses, blows from cool land to warmer water, and creates a "land breeze."

   c.   Figure 30 below diagrams land and sea breezes.

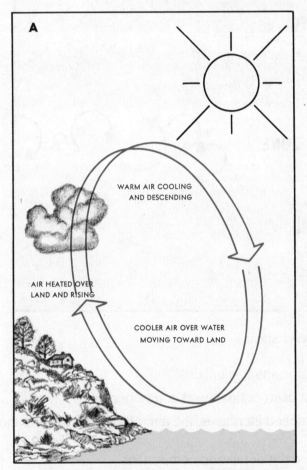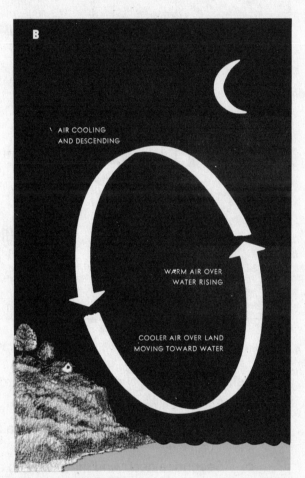

Figure 30.  Land and Sea Breezes

   d.   Land and sea breezes develop only when the overall pressure gradient is weak.

      1)   Wind occurring as a result of a strong pressure gradient mixes the air so rapidly that local temperature and pressure gradients do not develop along the shore line.

I. **Wind Shear**

  1.   Rubbing two objects against each other creates friction.

     a.   If the objects are solid, no exchange of mass occurs between the two.

     b.   However, if the objects are fluid currents, friction creates eddies along a common shallow mixing zone, and a mass transfer takes place in the shallow mixing layer.

        1)   This zone of induced eddies and mixing is a **shear zone**.

  2.   Figure 31 below shows two adjacent currents of air and their accompanying shear zone.

  3.   Part I, Chapter 9, Turbulence, beginning on page 93, relates wind shear to turbulence.

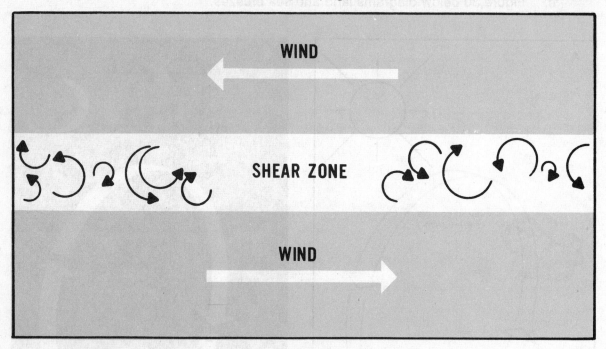

Figure 31.  Wind Shear

  4.   Basically, there are two potentially hazardous shear situations.

     a.   First, a tailwind may shear to either a calm or headwind component.

        1)   In this instance, initially the airspeed increases, the aircraft pitches up, and the altitude increases.

     b.   Second, a headwind may shear to a calm or tailwind component.

        1)   In this instance, initially the airspeed decreases, the aircraft pitches down, and the altitude decreases.

  5.   Aircraft speed, aerodynamic characteristics, power-to-weight ratio, powerplant response time, and pilot reactions along with other factors have a bearing on wind shear effects.

  6.   Remember that wind shear can cause problems for any aircraft and pilot.

J.  **In Closing**

1.  The horizontal air flow in a convective current is wind.

    a.  Wind speed is proportional to the spacing of isobars or contours on a weather map.

        1)  However, with the same spacing, wind speed at the surface will be less than aloft because of surface friction.

    b.  You can also determine wind direction from a weather map.

        1)  If you visualize standing with the wind aloft at your back (Northern Hemisphere), low pressure should be on your left and high pressure on your right.

            a)  If you visualize standing with the surface wind to your back, then turn clockwise about 30°, low pressure should be on your left and high pressure on your right.

        2)  On a surface map, wind will cross the isobar at an angle toward lower pressure; on an upper air chart, it will be nearly parallel to the contour.

2.  As the pressure gradient force drives the wind from high- to low-pressure areas, the Coriolis force deflects the wind to the right (in the Northern Hemisphere), resulting in the general circulation around high- and low-pressure systems.

    a.  In the Northern Hemisphere, wind blows counterclockwise around a low and clockwise around a high.

        1)  At the surface where winds cross the isobars at an angle, you can see a transport of air from high to low pressure.

        2)  Although winds are virtually parallel to contours on an upper air chart, there still is a slow transport of air from high to low pressure.

3.  At the surface when air converges into a low, it cannot go outward against the pressure gradient, nor can it go downward into the ground; it must go upward.

    a.  Thus, a low or trough is an area of rising air.

        1)  Rising air is conducive to cloudiness and precipitation; thus, we have the general association of low pressure and bad weather.

    b.  By similar reasoning, air moving out of a high or ridge depletes the quantity of air. Highs and ridges, therefore, are areas of descending air.

        1)  Descending air favors dissipation of cloudiness; hence the association of high pressure and good weather.

4.  Three exceptions to the low pressure -- bad weather, high pressure -- good weather rule:

    a.  Many times weather is more closely associated with an upper air pattern than with features shown by the surface map.  Although features on the two charts are related, they seldom are identical.

        1)  A weak surface system often loses its identity in the upper air pattern, while another system may be more evident on the upper air chart than on the surface map.

        2)  Widespread cloudiness and precipitation often develop in advance of an upper trough or low.

            a)  A line of showers and thunderstorms is not uncommon with a trough aloft even though the surface pressure pattern shows little or no cause for the development.

    b.   Downward motion in a high or ridge places a "cap" on convection, preventing any upward motion.

        1)   Air may become stagnant in a high, trap moisture and contamination in low levels, and restrict ceiling and visibility.

            a)   Low stratus, fog, haze, and smoke are not uncommon in high pressure areas.

        2)   However, a high or ridge aloft with moderate surface winds most often produces good flying weather.

    c.   A dry, sunny region becomes quite warm from intense surface heating, thus generating a surface low pressure area, called a thermal low.

        1)   The warm air is carried to high levels by convection, but cloudiness is scant because of lack of moisture.

            a)   Since in warm air, pressure decreases slowly with altitude, the warm surface low is not evident at upper levels.

        2)   The thermal low is relatively shallow with weak pressure gradients and no well-defined cyclonic circulation.

            a)   It generally supports good flying weather.

            b)   However, during the heat of the day, one must be alert for high density altitude and convective turbulence.

5.   Due to surface friction, highs and lows tend to *lean* from the surface into the upper atmosphere, i.e., the upper portion moves faster than the lower portion, which is slowed by friction with the surface.

    a.   Winds aloft often flow across the associated surface systems.

    b.   Upper winds tend to steer surface systems in the general direction of the upper wind flow.

    c.   An intense, cold, low pressure vortex *leans less* than does a weaker system.

        1)   The intense low becomes oriented almost vertically and is clearly evident on both surface and upper air charts.

            a)   Upper winds encircle the surface low and do not blow across it.

            b)   Thus, the storm moves very slowly and usually causes an extensive and persistent area of clouds, precipitation, strong winds, and generally adverse flying weather.

6.   There are three major wind belts in the Northern Hemisphere:

    a.   Subtropical northeasterly trade winds, which carry tropical storms from east to west

    b.   Midlatitude westerlies, which drive midlatitude storms generally from west to east (which encompasses the United States)

    c.   Polar easterlies, which primarily contribute to the development of midlatitude storms

7.   Local terrain features influence local winds and weather, e.g.,

    a.   Mountain and valley winds
    b.   Katabatic wind
    c.   Land and sea breezes

# END OF CHAPTER

# CHAPTER FIVE
# MOISTURE, CLOUD FORMATION, AND PRECIPITATION

Please take a few minutes to study each of the concepts listed above and anticipate/imagine what they are and how they relate to the other listed concepts.

## A. Water Vapor

1. Water evaporates into the air and becomes an ever-present but variable part of the atmosphere.

2. Water vapor is invisible, just as oxygen and other gases are invisible.

   a. We can readily measure water vapor and express the results in different ways. Two commonly used terms are:

      1) Relative humidity, and
      2) Dew point.

3. **Relative humidity** relates the actual water vapor present to that which could be present.

   a. Relative humidity is routinely expressed as a percentage.

   b. Temperature largely determines the maximum amount of water vapor a parcel of air can hold.

      1) As Figure 32 below shows, warm air can hold more water vapor than cool air.

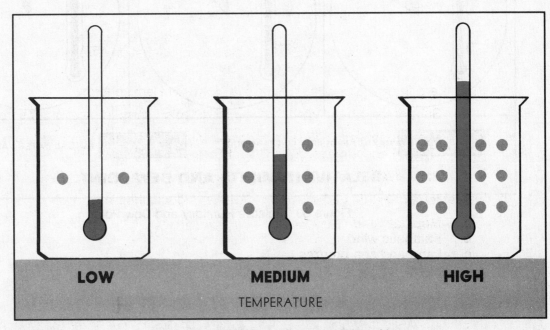

Figure 32. Air Temperature and Water Vapor

    c.    Relative humidity expresses the degree of saturation.  Air with 100% relative humidity is saturated; at less than 100%, unsaturated.

        1)    If a given volume of air is cooled to some specific temperature, it can hold no more water vapor than is actually present; relative humidity becomes 100%, and saturation occurs.

4.    **Dew point** is the temperature to which air must be cooled to become saturated by the water vapor already present in the air.

    a.    Aviation weather reports normally include the air temperature and dew point.

    b.    Comparing the dew point to the air temperature reveals how close the air is to saturation.

    c.    Figure 33 below relates water vapor, temperature, and relative humidity.

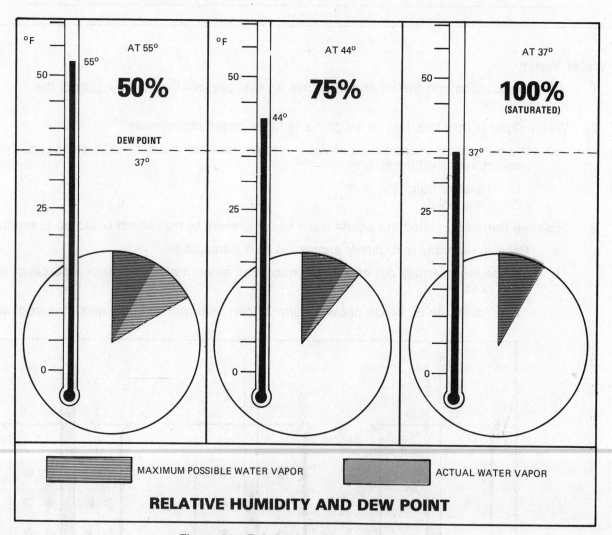

Figure 33.  Relative Humidity and Dew Point

5.    The difference between air temperature and dew point is commonly known as the **temperature-dew point spread**.

   a.    As this spread becomes smaller, relative humidity increases, and it reaches 100% when temperature and dew point are the same.

   b.    Surface temperature-dew point spread is important in anticipating fog, but has little bearing on precipitation.

   1)    To support precipitation, air must be saturated through thick layers aloft.

   c.    Sometimes the spread at ground level may be quite large, yet at higher altitudes the air is saturated and clouds form.

   1)    Some rain may reach the ground, or it may evaporate as it falls into the drier air.

   a)    Figure 34 below is a photograph of "virga" -- streamers of precipitation trailing beneath clouds but evaporating before reaching the ground.

Figure 34.  Virga

B.    **Change of State**

1.    Evaporation, condensation, freezing, melting, and sublimation are changes of state in water.

   a.    Evaporation is the changing of liquid water to invisible water vapor.

   1)    Condensation is the reverse process, i.e., water vapor into water.

   b.    Freezing is the changing of liquid water into solid water, or ice.

   1)    Melting is the reverse process, i.e., ice into water.

   c.    Sublimation is the changing of ice directly to water vapor, or water vapor to ice, bypassing the liquid state in each process.

   1)    Snow or ice crystals result from the sublimation of water vapor directly to the solid state.

2.    **Latent Heat**

   a.    The exchange of heat energy is required to change a substance from one state to another.

   1)    This energy is called "latent heat" and is significant, as we will learn in later chapters.

   2)    Figure 35 below diagrams the heat exchanges between the different states.

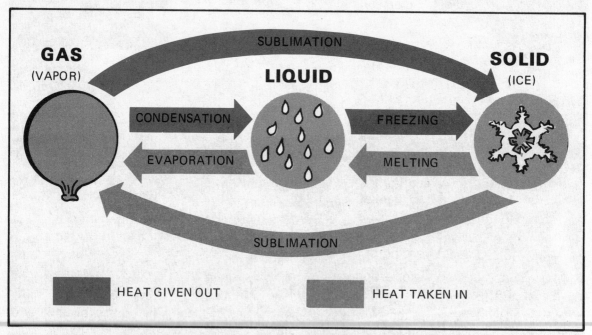

Figure 35.  Heat Transactions When Water Changes State

   b.    Evaporation of water, for example, requires heat energy that may come from the water itself, or some other source such as the surrounding air.

   1)    This energy is known as the "latent heat of vaporization," and its transference to the water vapor cools the source it comes from.

   a)    Evaporation is thus a cooling process.

   b)    When you step out of the shower, the evaporation of water from your body removes heat from your skin, making you feel cold.

2) The heat energy is stored or hidden in the water vapor, thus the term "latent (hidden) heat."

    a) It is considered hidden because the temperature of the evaporated water remains the same.

3) When the water vapor condenses back into liquid water, this energy is released to the atmosphere as the "latent heat of condensation."

    a) Thus, condensation is a warming process.

  c. Melting and freezing involve the exchange of "latent heat of fusion" in a similar manner.

    1) The latent heat of fusion is much less than that of condensation and evaporation; however, each plays an important role in aviation weather.

3. **Condensation Nuclei.** As air becomes saturated, water vapor begins to condense on the nearest available surface.

  a. The atmosphere is never completely clean; an abundance of microscopic solid particles suspended in the air are condensation surfaces.

    1) These particles, such as salt, dust, and combustion by-products are condensation nuclei.

    2) Some condensation nuclei have an affinity for water and can induce condensation or sublimation even when air is almost, but not completely, saturated.

  b. As water vapor condenses or sublimates on condensation nuclei, liquid or ice particles begin to grow.

    1) Whether the particles are liquid or ice does not depend entirely on temperature.

      a) Liquid water may be present at temperatures well below freezing.

4. **Supercooled Water**

  a. Freezing is complex, and liquid water droplets often condense or persist at temperatures colder than 0°C because the molecular motion of the droplet remains large enough to weaken any formation of an ice crystal within the droplet.

    1) Water droplets colder than 0°C are supercooled.
    2) When they strike an exposed object, the impact induces freezing.
    3) Impact freezing of supercooled water can result in aircraft icing.

  b. Supercooled water drops very often are in abundance in clouds at temperatures between 0°C and −15°C, with decreasing amounts at colder temperatures.

    1) Usually, at temperatures colder than −15°C, sublimation is prevalent, and clouds and fog may be mostly ice crystals with a lesser amount of supercooled water.

    2) Strong vertical currents may carry supercooled water to great heights where temperatures are much colder than −15°C.

      a) Supercooled water has been observed at temperatures colder than −40°C.

5.    **Dew and Frost**

    a.    During clear nights with little or no wind, vegetation often cools by radiation to a temperature at or below the dew point of the adjacent air.

        1)    Moisture then collects on the leaves just as it does on a pitcher of ice water in a warm room.

        2)    Heavy dew often collects on grass and plants while none collects on pavements or large solid objects.

            a)    These larger objects absorb abundant heat during the day, lose it slowly during the night, and cool below the dew point only in extreme cases.

    b.    Frost forms in much the same way as dew.  The difference is that the dew point of surrounding air must be colder than freezing.

        1)    Water vapor then sublimates directly as ice crystals or frost rather than condensing as dew.

        2)    Sometimes dew forms and later freezes; however, frozen dew is easily distinguished from frost.

            a)    Frozen dew is hard and transparent while frost is white and opaque.

## C.    Cloud Formation

1.    Normally, air must become saturated for condensation or sublimation to occur.

    a.    Saturation may result from cooling temperature, increasing dew point, or both.

        1)    Cooling is far more predominant.

2.    Three basic processes may cool air to saturation.

    a.    Air moving over a colder surface,
    b.    Stagnant air overlying a cooling surface,
    c.    Expansional (adiabatic) cooling in upward-moving air.

        1)    Expansional cooling is the major cause of cloud formation.
        2)    See Part I, Chapter 6, Stable and Unstable Air, beginning on page 55.

3.    A cloud is a visible collection of minute water or ice particles suspended in air.

    a.    If the cloud is near the ground, it is called fog.

    b.    When entire layers of air cool to saturation, fog or sheet-like clouds result.

    c.    Saturation of a localized updraft produces a towering cloud.

    d.    A cloud may be composed entirely of liquid water, ice crystals, or a mixture of the two.

D. **Precipitation**

  1. Precipitation is an all-inclusive term denoting drizzle, rain, snow, ice pellets, hail, and ice crystals.

      a. Precipitation occurs when these particles grow in size and weight until the atmosphere no longer can suspend them and they fall.

  2. These particles grow primarily in two ways.

      a. Once a water droplet or ice crystal forms, it continues to grow by added condensation or sublimation directly onto the particle.

          1) This is the slower of the two methods and usually results in drizzle or very light rain or snow.

      b. Cloud particles collide and merge into a larger drop in the more rapid growth process.  This process produces larger precipitation particles and does so more rapidly than the simple condensation growth process.

          1) Upward currents enhance the growth rate and also support larger drops, as shown in Figure 36 below.

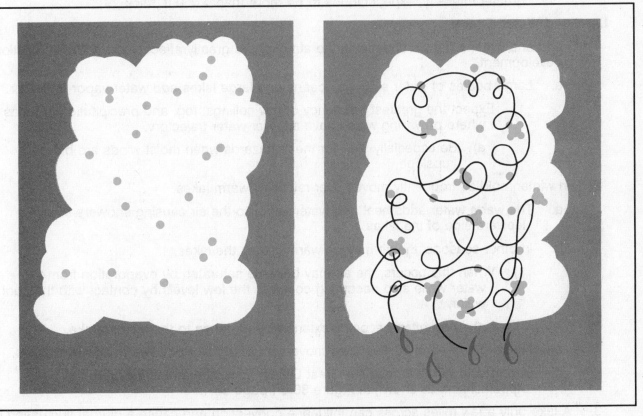

Figure 36.  Growth of Raindrops by Collision of Cloud Droplets

          2) Precipitation formed by merging drops with mild upward currents can produce light to moderate rain and snow.

          3) Strong upward currents support the largest drops and build clouds to great heights.

              a) They can produce heavy rain, heavy snow, and hail.

3.  **Liquid, Freezing, and Frozen Precipitation**

    a.  Precipitation that forms and remains in a liquid state is either rain or drizzle.

    b.  Sublimation forms snowflakes, and they reach the ground as snow if temperatures remain below freezing.

    c.  Precipitation can change its state as the temperature of its environment changes.

        1)  Falling snow may melt in warmer layers of air at lower altitudes to form rain.

        2)  Rain falling through colder air may become supercooled, freezing on impact as freezing rain; or it may freeze during its descent, falling as ice pellets.

            a)  Ice pellets always indicate freezing rain at higher altitude.

    d.  Sometimes strong upward currents sustain large supercooled water drops until some freeze; subsequently, other drops freeze to them forming hailstones.

4.  To produce significant precipitation, clouds usually must be 4,000 ft. thick or more.

    a.  The thicker the clouds, the heavier the precipitation is likely to be.

    b.  When arriving at or departing from an airport reporting precipitation of light or greater intensity, expect clouds to be more than 4,000 ft. thick.

E.  **Land and Water Effects**

1.  Land and water surfaces underlying the atmosphere greatly affect cloud and precipitation development.

    a.  Large bodies of water such as oceans and large lakes add water vapor to the air.

        1)  Expect the greatest frequency of low ceilings, fog, and precipitation in areas where prevailing winds have an over-water trajectory.

            a)  Be especially alert for these hazards when moist winds are blowing upslope.

2.  In winter, cold air frequently moves over relatively warm lakes.

    a.  The warm water adds heat and water vapor to the air causing showers to the leeward side of the lakes.

    b.  In other seasons, the air may be warmer than the lakes.

        1)  When this occurs, the air may become saturated by evaporation from the water while also becoming cooler in the low levels by contact with the cool water.

            a)  Fog often becomes extensive and dense to the lee of a lake.

    c.  Figure 37 on page 53 illustrates movement of air over both warm and cold lakes.

    d.  Strong cold winds across the Great Lakes often carry precipitation to the Appalachians as shown in Figure 38 on page 53.

3.  A lake only a few miles across can influence convection and cause a diurnal fluctuation in cloudiness.

    a.  During the day, cool air over the lake blows toward the land, and convective clouds form over the land.

    b.  At night, the pattern reverses; clouds tend to form over the lake as cool air from the land flows over the lake, creating convective clouds over the water.

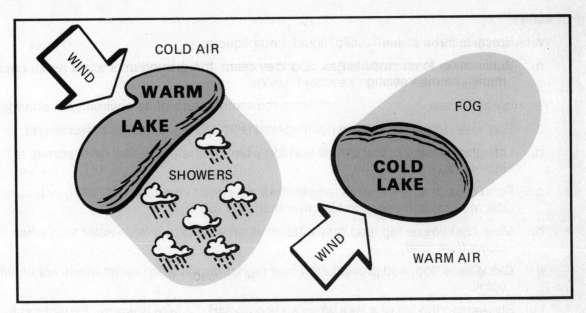

Figure 37. Lake Effects

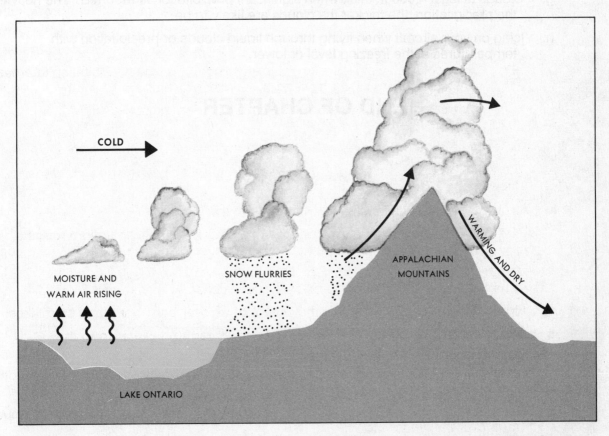

Figure 38. Strong Cold Winds across the Great Lakes Absorb Water Vapor and May Carry Showers as Far Eastward as the Appalachians

F.  **In Closing**

1.  Water exists in three states -- solid, liquid, and gaseous.

    a.  Water vapor is an invisible gas.  Condensation or sublimation of water vapor creates many common aviation weather hazards.

2.  You may anticipate

    a.  Fog when temperature-dew point spread is 3°C (5°F) or less and decreasing.

    b.  Lifting or clearing of low clouds and fog when temperature-dew point spread is increasing.

    c.  Frost on a clear night when temperature-dew point spread is 3°C (5°F) or less and is decreasing, and dew point is lower than 0°C (32°F).

    d.  More cloudiness, fog, and precipitation when wind blows from water than when it blows from land.

    e.  Cloudiness, fog, and precipitation over higher terrain when moist winds are blowing uphill.

    f.  Showers to the lee of a lake when air is cold and the lake is warm.  Expect fog to the lee of the lake when the air is warm and the lake is cold.

    g.  Clouds at least 4,000 ft. thick when significant precipitation is reported.  The heavier the precipitation, the thicker the clouds are likely to be.

    h.  Icing on your aircraft when flying through liquid clouds or precipitation with temperatures at the freezing level or lower.

# END OF CHAPTER

# CHAPTER SIX
# STABLE AND UNSTABLE AIR

Please take a few minutes to study each of the concepts listed above and anticipate/imagine what they are and how they relate to the other listed concepts.

## A. Definitions

1. A *stable* atmosphere resists any upward or downward displacement.

2. An *unstable* atmosphere allows an upward or downward disturbance to grow into a vertical, or convective, current.

## B. Adiabatic Cooling and Heating

1. Anytime air moves upward, it expands because of decreasing atmospheric pressure.

   a. Conversely, downward-moving air is compressed by increasing pressure.
   b. But as pressure and volume change, temperature also changes.

2. When air expands, it cools, and when compressed, it warms.

   a. These changes are **adiabatic**, i.e., no heat is removed from or added to the air.

   b. We frequently use the terms **expansional**, or **adiabatic**, **cooling** and **compressional**, or **adiabatic**, **heating**.

   c. The adiabatic rate of change of temperature is constant in unsaturated air but varies with the amount of moisture in saturated air.

3. **Unsaturated air** moving upward and downward cools and warms at about 3°C per 1,000 ft. (5.4°F per 1,000 ft.).

   a. This rate is the dry adiabatic rate of temperature change and is independent of the temperature of the mass of air through which the vertical movements occur.

   b. Figure 41, on page 56, illustrates a "chinook wind," which is a warm wind blowing down the side of a mountain -- an excellent example of dry adiabatic warming.

      1) As the cold, relatively dry air moves, it sinks down the side of the mountain.

      2) As it is compressed by the increasing atmospheric pressure, it becomes warmer through adiabatic heating.

4. When **saturated air** moves upward and cools, condensation occurs.

   a. Latent heat released through condensation (see Part I, Chapter 5, Moisture, Cloud Formation, and Precipitation, beginning on page 45) partially offsets the expansional cooling.

      1) Thus, the saturated adiabatic rate of cooling is slower than the dry rate.

      2) The saturated cooling rate depends on saturation temperature (dew point) of the air, i.e., the relative humidity.

   b. Because warmer air can hold more water before becoming saturated, more latent heat is released through condensation in saturated warm air than in cold.

      1) Thus, the saturated adiabatic rate of cooling is less in warm air than in cold.

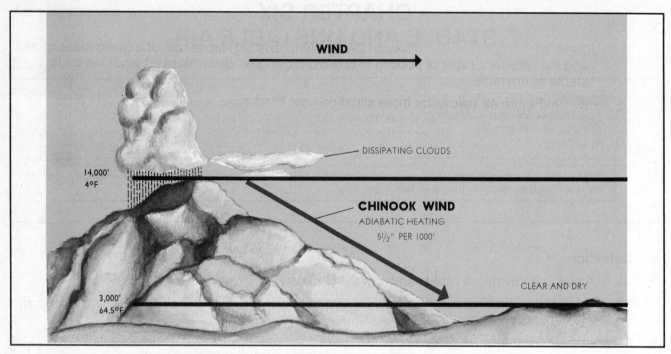

WIND

DISSIPATING CLOUDS

14,000'
4°F

CHINOOK WIND
ADIABATIC HEATING
5¹/₂° PER 1000'

CLEAR AND DRY

3,000'
64.5°F

Figure 41.  Adiabatic Warming of Downward Moving Air

c.    When saturated air moves downward, it heats at the same rate as it cools on ascent *provided* liquid water evaporates rapidly enough to maintain saturation.

    1)    Minute water droplets in the air evaporate rapidly enough.

    2)    Larger drops evaporate more slowly and complicate the moist adiabatic process in downward moving air.

5.    **Adiabatic Cooling and Vertical Air Movement**

    a.    At this point we should clarify the terms "ambient, or existing, lapse rate" and "adiabatic rates of cooling."

        1)    As you move upward through the atmosphere, the temperature of the air around you generally changes with altitude.

            a)    This change is the ambient (existing) lapse rate.

            b)    In the standard atmosphere, the lapse rate is 2°C per 1,000 ft. (see Part I, Chapter 2, Temperature, beginning on page 7).

        2)    As a parcel of air is forced upward, it expands, and therefore cools, at a given rate.

            a)    This is the adiabatic rate of cooling, or adiabatic lapse rate.

            b)    This rate depends on the amount of moisture present in the parcel of air that is forced upward.

    b.    If a parcel of air is forced upward into the atmosphere, we must consider two possibilities:

        1)    As the sample cools, it may become colder than the surrounding air, or
        2)    Even though it cools, the air may remain warmer than the surrounding air.

    c.    If the upward moving air becomes colder than the surrounding air, the adiabatic lapse rate is greater than the ambient (existing) lapse rate, and the parcel of air sinks.

        1)    If it remains warmer, the adiabatic lapse rate is less than the ambient (existing) lapse rate, and the parcel of air continues to rise as a convective current.

## C. **Stability and Instability**

1.  EXAMPLE:  The difference between the ambient (existing) lapse rate of a given mass of air and the adiabatic rates of cooling in upward-moving air determines whether the air is stable or unstable.

    a.  In Figure 42 below, for three situations, we filled a balloon at sea level with air at 31°C -- the same as the ambient temperature.  We carried the balloon to 5,000 ft.  In each situation, the air in the balloon expanded and cooled at the dry adiabatic rate of 3°C for each 1,000 ft. to a temperature of 16°C at 5,000 ft.

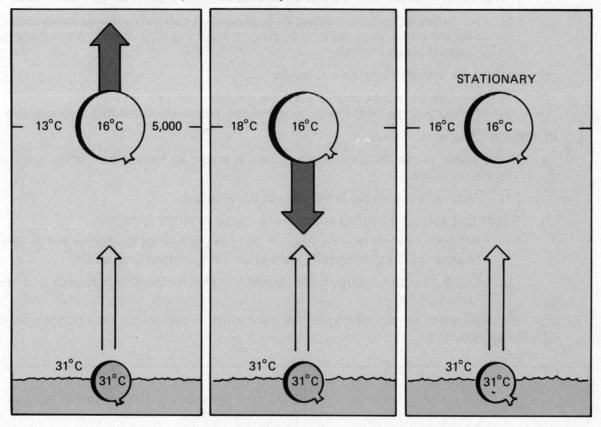

Figure 42.  Stability Related to Temperatures Aloft and Adiabatic Cooling

1)  In the first situation (left), air inside the balloon, even though cooled adiabatically, remained warmer than surrounding air.

    a)  Vertical motion was favored.
    b)  The colder, more dense surrounding air forced the balloon further upward.
    c)  The air was unstable, and a convective current developed.

2)  In the second situation (center), the air aloft was warmer.

    a)  Air inside the balloon, cooled adiabatically, became colder than the surrounding air.

    b)  The balloon sank under its own weight, returning to its original position when the lifting force was removed.

    c)  The air was stable, and spontaneous convection was impossible.

3)  In the third situation (right), the temperature of the air inside the balloon was the same as that of the surrounding air.  The balloon remained at rest.

    a)  This condition was neutrally stable; that is, the air was neither stable nor unstable.

4)   Note that, in all three situations, temperature of air in the expanding balloon cooled at a fixed rate.

a)   The differences in the three conditions depend, therefore, on the temperature differences between the surface and 5,000 ft., that is, on the ambient (existing) lapse rates.

2.   Stability runs the gamut from absolutely stable to absolutely unstable, and the atmosphere usually is in a delicate balance somewhere in between.

a.   A change in ambient temperature lapse rate of an air mass can tip this balance.

1)   For example, surface heating or cooling aloft can make the air more unstable; on the other hand, surface cooling or warming aloft often tips the balance toward greater stability.

b.   Air may be stable or unstable in layers.

1)   A stable layer may overlie and cap unstable air; or
2)   Conversely, air near the surface may be stable with unstable layers above.

3.   **Stratiform Clouds**

a.   Since stable air resists convection, clouds in stable air form in horizontal, sheet-like layers or "strata."

1)   Thus, within a stable layer, clouds are *stratiform*.

b.   Recall that adiabatic cooling is the major cause of cloud formation.

1)   Adiabatic cooling may be caused by upslope flow as illustrated in Figure 43 below, by lifting over cold, denser air, or by converging winds.

2)   Cooling by an underlying cold surface is also a stabilizing process and may produce fog.

c.   If clouds are to remain stratiform, the layer must remain stable after condensation has occurred.

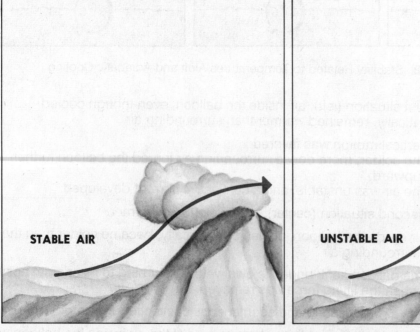
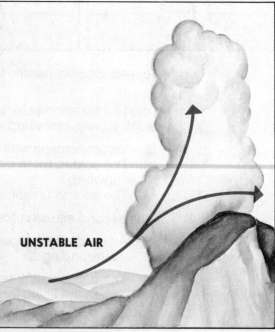

STABLE AIR                   UNSTABLE AIR

Figure 43.  Stable and Unstable Air Forced Upward

4.  **Cumuliform Clouds**

    a.  Unstable air favors convection.  A cumulus (heaping) cloud forms in a convective updraft and builds upward, also shown in Figure 43 on the previous page.

        1)  Thus, within an unstable layer, clouds are **cumuliform**, and the vertical extent of the cloud depends on the depth of the unstable layer.

    b.  Initial lifting to trigger a cumuliform cloud may be the same as that for lifting stable air.

        1)  In addition, convection may be set off by surface heating (see Part I, Chapter 4, Wind, beginning on page 29).

    c.  Air may be unstable or slightly stable before condensation occurs, but for convective cumuliform clouds to develop, it must be unstable after saturation.

        1)  Cooling in the updraft is now at the slower moist adiabatic rate because of the release of latent heat of condensation.

            a)  Temperature in the saturated updraft is warmer than ambient temperature, and convection is spontaneous.

        2)  Updrafts accelerate until temperature within the cloud cools below the ambient temperature.

            a)  This condition occurs where the unstable layer is capped by a stable layer often marked by a temperature inversion.

        3)  Vertical heights range from the shallow fair weather cumulus to the giant thunderstorm cumulonimbus -- the ultimate in atmospheric instability capped by the tropopause.

    d.  You can estimate the heights of cumuliform cloud bases using surface temperature-dew point spread.

        1)  Unsaturated air in a convective current cools at about 3°C (5.4°F) per 1,000 ft.; dew point decreases at about 5/9°C (1°F).

            a)  Thus, in a convective current, temperature and dew point converge at about 2.5°C (4.4°F) per 1,000 ft. as illustrated in Figure 44 on page 60.

            b)  The point at which they converge is the base of the clouds.

        2)  We can get a quick estimate of a convective cloud base in thousands of feet by rounding these values and dividing into the spread.

            a)  When using Celsius, divide by 2; when using Fahrenheit, divide by 4.

            b)  This method of estimating is reliable only with instability clouds and during the warmer part of the day.

            c)  EXAMPLE:  If the surface temperature is 30°C and the dew point is 10°C, the spread is 20°C.  Dividing by 2 yields 10.  Thus, the convective cloud base is approximately 10,000 ft.

    e.  When unstable air lies above stable air, convective currents aloft sometimes form middle and high level cumuliform clouds.

        1)  In relatively shallow layers, they occur as altocumulus and ice crystal cirrocumulus clouds.

        2)  Altocumulus castellanus clouds develop in deeper midlevel unstable layers.

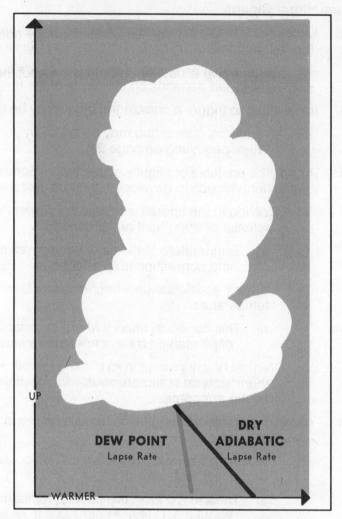

UP

DEW POINT
Lapse Rate

DRY
ADIABATIC
Lapse Rate

WARMER

Figure 44.  Cloud Base Determination

5. **Merging Stratiform and Cumuliform**

   a. A layer of stratiform clouds may sometimes form in a mildly stable layer while a few ambitious convective clouds penetrate the layer, thus merging stratiform with cumuliform.

   b. Convective clouds may be almost or entirely embedded in a massive stratiform layer and pose an unseen threat to instrument flight.

## D. **In Closing**

1. The usual convection in unstable air gives a bumpy ride; only at times is it violent enough to be hazardous.

   a. In stable air, flying is usually smooth but sometimes can be plagued by low ceiling and visibility.

   b. In preflight planning, you need to take into account stability or instability and any associated hazards.

2. Certain observations you can make on your own:

   a. Thunderstorms are sure signs of violently unstable air. Give these storms a wide berth.

   b. Showers and clouds towering upward indicate strong updrafts and rough (turbulent) air. Stay clear of these clouds.

   c. Fair weather cumulus clouds often indicate bumpy turbulence beneath and in the clouds. The cloud tops indicate the approximate upper limit of convection; flight above is usually smooth.

   d. Dust devils are a sign of dry, unstable air, usually to considerable height. Your ride may be fairly rough unless you can get above the instability.

   e. Stratiform clouds indicate stable air. Flight generally will be smooth, but low ceiling and visibility might require IFR.

   f. Restricted visibility at or near the surface over large areas usually indicates stable air. Expect a smooth ride, but poor visibility may require IFR.

   g. Thunderstorms may be embedded in stratiform clouds posing an unseen threat to instrument flight.

   h. Even in clear weather, you have some clues to stability, viz.:

      1) When temperature decreases uniformly and rapidly as you climb (approaching 3°C per 1,000 ft.), you have an indication of unstable air.

      2) If temperature remains unchanged or decreases only slightly with altitude, the air tends to be stable.

      3) If the temperature increases with altitude through a layer -- an inversion -- the layer is stable and convection is suppressed. Air may be unstable beneath the inversion.

      4) When air near the surface is warm and moist, suspect instability. Surface heating, cooling aloft, converging or upslope winds, or an invading mass of colder air may lead to instability and cumuliform clouds.

# END OF CHAPTER

# CHAPTER SEVEN
# CLOUDS

A. **Introduction** -- To you as a pilot, clouds are your weather "signposts in the sky."

   1.   They give you an indication of air motion, stability, and moisture.

   2.   Clouds help you visualize weather conditions and potential weather hazards you might encounter in flight.

   3.   The photographs on pages 64 through 71 illustrate some of the basic cloud types discussed below.

      a.   The caption with each photograph describes the cloud type and its significance to flight.

      b.   Study the descriptions and potential hazards posed by each type.

B. **Classification** -- Clouds are classified according to the way they are formed.

   1.   Clouds formed by vertical currents in unstable air are *cumulus*, meaning *accumulation* or *heap*.

      a.   They are characterized by their lumpy, billowy appearance.

   2.   Clouds formed by the cooling of a stable layer are *stratus*, meaning *stratified* or *layered*.

      a.   They are characterized by their uniform, sheet-like appearance.

   3.   In addition to the above, the prefix *nimbo* or the suffix *nimbus* means raincloud.

      a.   Thus, stratified clouds from which rain is falling are *nimbostratus*.

      b.   A heavy, swelling cumulus type cloud which produces precipitation is a *cumulonimbus*.

   4.   Clouds broken into fragments are often identified by adding the suffix *fractus*.

      a.   EXAMPLE:  Fragmentary cumulus is *cumulus fractus*.

C. **Identification** -- For identification purposes, you must also be concerned with the four families of clouds.

   1.   The **high cloud** family is cirriform and includes cirrus, cirrocumulus, and cirrostratus.

      a.   They are composed almost entirely of ice crystals.

      b.   The height of the bases of these clouds ranges from about 16,500 to 45,000 ft. in middle latitudes.

      c.   Figures 45 through 47 on pages 64 and 65 are photographs of high clouds.

2. In the **middle cloud** family are the altostratus, altocumulus, and nimbostratus clouds.

   a. These clouds are primarily water, much of which may be supercooled.

   b. The height of the bases of these clouds ranges from about 6,500 to 23,000 ft. in middle latitudes.

   c. Figures 48 through 52 on pages 66 through 69 are photographs of middle clouds.

3. In the **low cloud** family are the stratus, stratocumulus, and fair weather cumulus clouds.

   a. Low clouds are almost entirely water, but at times the water may be supercooled.

      1) Low clouds at subfreezing temperatures can also contain snow and ice particles.

   b. The bases of these clouds range from near the surface to about 6,500 ft. in middle latitudes.

   c. Figures 53 through 55 on pages 69 and 70 are photographs of low clouds.

4. The **clouds with extensive vertical development** family includes towering cumulus and cumulonimbus.

   a. These clouds usually contain supercooled water above the freezing level.

      1) But when a cumulus grows to great heights, water in the upper part of the cloud freezes into ice crystals, forming a cumulonimbus.

   b. The heights of cumuliform cloud bases range from 1,000 ft. or less to above 10,000 ft.

   c. Figures 56 and 57 on page 71 are photographs of clouds with extensive vertical development.

FIGURE 45.   CIRRUS. Cirrus are thin, feather-like ice crystal clouds in patches or narrow bands. Larger ice crystals often trail downward in well-defined wisps called "mares' tails." Wispy, cirrus-like, these contain no significant icing or turbulence. Dense, banded cirrus, which often are turbulent, are discussed in chapter 13.

FIGURE 46.   CIRROCUMULUS. Cirrocumulus are thin clouds, the individual elements appearing as small white flakes or patches of cotton. May contain highly supercooled water droplets. Some turbulence and icing.

FIGURE 47.   CIRROSTRATUS. Cirrostratus is a thin whitish cloud layer appearing like a sheet or veil. Cloud elements are diffuse, sometimes partially striated or fibrous. Due to their ice crystal makeup, these clouds are associated with halos— large luminous circles surrounding the sun or moon. No turbulence and little if any icing. The greatest problem flying in cirriform clouds is restriction to visibility. They can make the strict use of instruments mandatory.

FIGURE 48.   ALTOCUMULUS. Altocumulus are composed of white or gray colored layers or patches of solid cloud. The cloud elements may have a waved or roll-like appearance. Some turbulence and small amounts of icing.

FIGURE 49.   ALTOSTRATUS. Altostratus is a bluish veil or layer of clouds. It is often associated with altocumulus and sometimes gradually merges into cirrostratus. The sun may be dimly visible through it. Little or no turbulence with moderate amounts of ice.

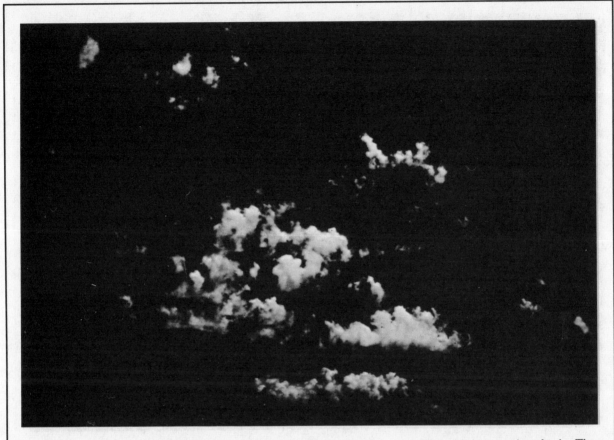

FIGURE 50. ALTOCUMULUS CASTELLANUS. Altocumulus castellanus are middle level convective clouds. They are characterized by their billowing tops and comparatively high bases. They are a good indication of mid-level instability. Rough turbulence with some icing.

FIGURE 51.     STANDING LENTICULAR ALTOCUMULUS CLOUDS. Standing lenticular altocumulus clouds are formed on the crests of waves created by barriers in the wind flow. The clouds show little movement, hence the name *standing*. Wind, however, can be quite strong blowing through such clouds. They are characterized by their smooth, polished edges. The presence of these clouds is a good indication of very strong turbulence and should be avoided. Chapter 9, "Turbulence," further explains the significance of this cloud.

FIGURE 52. NIMBOSTRATUS. Nimbostratus is a gray or dark massive cloud layer, diffused by more or less continuous rain, snow, or ice pellets. This type is classified as a middle cloud although it may merge into very low stratus or strato-cumulus. Very little turbulence, but can pose a serious icing problem if temperatures are near or below freezing.

FIGURE 53. STRATUS. Stratus is a gray, uniform, sheet-like cloud with relatively low bases. When associated with fog or precipitation, the combination can become troublesome for visual flying. Little or no turbulence, but temperatures near or below freezing can create hazardous icing conditions.

FIGURE 54.  STRATOCUMULUS. Stratocumulus bases are globular masses or rolls unlike the flat, sometimes indefinite, bases of stratus. They usually form at the top of a layer mixed by moderate surface winds. Sometimes, they form from the breaking up of stratus or the spreading out of cumulus. Some turbulence, and possible icing at subfreezing temperatures. Ceiling and visibility usually better than with low stratus.

FIGURE 55.  CUMULUS. Fair weather cumulus clouds form in convective currents and are characterized by relatively flat bases and dome-shaped tops. Fair weather cumulus do not show extensive vertical development and do not produce precipitation. More often, fair weather cumulus indicates a shallow layer of instability. Some turbulence and no significant icing.

FIGURE 56.   TOWERING CUMULUS. Towering cumulus signifies a relatively deep layer of unstable air. It shows considerable vertical development and has billowing *cauliflower* tops. Showers can result from these clouds. Very strong turbulence; some clear icing above the freezing level.

FIGURE 57.   CUMULONIMBUS. Cumulonimbus are the ultimate manifestation of instability. They are vertically developed clouds of large dimensions with dense *boiling* tops often crowned with thick veils of dense cirrus (the anvil). Nearly the entire spectrum of flying hazards are contained in these clouds including violent turbulence. They should be avoided at all times! This cloud is the thunderstorm cloud and is discussed in detail in chapter 11, "Thunderstorms."

D.  **In Closing**

1.  Clouds give an indication of weather conditions and weather hazards you may encounter. Remember there are four families of clouds:

    a.  The high cloud family, including cirrus, cirrocumulus, and cirrostratus

    b.  The middle cloud family, including altostratus, altocumulus, and nimbostratus

    c.  The low cloud family, including stratus, stratocumulus, and fair weather cumulus clouds

    d.  The clouds with extensive vertical development family, including towering cumulus and cumulonimbus

2.  Study the photographs and descriptions on the preceding pages so you will be familiar with each cloud type and its potential hazards.

# END OF CHAPTER

# CHAPTER EIGHT
# AIR MASSES AND FRONTS

## A.  Air Masses

1.   When a body of air comes to rest or moves slowly over an extensive area that has fairly uniform properties of temperature and moisture, the body of air takes on those properties.

   a.   Thus, the air over the area becomes a kind of entity as illustrated in Figure 58 below and has a fairly uniform horizontal distribution of its properties.

   b.   The area over which the air mass acquires its identifying distribution of moisture and temperature is known as its **source region**.

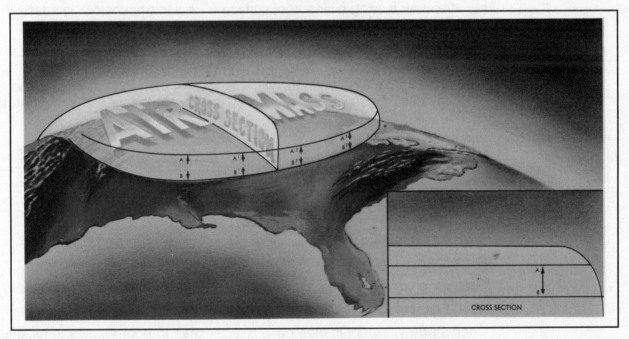

Figure 58.  Horizontal Uniformity of an Air Mass

2.    Source regions are many and varied, but the best source regions for air masses are large snow or ice-covered polar regions, cold northern oceans, tropical oceans, and large desert areas.

    a.    Mid-latitudes are poor source regions because air movement is so constantly varied that air masses have little opportunity to stagnate and take on the properties of the underlying region.

3.    Just as an air mass took on the properties of its source region, it will tend to take on properties of a new underlying surface when it moves away from its source region, thus becoming modified.

    a.    The degree of modification depends on the speed with which the air mass moves, the nature of the region over which it moves, and the temperature difference between the new surface and the air mass.

    b.    Some ways air masses are modified are

        1)    Warming from below.

            a)    Cool air moving over a warm surface is heated from below, generating instability and increasing the possibility of showers.

        2)    Cooling from below.

            a)    Warm air moving over a cool surface is cooled from below, increasing stability.  If air is cooled to its dew point, stratus and/or fog forms.

        3)    Addition of water vapor.

            a)    Evaporation from water surfaces and falling precipitation adds water vapor to the air.  When the water is warmer than the air, evaporation can raise the dew point sufficiently to saturate the air and form stratus or fog.

        4)    Subtraction of water vapor.

            a)    Water vapor is removed by condensation and precipitation.

4.    The **stability** of an air mass determines its typical weather characteristics.

    a.    Characteristics typical of an unstable and a stable air mass are as follows:

| Unstable Air | Stable Air |
| --- | --- |
| Cumuliform clouds | Stratiform clouds and fog |
| Showery precipitation | Continuous precipitation |
| Rough air (turbulence) | Smooth air |
| Good visibility, except in blowing obstructions | Fair to poor visibility in haze and smoke |

B.  **Fronts**

1.  As air masses move out of their source regions, they come in contact with other air masses of different properties.  The zone between two different air masses is a *frontal zone*, or *front*.

    a.  Across this zone, temperature, humidity, and wind often change rapidly over short distances.

2.  **Discontinuities**.  When you pass through a front, the change from the properties of one air mass to those of the other is sometimes quite abrupt.  Abrupt changes indicate a narrow frontal zone.  At other times, the change of properties is very gradual, indicating a broad and diffuse frontal zone.

    a.  **Temperature** is one of the most easily recognized discontinuities across a front.

        1)  At the surface, the passage of a front usually causes a noticeable temperature change.

        2)  When flying through a front, you note a significant change in temperature, especially at low altitudes.

        3)  Remember that the temperature change, even when gradual, is faster and more pronounced than a change during a flight wholly within one air mass.

            a)  Thus, for safety, obtain a new altimeter setting after flying through a front.

    b.  **Dew point** and temperature-dew point spread usually differ across a front.

        1)  This difference helps identify the front and may give a clue to changes in cloudiness and/or fog.

    c.  **Wind** always changes across a front.

        1)  Wind discontinuity may be in direction, in speed, or in both.
        2)  Be alert for a wind shift when flying in the vicinity of a frontal surface.
        3)  The relatively sudden change in wind also creates wind shear.

    d.  **Pressure**.  A front lies in a pressure trough, and pressure generally is higher in the cold air.

        1)  Thus, when you cross a front directly into colder air, pressure usually rises abruptly.

        2)  When you approach a front toward warm air, pressure generally falls until you cross the front and then remains steady or falls slightly in the warm air.

        3)  However, pressure patterns vary widely across fronts, and your course may not be directly across a front.

        4)  The important thing to remember is that when crossing a front, you will encounter a difference in the rate of pressure change; be especially alert in keeping your altimeter setting current.

3.   The three principal types of fronts are the **cold front**, the **warm front**, and the **stationary front**.

   a.   The leading edge of an advancing cold air mass is a **cold front**.

   1)   At the surface, cold air is overtaking and replacing warmer air.

   2)   Cold fronts move at about the speed of the wind component perpendicular to the front just above the frictional layer.

   3)   Figure 59 on page 77 shows the vertical cross section of a cold front and the symbol depicting it on a surface weather chart.

   a)   The vertical cross section in the top illustration shows the frontal slope.

   i)   The frontal slope is steep near the leading edge as cold air replaces warm air.

   ii)   The solid heavy arrow shows movement of the front.

   iii)   Warm air may descend over the front as indicated by the dashed arrows, but more commonly, the cold air forces warm air upward over the frontal surface as shown by the solid arrows.

   b)   The symbol in the bottom illustration is a line with barbs pointing in the direction of movement.

   i)   If a map is in color, a blue line represents the cold front.

   4)   A shallow cold air mass or a slow moving cold front may have a frontal slope more like a warm front shown in Figure 60 on page 79.

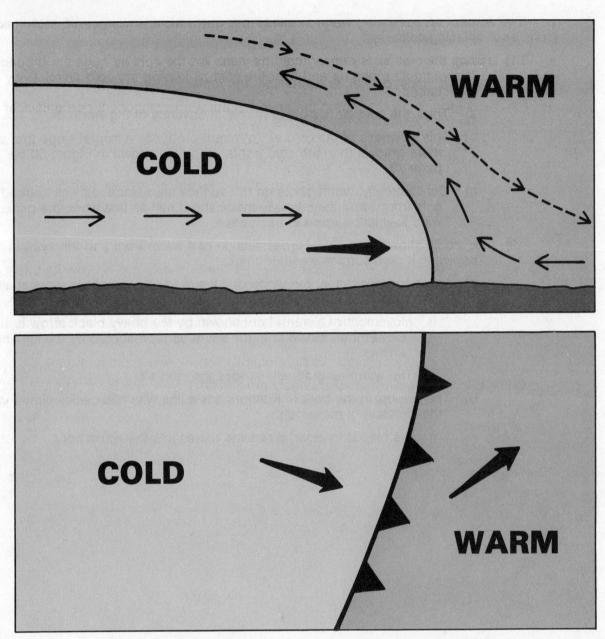

Figure 59.  Cross Section and Weather Map Symbol of a Cold Front

b.   The edge of an advancing warm air mass is a **warm front** -- warmer air is overtaking and replacing colder air.

1)   Since the cold air is denser than the warm air, the cold air hugs the ground. The warm air slides up and over the cold air leaving the cold air relatively undisturbed.

   a)   Thus, the cold air is slow to retreat in advance of the warm air.

   b)   This slowness of the cold air to retreat produces a frontal slope that is more gradual than the cold frontal slope, as shown in Figure 60 on page 79.

   c)   Consequently, warm fronts on the surface are seldom as well marked as cold fronts, and they usually move about half as fast when the general wind flow is the same in each case.

2)   Figure 60 shows the vertical cross section of a warm front and the symbol depicting it on a surface weather chart.

   a)   In the top illustration, the slope of a warm front generally is more shallow than slope of a cold front.

      i)   Movement of a warm front shown by the heavy black arrow is slower than the wind in the warm air represented by the light solid arrows.

      ii)   The warm air gradually erodes the cold air.

   b)   The symbol in the bottom illustration is a line with half circles aimed in the direction of movement.

      i)   If a map is in color, a red line represents the warm front.

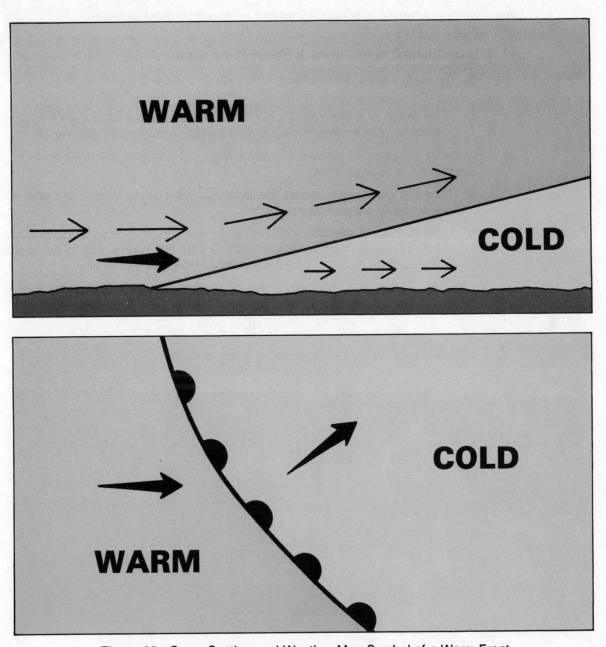

Figure 60.  Cross Section and Weather Map Symbol of a Warm Front

   c.   When neither air mass is replacing the other, the front is **stationary**.

      1)   The opposing forces exerted by adjacent air masses of different densities are such that the frontal surface between them shows little or no movement.

         a)   In such cases, the surface winds tend to blow parallel to the frontal zone.

         b)   The slope of a stationary front is normally shallow, although it may be steeper, depending on wind distribution and density difference.

      2)   Figure 61 on page 81 shows a cross section of a stationary front and its symbol on a surface chart.

         a)   The top illustration shows the front has little or no movement and winds are nearly parallel to the front.

         b)   The symbol in the bottom illustration is a line with alternating barbs and half circles on opposite sides of the line, the barbs aiming away from the cold air and the half circles away from the warm air.

            i)   If a map is in color, a line of alternating red and blue segments represents the stationary front.

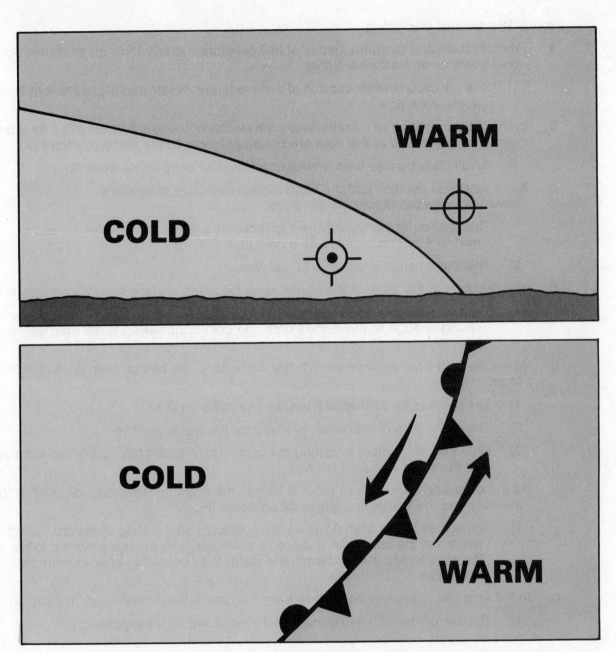

Figure 61.  Cross Section and Weather Map Symbol of a Stationary Front

4.    **Frontal Waves and Occlusion**

    a.    Frontal waves and cyclones (areas of low pressure) usually form on slow-moving cold fronts or on stationary fronts.

        1)    The life cycle and movement of a cyclone are dictated to a great extent by the upper wind flow.

    b.    In the initial condition of a frontal wave development in Figure 62 on page 83, the winds on both sides of the front are blowing parallel to the stationary front (A).

        1)    Small disturbances then may start a wavelike bend in the front (B).

    c.    If this tendency persists and the wave increases in size, a cyclonic (counterclockwise) circulation develops.

        1)    One section of the front begins to move as a warm front, while the section next to it begins to move as a cold front (C).

        2)    This deformation is called a *frontal wave*.

    d.    The pressure at the peak of the frontal wave falls, and a low-pressure center forms.

        1)    The cyclonic circulation becomes stronger, and the surface winds are now strong enough to move the fronts; the cold front moves faster than the warm front (D).

    e.    When the cold front catches up with the warm front, the two of them *occlude* (close together).

        1)    The result is an **occluded front**, or an *occlusion* (E).

        2)    This is the time of maximum intensity for the wave cyclone.

        3)    Note that the symbol depicting the occlusion is a combination of the symbols for the warm and cold fronts.

    f.    As the occlusion continues to grow in length, the cyclonic circulation diminishes in intensity and the frontal movement slows down (F).

        1)    Sometimes a new frontal wave begins to form on the long westward-trailing portion of the cold front (F,G), or a secondary low pressure system forms at the apex where the cold front and warm front come together to form the occlusion.

    g.    In the final stage, the two fronts may have become a single stationary front again.

        1)    The low center with its remnant of the occlusion is disappearing (G).

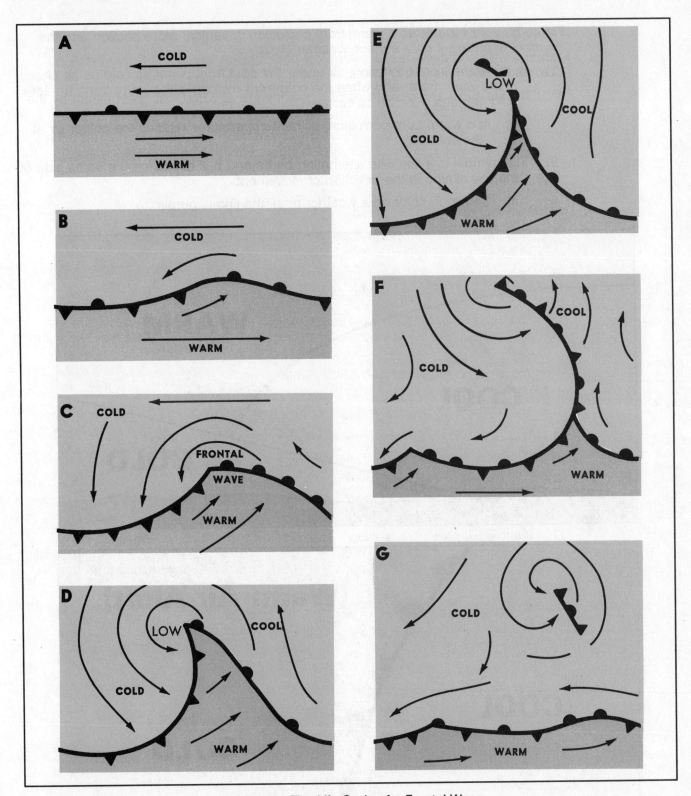

Figure 62.  The Life Cycle of a Frontal Wave

h.   Figure 63 below indicates a warm-front occlusion in vertical cross section and the symbol depicting it on a surface weather chart.

    1)   In the warm front occlusion, air under the cold front is not as cold as air ahead of the warm front; and when the cold front overtakes the warm front, the less cold air rides over the colder air.

        a)   In a warm front occlusion, milder temperatures replace the colder air at the surface.

    2)   The symbol is a line with alternating barbs and half circles on the same side of the line aiming in the direction of movement.

        a)   Shown in color on a weather map, the line is purple.

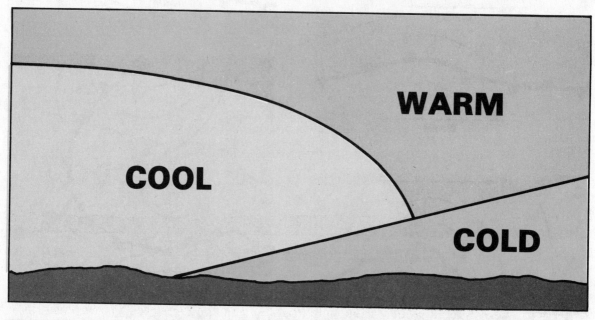

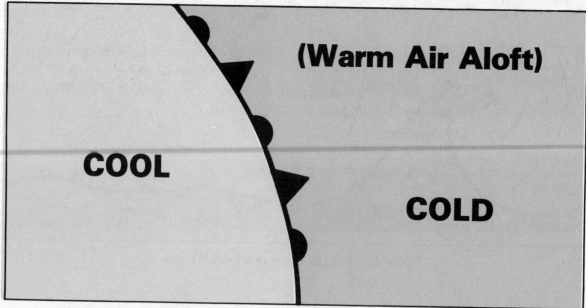

Figure 63.  Cross Section and Weather Map Symbol for a Warm-Front Occlusion

i.    Figure 64 below indicates a cold-front occlusion in vertical cross section.

1)    In the cold-front occlusion, the coldest air is under the cold front.

2)    When it overtakes the warm front, it lifts the warm front aloft; and cold air replaces cool air at the surface.

3)    The weather map symbol is the same as used for a warm-front occlusion.

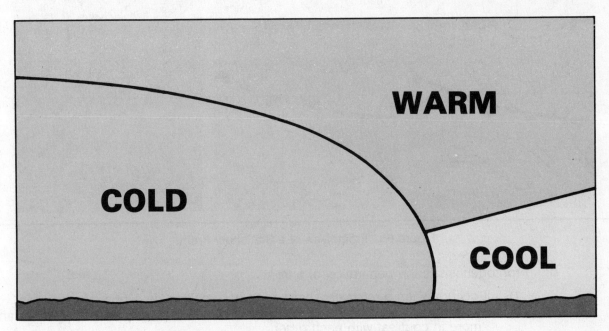

Figure 64.  Cross Section of a Cold-Front Occlusion

## 5.    Nonfrontal Lows

a.    As we have learned, fronts lie in troughs, or elongated areas of low pressure.

1)    These frontal troughs mark the boundaries between air masses of different properties.

2)    Low-pressure areas lying solely in a homogeneous air mass are called nonfrontal lows.

b.    Nonfrontal lows are infrequent east of the Rocky Mountains in mid-latitudes but do occur occasionally during the warmer months.

1)    Small nonfrontal lows over the western mountains are common as is the semistationary thermal low in the extreme southwestern United States.

c.    Tropical lows are also nonfrontal.

6.    **Frontolysis and Frontogenesis**

a.    As adjacent air masses modify and as temperature and pressure differences equalize across a front, the front dissipates.

1)    This process, frontolysis, is illustrated in Figure 65 below.

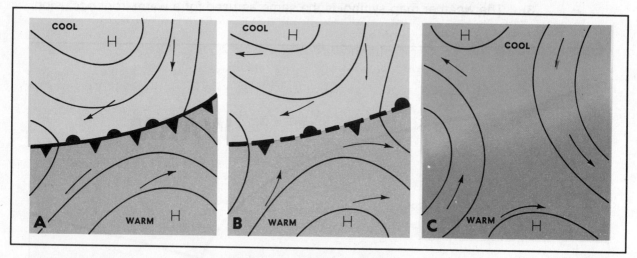

Figure 65.  Frontolysis of a Stationary Front

b.    Frontogenesis is the generation of a front.

1)    It occurs when a relatively sharp zone of transition develops over an area between two air masses which have densities gradually becoming more and more in contrast with each other.

a)    The necessary wind flow pattern develops at the same time.

2)    Figure 66 below shows an example of frontogenesis with the symbol.

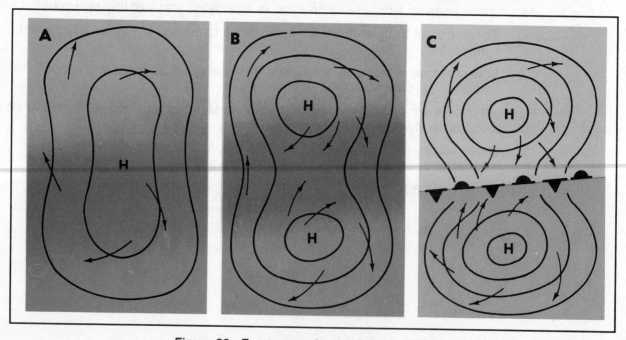

Figure 66.  Frontogenesis of a Stationary Front

### 7. Frontal Weather

    a.    In fronts, flying weather varies from virtually clear skies to extreme hazards including hail, turbulence, icing, low clouds, and poor visibility.

        1)   Weather occurring with a front depends on

            a)   The amount of moisture available,
            b)   The degree of stability of the air that is forced upward,
            c)   The slope of the front,
            d)   The speed of frontal movement, and
            e)   The upper wind flow.

    b.    Sufficient moisture must be available for clouds to form, or there will be no clouds.

        1)   As an inactive front (i.e., no precipitation) comes into an area of moisture, clouds and precipitation may develop rapidly.

        2)   A good example of this is a cold front moving eastward from the dry slopes of the Rocky Mountains into a tongue of moist air from the Gulf of Mexico over the Plains States.

            a)   Thunderstorms may build rapidly and catch a pilot unaware.

    c.    The degree of stability of the lifted air determines whether cloudiness will be predominately stratiform or cumuliform.

        1)   If the warm air overriding the front is stable, stratiform clouds develop.

            a)   Precipitation from stratiform clouds is usually steady, as illustrated in Figure 67 below, and there is little or no turbulence.

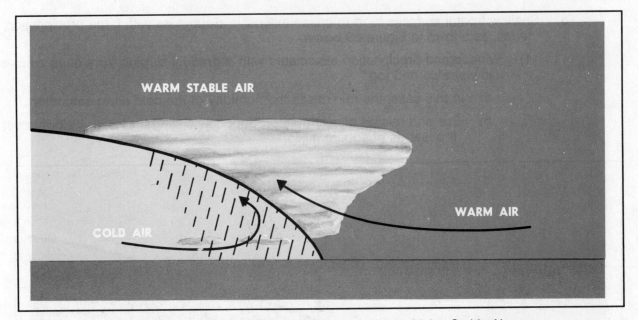

Figure 67. A Cold Front Underrunning Warm, Moist, Stable Air

2)    If the warm air is unstable, cumuliform clouds develop.

    a)    Precipitation from cumuliform clouds is showery, as in Figure 68 below, and the clouds are turbulent.

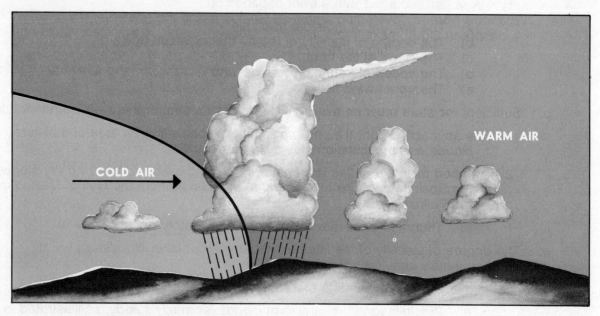

Figure 68.  A Cold Front Underrunning Warm, Moist, Unstable Air

d.    Shallow frontal surfaces tend to cause extensive cloudiness with large precipitation areas, as shown in Figure 69 below.

1)    Widespread precipitation associated with a gradual sloping front often causes low stratus and fog.

    a)    In this case, the rain raises the humidity of the cold air to saturation.

    b)    This and related effects may produce low ceiling and poor visibility over thousands of square miles.

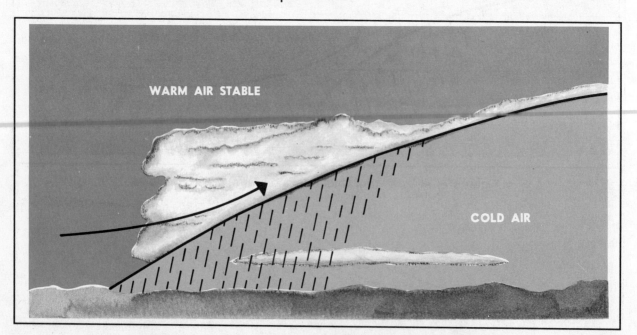

Figure 69.  A Warm Front with Overrunning Moist, Stable Air

2)   If temperature of the cold air near the surface is below freezing but the warmer air aloft is above freezing, precipitation falls as freezing rain or ice pellets.

   a)   However, if temperature of the warmer air aloft is well below freezing, precipitation forms as snow.

e.   When the warm air overriding a shallow front is moist and unstable, the usual widespread cloud mass forms; but embedded in the cloud mass are altocumulus, cumulus, and even thunderstorms as in Figures 70 and 71 below.

   1)   These embedded storms are more common with warm and stationary fronts but may occur with a slow moving, shallow cold front.

      a)   A good preflight briefing helps you to foresee the presence of these hidden thunderstorms.

      b)   Radar also helps in this situation and is discussed in Part I, Chapter 11, Thunderstorms, beginning on page 115.

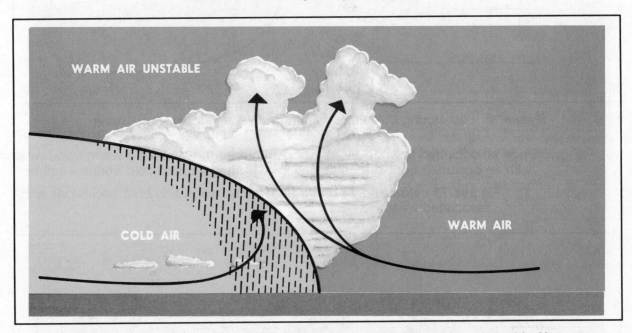

Figure 70.  A Slow-Moving Cold Front Underrunning Warm, Moist, Unstable Air

Figure 71.  Warm Front with Overrunning Warm, Moist, Unstable Air

f.    A fast moving, steep cold front forces upward motion of the warm air along its leading edge.

  1)    If the warm air is moist, precipitation occurs immediately along the surface position of the front as shown in Figure 72 below.

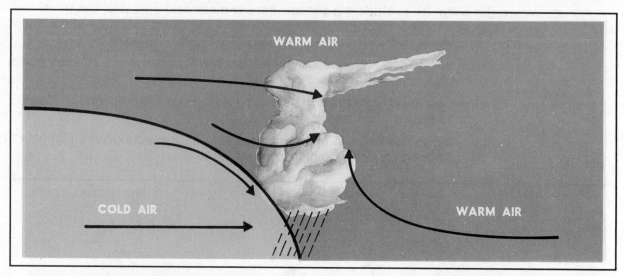

Figure 72.  Fast Moving Cold Front Underrunning Warm, Moist, Unstable Air

g.    Since an occluded front develops when a cold front overtakes a warm front, weather with an occluded front is a combination of both warm and cold frontal weather.

  1)    Figures 73 below and 74 on page 91 show warm and cold occlusions and associated weather.

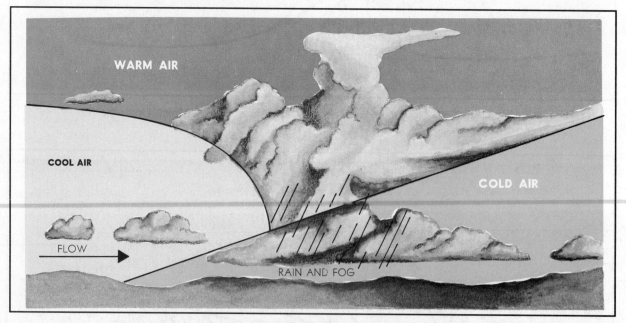

Figure 73.  Warm Front Occlusion Lifting Warm, Moist, Unstable Air

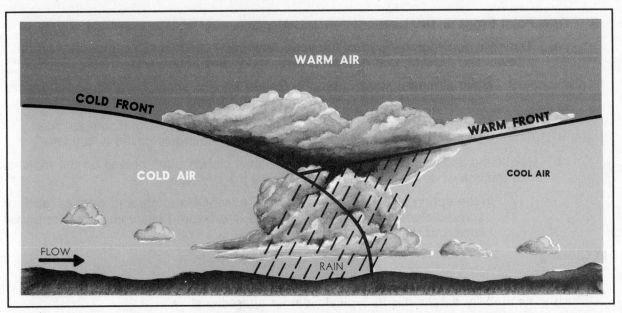

Figure 74.  Cold Front Occlusion Lifting Warm, Moist, Stable Air

h.   A front may have little or no cloudiness associated with it.

   1)   *Dry fronts* occur when the warm air aloft is flowing down the frontal slope or the air is so dry that any cloudiness that occurs is at high levels.

i.   The upper wind flow dictates to a great extent the amount of cloudiness and rain accompanying a frontal system as well as movement of the front itself.

   1)   When winds aloft blow across a front, it tends to move with the wind.

      a)   When winds aloft parallel a front, the front moves slowly, if at all.

   2)   A deep, slow moving trough aloft forms extensive cloudiness and precipitation, while a rapid moving minor trough more often restricts weather to a rather narrow band.

      a)   However, the latter often breeds severe, fast moving, turbulent spring weather.

8.   **Instability Line**

   a.   An instability line is a narrow, nonfrontal line or band of convective activity.

      1)   If the activity consists of fully developed thunderstorms, the line is a *squall line* (see Part I, Chapter 11, Thunderstorms, beginning on page 115).

   b.   Instability lines form in moist, unstable air.

      1)   An instability line may develop far from any front.

      2)   More often, it develops ahead of a cold front, and sometimes a series of these lines move out ahead of the front.

      3)   A favored location for instability lines which frequently erupt into severe thunderstorms is a dew point front or dry line.

9. **Dew Point Front or Dry Line**

    a. During a considerable part of the year, dew point fronts are common in western Texas and New Mexico northward over the Plains States.

        1) Moist air flowing north from the Gulf of Mexico abuts the drier and therefore slightly denser air flowing from the southwest.

        2) Except for moisture differences, there is seldom any significant air mass contrast across this front, and thus, it is commonly called a "dry line."

    b. Nighttime and early morning fog and low-level clouds often prevail on the moist side of the line while generally clear skies mark the dry side.

        1) In the spring and early summer over Texas, Oklahoma, and Kansas, and for some distance eastward, the dry line is a favored spawning area for squall lines and tornadoes.

## C. **In Closing**

1. An air mass is a body of air with fairly uniform properties of temperature and moisture.

    a. Its stability determines its typical weather characteristics.

2. A front is the zone between two different air masses.

    a. There are three principal types of fronts:  cold, warm, and stationary.

3. Frontal waves usually form on slow moving cold fronts or on stationary fronts.

    a. One section of the front begins to move as a warm front, and the section beside it begins to move as a cold front.

    b. The cold front moves faster than the warm front, and the two close together to form an occlusion.

4. Nonfrontal lows are low-pressure areas lying solely in a homogeneous air mass.
5. Frontolysis is the dissipation of a front.
6. Frontogenesis is the development of a front.
7. In fronts, flying weather varies from virtually clear skies to extremely hazardous conditions.

    a. Surface weather charts pictorially portray fronts and, in conjunction with other forecast charts and special analyses, help you in determining expected weather conditions along your proposed route.

        1) Knowing the locations of fronts and associated weather helps you determine whether you can proceed as planned.

        2) Often you can change your route to avoid adverse weather.

    b. Frontal weather may change rapidly.

        1) For example, there may be only cloudiness associated with a cold front over northern Illinois during the morning but with a strong squall line forecast by afternoon.

    c. A mental picture of what is happening and what is forecast should greatly help you in avoiding adverse weather conditions.

        1) If unexpected adverse weather develops en route, your mental picture helps you in planning the best diversion.

        2) *Always obtain a good preflight weather briefing.*

# END OF CHAPTER

# CHAPTER NINE
# TURBULENCE

Please take a few minutes to study each of the concepts listed above and anticipate/imagine what they are and how they relate to the other listed concepts.

A. **Introduction** -- A turbulent atmosphere is one in which air currents vary greatly over short distances.

1. These currents may range from rather mild eddies to strong currents of relatively large dimensions.

2. As an aircraft moves through these currents, it undergoes changing accelerations which jostle it from its smooth flight path.

   a. This jostling is turbulence.

      1) Turbulence ranges from bumpiness which can annoy crew and passengers to severe jolts which can structurally damage the aircraft or injure passengers.

   b. Aircraft reaction to turbulence varies with the difference in wind speed in adjacent currents, size of the aircraft, wing loading, airspeed, and aircraft attitude.

      1) When an aircraft travels rapidly from one current to another, it undergoes abrupt changes in acceleration.

      2) Obviously, if the aircraft were to move more slowly, the changes in acceleration would be less.

      3) The first rule in flying through turbulence, then, is to reduce airspeed.

         a) Your aircraft manual most likely lists recommended airspeed for penetrating turbulence.

3. The main causes of turbulence are

   a. Convective currents,
   b. Obstructions to wind flow, and
   c. Wind shear.

4. Turbulence also occurs in the wake of moving aircraft whenever the airfoils exert lift, i.e., wake turbulence.

5. Any combination of causes may occur at one time.

B. **Convective Currents**

1. Convective currents are a common cause of turbulence, especially at low altitudes. These currents are localized vertical air movements, both ascending and descending.

   a. For every rising current, there is a compensating downward current.

   b. The downward currents frequently occur over broader areas than do the upward currents. Therefore, they have a slower vertical speed than do the rising currents.

2.    Convective currents are most active on warm summer afternoons when winds are light.

    a.    Heated air at the surface creates a shallow, unstable layer because the warm air is forced upward.

        1)    Convection increases in strength and to greater heights as surface heating increases.

    b.    Barren surfaces such as sandy or rocky wastelands and plowed fields become hotter than open water or ground covered by vegetation.

        a)    Thus, air at and near the surface heats unevenly.

        1)    Because of this uneven heating, the strength of convective currents can vary considerably within short distances.

3.    When cold air moves over a warm surface, it becomes unstable in lower levels.

    a.    Convective currents may extend several thousand feet above the surface, resulting in rough, choppy turbulence when you are flying through the cold air.

        1)    This condition often occurs in any season after the passage of a cold front.

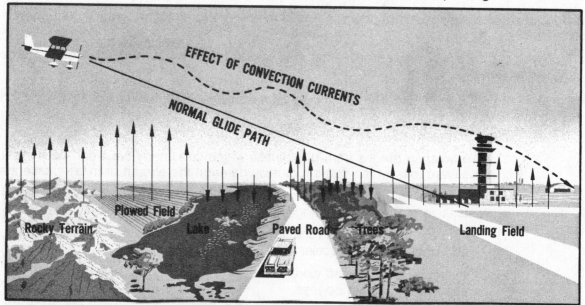

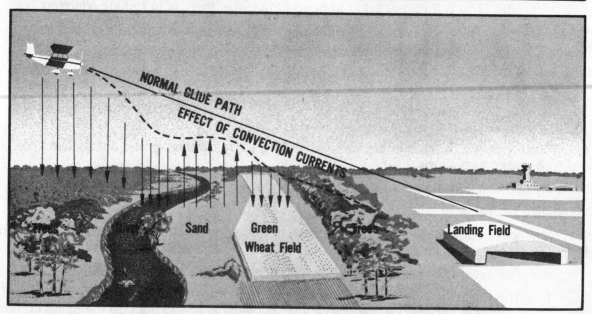

Figure 76.  Effect of Convective Currents on Final Approach

b. Figure 76 on page 94 illustrates the effect of low-level convective turbulence on an aircraft approaching a landing field.

1) Predominantly upward currents (top) tend to cause the aircraft to overshoot the intended touchdown point.

2) Predominantly downward currents (bottom) tend to cause the aircraft to undershoot the intended touchdown point.

4. Turbulence on approach can cause abrupt changes in airspeed and may even result in a stall at a dangerously low altitude.

a. To guard against this danger, increase airspeed slightly over normal approach speed.

1) This procedure may appear to conflict with the rule of reducing airspeed for turbulence penetration, but remember, the approach speed for your aircraft is well below the recommended turbulence penetration speed.

5. As air moves upward, it cools by expansion.

a. A convective current continues upward until it reaches a level where its temperature cools to that of the surrounding air.

1) If it cools to saturation, a cloud forms.

b. Billowy fair weather cumulus clouds, usually seen on sunny afternoons, are signposts in the sky indicating convective turbulence.

1) The cloud top usually marks the approximate upper limit of the convective current.

2) You can expect to encounter turbulence beneath or in the clouds. While above the clouds, air generally is smooth.

3) You will find flight more comfortable above the cumulus as illustrated in Figure 77 below.

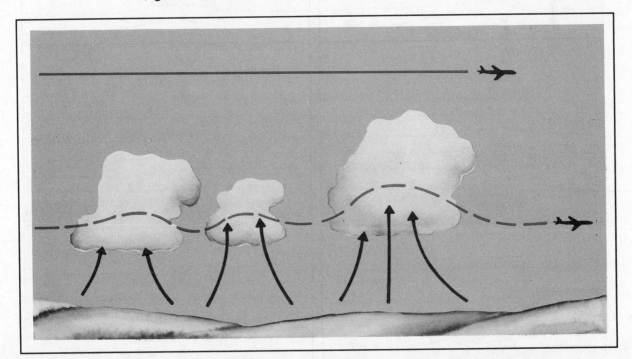

Figure 77. Avoiding Turbulence by Flying above Convective Clouds

   c.  When convection extends to greater heights, it develops larger towering cumulus
       clouds and cumulonimbus with anvil-like tops.

       1)  The cumulonimbus (i.e., thunderstorm) provides visual warning of violent
           convective turbulence.  See Part I, Chapter 11, Thunderstorms, beginning on
           page 115.

   d.  You should also know that when air is too dry for cumulus to form, convective currents
       still can be active.

       1)  There is little indication of their presence until you encounter turbulence.
       2)  This is sometimes referred to as clear air turbulence (CAT).

## C.  Obstructions to Wind Flow

   1.  Obstructions such as buildings, trees, and rough or mountainous terrain disrupt smooth
       wind flow into a complex snarl of eddies as diagramed in Figure 78 below.

       a.  An aircraft flying through these eddies experiences turbulence.

       b.  This turbulence is classified as "mechanical" since it results from mechanical
           disruption of the ambient wind flow.

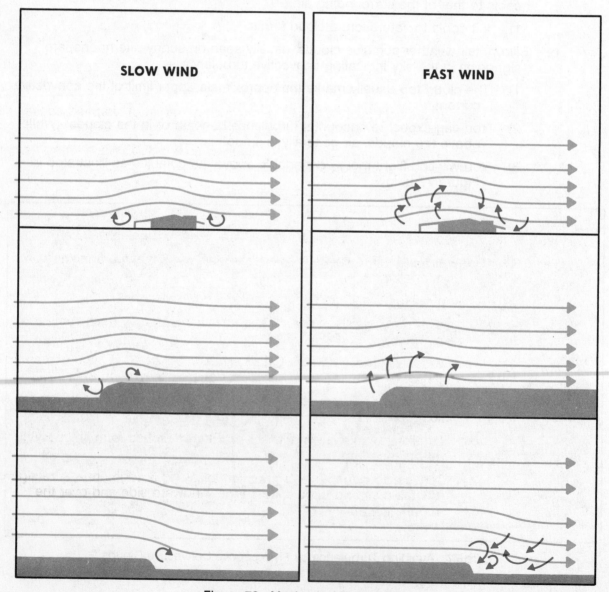

Figure 78.  Mechanical Turbulence

2.  The degree of mechanical turbulence depends on wind speed and roughness of the obstructions.

   a.  The higher the speed and/or the rougher the surface, the greater is the turbulence.

   b.  The wind carries the turbulent eddies downstream -- how far depends on wind speed and stability of the air.

      1)  Unstable air allows larger eddies to form than those that form in stable air.

      2)  However, the instability breaks up the eddies quickly, whereas in stable air they dissipate slowly.

3.  Mechanical turbulence can cause cloudiness near the top of the mechanically disturbed layer, just as convective turbulence can.

   a.  The type of cloudiness tells you whether it is from mechanical or convective mixing.

      1)  Mechanical mixing produces stratocumulus clouds in rows or bands while convective clouds form a random pattern.

      2)  The cloud rows developed by mechanical mixing may be parallel to or perpendicular to the wind depending on meteorological factors which we do not discuss here.

4.  The airport area is especially vulnerable to mechanical turbulence which invariably causes gusty surface winds.

   a.  When an aircraft is in a low-level approach or a climb, airspeed fluctuates in the gusts, and the aircraft may even stall.

      1)  During extremely gusty conditions, maintain a margin of airspeed above normal approach or climb speed to allow for changes in airspeed.

   b.  When landing with a gusty crosswind, be alert for control problems in mechanical turbulence caused by airport structures upwind.

      1)  Surface gusts can also create taxi problems.

5.  Mechanical turbulence can affect low-level cross-country flight almost anywhere.

   a.  When flying over rolling hills, you may experience mechanical turbulence.

      1)  Generally, such turbulence is not hazardous, but it may be annoying or uncomfortable.

      2)  A climb to higher altitude should reduce the turbulence.

   b.  When flying over rugged hills or mountains, however, you may have some real turbulence problems.

      1)  When wind speed across a mountain exceeds about 40 kt., you can anticipate turbulence.  Where and to what extent depends largely on air stability.

      2)  If the air crossing the mountains is unstable, turbulence on the windward side is almost certain.

         a)  If sufficient moisture is present, convective clouds form, intensifying the turbulence.

         b)  Convective clouds over a mountain or along a ridge are a sure sign of unstable air and turbulence on the windward side and over the mountain crest.

3)    As the unstable air crosses the barrier, it spills down the leeward slope often as a violent downdraft.

a)    Sometimes the downward speed exceeds the maximum climb rate for your aircraft and may drive the craft into the mountainside as shown in Figure 80 below.

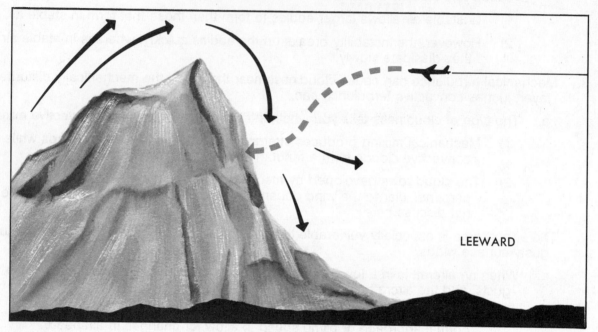

LEEWARD

Figure 80.  Wind Flow in Mountain Areas

4)    In the process of crossing the mountains, mixing reduces the instability to some extent.

a)    Thus, hazardous turbulence in unstable air generally does not extend a great distance downwind from the barrier.

6.    **Mountain Wave**

a.    When stable air crosses a mountain barrier, the turbulent situation is somewhat reversed.

1)    Air flowing up the windward side is relatively smooth.
2)    Wind flow across the barrier is laminar -- that is, it tends to flow in layers.

b.    The barrier may set up waves in these layers much as waves develop on a disturbed water surface.

1)    The waves remain nearly stationary while the wind blows rapidly through them.

2)    The wave pattern, diagramed in Figure 81 on the opposite page, is a "standing" or "mountain" wave, so named because it remains essentially stationary and is associated with the mountains.

c.    Wave crests extend well above the highest mountains, sometimes into the lower stratosphere.

1)    Under each wave crest is a rotary circulation also diagramed in Figure 81.

a)    The "rotor" forms below the elevation of the mountain peaks.

b)    One of the most dangerous features of a mountain wave is the turbulent areas in and below rotor clouds.

2)    Updrafts and downdrafts in the waves can also create violent turbulence.

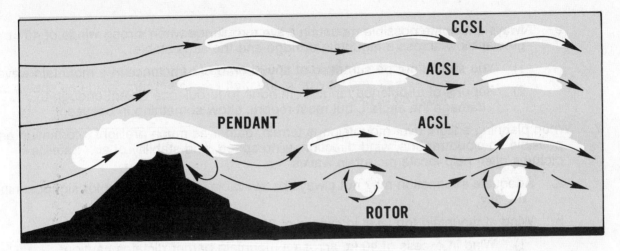

Figure 81.  Cross Section of a Mountain Wave

d.  Figure 81 above further illustrates clouds often associated with a mountain wave.

   1)  When moisture is sufficient to produce clouds on the windward side, they are stratified.

   2)  Crests of the standing waves may be marked by stationary, lens-shaped clouds known as "standing lenticular" clouds.  (See Figure 51 on page 68.)

      a)  They form in the updraft and dissipate in the downdraft, so they do not move as the wind blows through them.

   3)  The rotor may also be marked by a "rotor" cloud.

      a)  Figure 83 below is a photograph of a series of rotor clouds, each under the crest of a wave.

Figure 83.  Standing Wave Rotor Clouds

    e.    Always anticipate possible mountain wave turbulence when strong winds of 40 kt. or greater blow across a mountain or ridge and the air is stable.

        1)    You should not be surprised at any degree of turbulence in a mountain wave.

        2)    Reports of turbulence range from none to turbulence violent enough to damage the aircraft, but most reports show something in between.

7.    When planning a flight over mountainous terrain, gather as much preflight information as possible on cloud reports, wind direction, wind speed, and stability of air.  Satellite pictures often help locate mountain waves.

    a.    Adequate information may not always be available, so remain alert for signposts in the sky.

    b.    Wind at mountain top level in excess of 25 kt. suggests some turbulence.

        1)    Wind in excess of 40 kt. across a mountain barrier dictates caution.

    c.    Stratified clouds mean stable air.

        1)    Standing lenticular and/or rotor clouds suggest a mountain wave.

            a)    Expect turbulence many miles to the lee of mountains and relatively smooth flight on the windward side.

    d.    Convective clouds on the windward side of mountains mean unstable air.

        1)    Expect turbulence in close proximity to and on either side of the mountain.

    e.    When approaching mountains from the leeward side during strong winds, begin your climb well away from the mountains -- 100 miles in a mountain wave and 30 to 50 miles otherwise.

        1)    Climb to an altitude 3,000 to 5,000 ft. above mountain tops before attempting to cross.

        2)    It is recommended that you approach a ridge at a 45° angle to enable a rapid retreat to calmer air.

            a)    If unable to make good on your first attempt and you have higher altitude capabilities, you may choose to back off and make another attempt at higher altitude.

            b)    Sometimes you may have to choose between turning back or detouring the area.

    f.    Flying through mountain passes and valleys is not a safe procedure in high winds.

        1)    The mountains funnel the wind into passes and valleys, thus increasing wind speed and intensifying turbulence.

        2)    If winds at mountain top level are strong, go high or go around.

    g.    Surface wind may be relatively calm in a valley surrounded by mountains when wind aloft is strong.

        1)    If taking off in the valley, climb above mountain top level before leaving the valley.

        2)    Maintain lateral clearance from the mountains sufficient to allow recovery if caught in a downdraft.

## D.  Wind Shear

1.    Wind shear generates eddies between two wind currents of differing velocities.

    a.    The differences may be in wind speed, wind direction, or in both.

        1)    Wind shear may be associated with either a wind shift or a wind speed gradient at any level in the atmosphere.

b.  Three conditions are of special interest.

1)  Wind shear with a low-level temperature inversion,

2)  Wind shear in a frontal zone, and

3)  Clear air turbulence (CAT) at high levels associated with a jet stream or strong circulation.

a)  High-level CAT is discussed in detail in Part II, Chapter 13, High Altitude Weather, beginning on page 144.

2.  Wind shear with a low-level temperature inversion

a.  A temperature inversion forms near the surface on a clear night with calm or light surface wind.

b.  Wind just above the inversion may be relatively strong.

1)  As illustrated in Figure 86 below, a wind shear zone develops between the calm and the stronger winds above.

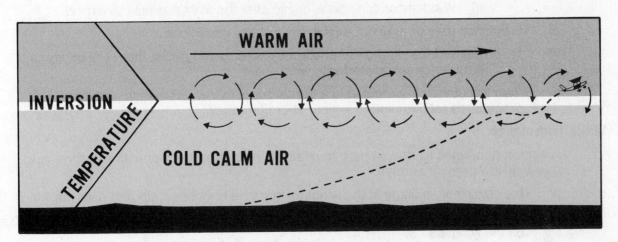

Figure 86.  Wind Shear in a Low-level Temperature Inversion

c.  Eddies in the shear zone cause airspeed fluctuations as an aircraft climbs or descends through the inversion.

1)  An aircraft most likely is either climbing from takeoff or approaching to land when passing through the inversion; therefore, airspeed is relatively slow -- perhaps only a few knots greater than stall speed.

2)  The fluctuation in airspeed can induce a stall precariously close to the ground.

d.  Since surface wind is calm or very light, takeoff or landing can be in any direction.

1)  Takeoff may be in the direction of the wind above the inversion.

a)  If so, the aircraft encounters a sudden tailwind and a corresponding loss of airspeed when climbing through the inversion.  Stall is possible.

2)  If the landing approach is into the wind above the inversion, the headwind is suddenly lost when descending through the inversion.  Again, a sudden loss in airspeed may induce a stall.

e.  When taking off or landing in calm wind under clear skies within a few hours before or after sunrise, be prepared for a temperature inversion near the ground.

1)  You can be relatively certain of a shear zone in the inversion if you know the wind at 2,000 to 4,000 ft. is 25 kt. or more.

2)  Allow a margin of airspeed above normal climb or approach speed to alleviate danger of stall in event of turbulence or sudden change in wind velocity.

3.    Wind shear in a frontal zone

    a.    While wind changes abruptly in the frontal zone, not all fronts have associated wind shear.

    b.    The following is a method of determining the approximate height of a front above an airport, with the consideration that wind shear is most critical when it occurs close to the ground.

        1)    A cold front wind shear occurs just after the front passes the airport and for a short period afterward.

            a)    If the front is moving 30 kt. or more, the frontal surface (wind shear zone) will usually be 5,000 ft. above the airport about 3 hr. after the frontal passage.

        2)    With a warm front, the most critical period is approximately 6 hr. before the front passes the airport.  Warm front shear may exist from the surface to 5,000 ft. AGL.

            a)    Wind shear conditions cease after the front passes the airport.

    c.    Turbulence may or may not exist in wind shear conditions.

        1)    If the surface wind under the front is strong and gusty, there will be some turbulence associated with wind shear.

    d.    When turbulence is expected in a frontal zone, follow turbulence penetration procedures recommended in your aircraft manual.

E.    **Wake Turbulence**

1.    An aircraft generates lift due, in part, to relatively high pressure beneath its wings, and relatively low pressure above.

    a.    This causes air spillage at the wingtips from the underside, up and over the wings.

    b.    Thus, whenever the wings are producing lift, they generate rotary motions or vortices off the wingtips.

2.    When the landing gear bears the entire weight of the aircraft, no wingtip vortices develop.

    a.    But the instant the aircraft rotates on takeoff, these vortices begin.

        1)    Figure 87 below illustrates how they might appear if visible behind the plane as it breaks ground.

Figure 87.  Wingtip Vortices

    b.    These vortices continue throughout the flight until the craft again settles firmly on its landing gear (i.e., no lift is produced).

3.    These vortices spread downward and outward from the flight path. They also drift with the wind. Avoid flying through these vortices.

    a.    The strength of the vortices is proportional to the weight of the aircraft as well as other factors.

        1)    Therefore, wake turbulence is more intense behind large, transport category aircraft than behind small aircraft.

        2)    Generally, it is a problem only when following the larger aircraft.

4.    The turbulence persists several minutes and may linger after the aircraft is out of sight.

    a.    Most jets when taking off lift the nose wheel about midpoint in the takeoff roll; therefore, vortices begin at approximately the middle of the takeoff roll.

    b.    Vortices behind propeller aircraft begin only a short distance behind liftoff.

    c.    Following a landing of either type of aircraft, vortices end at approximately the point where the nose wheel touches down.

5.    When using the same runway as a heavier aircraft:

    a.    If landing behind another aircraft, keep your approach above its approach and keep your touchdown beyond the point where its nose wheel touched the runway, as in Figure 88 (A);

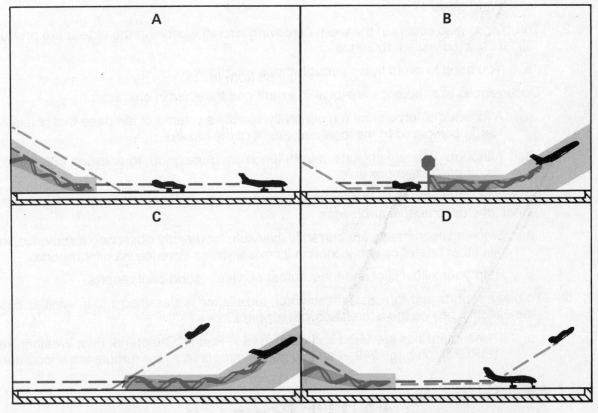

Figure 88. Planning Landing or Takeoff to Avoid Wake Turbulence

    b.    If landing behind a departing aircraft, land only if you can complete your landing roll before reaching the midpoint of its takeoff roll, as in Figure 88 (B).

c.   If departing behind another departing aircraft, take off only if you can become airborne before reaching the midpoint of its takeoff roll and only if you can climb fast enough to stay above its flight path, as in Figure 88 (C); and

d.   Do not depart behind a landing aircraft unless you can taxi onto the runway beyond the point at which its nosewheel touched down and have sufficient runway left for safe takeoff, as in Figure 88 (D).

e.   If parallel runways are available and the heavier aircraft takes off with a crosswind on the downwind runway, you may safely use the upwind runway.

f.   Never land or take off downwind from the heavier aircraft.

g.   When using a runway crossing the heavier aircraft's runway, you may safely use the upwind portion of your runway.

h.   You may cross behind a departing aircraft behind the midpoint of its takeoff roll.

i.   If none of these procedures is possible, wait 5 minutes or so for the vortices to dissipate or to blow off the runway.

6.   The problem of wake turbulence is more operational than meteorological.

## F.   In Closing

1.   The main causes of turbulence are

   a.   Convective currents,
   b.   Obstructions to wind flow, and
   c.   Wind shear.

2.   Turbulence also occurs in the wake of moving aircraft whenever the airfoils are producing lift.  It is called wake turbulence.

   a.   You need to avoid flying through these wingtip vortices.

3.   Occurrences of turbulence are local in extent and transient in character.

   a.   A forecast of turbulence will generally specify a volume of airspace that is relatively large compared to the localized extent of the hazard.

   b.   Although general forecasts of turbulence are quite good, forecasting precise locations is, at present, impossible.

4.   Generally, when you receive a forecast of turbulence, you should plan your flight to avoid areas of *most probable turbulence*.

   a.   Since no instruments are currently available for directly observing turbulence, the weather briefer can only confirm its existence or absence via pilot reports.

   b.   Help your fellow pilot and the weather service -- send pilot reports.

5.   To make reports and forecasts meaningful, turbulence is classified into intensities based on the effects it has on the aircraft and passengers.

   a.   These intensities are listed and described in Part III, Chapter 3, Pilot Weather Reports (PIREPs), on page 249.  Use this guide in reporting your turbulence encounters.

# END OF CHAPTER

# CHAPTER TEN
# ICING

Please take a few minutes to study each of the concepts listed above and anticipate/imagine what they are and how they relate to the other listed concepts.

A.  **Introduction** -- Aircraft icing is one of the major weather hazards to aviation.  It is a cumulative hazard.

1.  Icing reduces aircraft efficiency by increasing weight, reducing lift, decreasing thrust, and increasing drag.

    a.  As shown in Figure 89 below, each effect tends either to slow the aircraft or to force it downward.

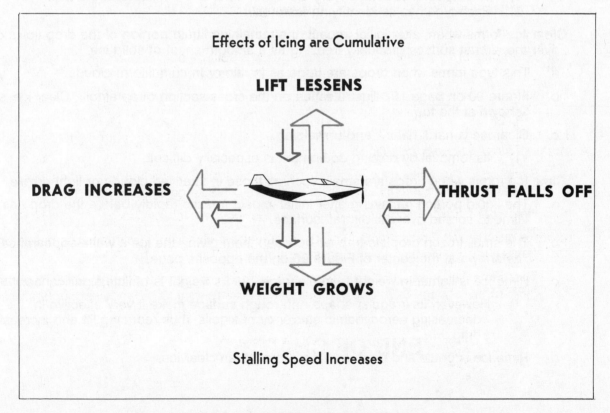

Figure 89.  Effects of Structural Icing

2.    Icing also seriously impairs aircraft engine performance.

3.    Other icing effects include false indications on flight instruments, loss of radio communications, and loss of operation of control surfaces, brakes, and landing gear.

4.    In this chapter we discuss the principles of structural, induction system, and instrument icing and relate icing to cloud types and other factors.

    a.    Although ground icing and frost are structural icing, we discuss them separately because of their different effect on an aircraft.

B.    **Structural Icing**

1.    Two conditions are necessary for structural icing in flight:

    a.    The aircraft must be flying through visible moisture such as rain droplets or clouds.

    b.    The temperature at the point where the moisture strikes the aircraft must be 0°C or colder.

        1)    Note that aerodynamic cooling can lower temperature of an airfoil to 0°C even though the ambient temperature is a few degrees warmer.

2.    Supercooled water increases the rate of icing and is essential to rapid accumulation.

    a.    Supercooled water is in an unstable liquid state.  When an aircraft strikes a supercooled drop, part of the drop freezes instantaneously.

    b.    The latent heat of fusion released by the freezing portion raises the temperature of the remaining portion to the melting point.

        1)    Aerodynamic effects may cause the remaining portion to freeze.

    c.    The way in which the remaining portion freezes determines the type of icing.

        1)    The types of structural icing are clear, rime, and a mixture of the two.
        2)    Each type has its identifying features.

3.    **Clear ice** forms when, after initial impact, the remaining liquid portion of the drop flows out over the aircraft surface gradually freezing as a smooth sheet of solid ice.

    a.    This type forms when drops are large as in rain or in cumuliform clouds.

    b.    Figure 90 on page 107 illustrates ice on the cross-section of an airfoil.  Clear ice is shown at the top.

    c.    Clear ice is hard, heavy, and unyielding.

        1)    Its removal by deicing equipment is especially difficult.

4.    **Rime ice** forms when drops are small, such as those in stratified clouds or light drizzle.

    a.    The liquid portion remaining after initial impact freezes rapidly before the drop has time to spread over the aircraft surface.

    b.    The small frozen droplets trap air between them giving the ice a white appearance as shown at the center of Figure 90 on the opposite page.

    c.    Rime ice is lighter in weight than clear ice, but its weight is of little significance.

        1)    However, its irregular shape and rough surface make it very effective in decreasing aerodynamic efficiency of airfoils, thus reducing lift and increasing drag.

    d.    Rime ice is brittle and more easily removed than clear ice.

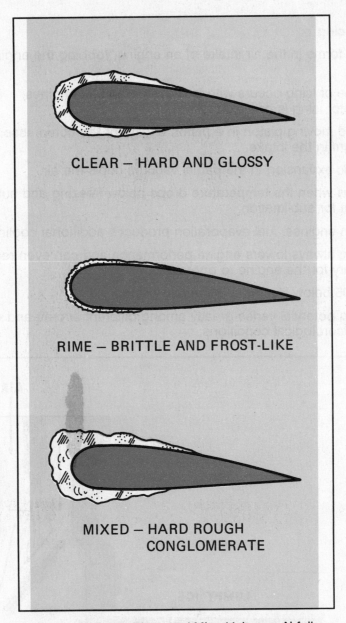

CLEAR — HARD AND GLOSSY

RIME — BRITTLE AND FROST-LIKE

MIXED — HARD ROUGH
CONGLOMERATE

Figure 90.  Clear, Rime, and Mixed Icing on Airfoils

5. **Mixed ice** forms when drops vary in size or when liquid drops are intermingled with snow or ice particles.  It can form rapidly.

   a. Ice particles become imbedded in clear ice, building a very rough accumulation sometimes in a mushroom shape on leading edges as shown at the bottom of Figure 90 above.

6. The FAA, National Weather Service, the military aviation weather services, and aircraft operating organizations have classified aircraft structural icing into intensity categories.

   a. Part III, Chapter 3, Pilot Weather Reports (PIREPs), on page 251, contains a table listing these intensities.

      1) The table is your guide in estimating how ice of a specific intensity will affect your aircraft.

      2) Use the table also in reporting ice when you encounter it.

## C.  Induction System Icing

1.   Ice frequently forms in the air intake of an engine, robbing the engine of air to support combustion.

    a.   This type of icing occurs with both piston and jet engines.

    b.   Carburetor icing is one example.

2.   The downward moving piston in a piston engine or the compressor in a jet engine forms a partial vacuum in the intake.

    a.   Adiabatic expansion in the partial vacuum cools the air.

    b.   Ice forms when the temperature drops below freezing and sufficient moisture is present for sublimation.

    c.   In piston engines, fuel evaporation produces additional cooling.

3.   Induction icing always lowers engine performance and can even reduce intake flow below that necessary for the engine to operate.

    a.   Figure 95 below illustrates carburetor icing.

4.   Induction icing potential varies greatly among different aircraft and occurs under a wide range of meteorological conditions.

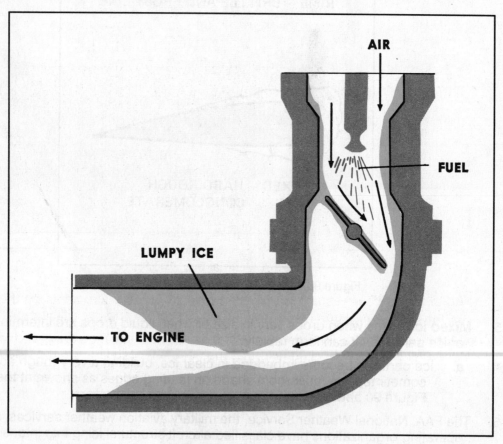

Figure 95.  Carburetor Icing

## D. Instrument Icing

1. Icing of the pitot tube as seen in Figure 96 below reduces ram air pressure to the airspeed indicator and renders the instrument unreliable.

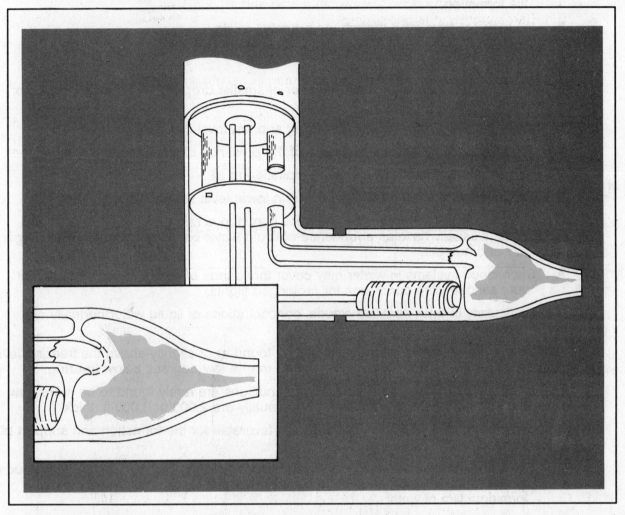

Figure 96. Internal Pitot Tube Icing

2. Icing of the static pressure port reduces reliability of all instruments on the pitot-static system -- the airspeed and vertical speed indicators and the altimeter.

3. Ice forming on the radio antenna distorts its shape, increases drag, and imposes vibrations that may result in failure in the communications system of the aircraft.

    a. The severity of this icing depends upon the shape, location, and orientation of the antenna.

E.    **Icing and Cloud Types**

1.    All clouds at subfreezing temperatures have icing potential.

    a.    However, drop size and distribution and aerodynamic effects of the aircraft influence ice formation.

    b.    Ice may not form even though the potential exists.

2.    The condition most favorable for very hazardous icing is the presence of many large, supercooled water drops.

    a.    Conversely, an equal or lesser number of smaller droplets favors a slower rate of icing.

3.    Small water droplets occur most often in fog and low-level clouds.

    a.    Drizzle or very light rain is evidence of the presence of small drops in such clouds.

        1)    In many cases there is no precipitation at all.

    b.    The most common type of icing found in lower-level stratus clouds is rime.

4.    Thick extensive stratified clouds that produce continuous rain such as altostratus and nimbostratus usually have an abundance of liquid water because of the relatively larger drop size and number.

    a.    Such cloud systems in winter may cover thousands of square miles and present very serious icing conditions for prolonged flights.

    b.    Particularly in thick stratified clouds, concentrations of liquid water normally are greater with warmer temperatures.

        1)    Thus, heaviest icing usually will be found at or slightly above the freezing level where temperature is never more than a few degrees below freezing.

    c.    In layer type clouds, continuous icing conditions are rarely found to be more than 5,000 ft. above the freezing level, and usually are 2,000 or 3,000 ft. thick.

5.    The upward currents in cumuliform clouds are favorable for the formation and support of many large water drops.

    a.    When an aircraft enters the heavy water concentrations found in cumuliform clouds, the large drops break and spread rapidly over the leading edge of the airfoil, forming a film of water.

        1)    If temperatures are freezing or colder, the water freezes quickly to form a solid sheet of clear ice.

    b.    You should avoid cumuliform clouds when possible.

    c.    The updrafts in cumuliform clouds lift large amounts of liquid water far above the freezing level.

        1)    On rare occasions icing has been encountered in thunderstorm clouds at altitudes of 30,000 to 40,000 ft. where the free air temperature was colder than −40°C.

    d.    While the vertical extent of critical icing potential cannot be specified in cumuliform clouds, their individual cell-like distribution usually limits the horizontal extent of icing conditions.

        1)    An exception, of course, may be found in a prolonged flight through a broad zone of thunderstorms or heavy showers.

F.  **Other Factors in Icing**

1.  **Fronts.**  A condition favorable for rapid accumulation of clear icing is freezing rain below a frontal surface.

    a.  Rain forms above the frontal surface at temperatures warmer than freezing.

        1)  Subsequently, it falls through air at temperatures below freezing and becomes supercooled.

        2)  The supercooled drops freeze on impact with an aircraft surface.

    b.  Figure 98 below diagrams this type of icing.  It may occur with either a warm front (top) or a cold front.

        1)  The icing can be critical because of the large amount of supercooled water.

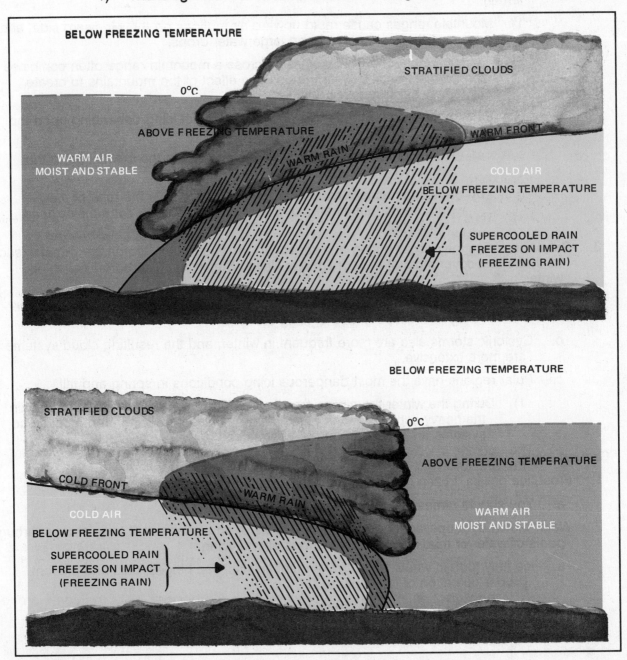

Figure 98.  Freezing Rain with a Warm Front and a Cold Front

    c.    Icing can also become serious in cumulonimbus clouds along a surface cold front, along a squall line, or embedded in the cloud shield of a warm front.

2.    **Terrain.**  Air blowing upslope is cooled adiabatically.  When the air is cooled below the freezing point, any water it contains becomes supercooled.

    a.    In stable air blowing up a gradual slope, the cloud drops generally remain comparatively small since larger drops fall out as rain.

        1)    Ice accumulation is rather slow and you should have ample time to get out of it before the accumulation becomes extremely dangerous.

    b.    When air is unstable, convective clouds develop a more serious hazard.

    c.    Icing is more probable and more hazardous in mountainous regions than over other terrain.

        1)    Mountain ranges cause rapid upward air motions on the windward side, and these vertical currents support large water drops.

        2)    The movement of a frontal system across a mountain range often combines the normal frontal lift with the upslope effect of the mountains to create extremely hazardous icing zones.

        3)    Each mountainous region has preferred areas of icing depending upon the orientation of mountain ranges to the wind flow.

            a)    The most dangerous icing takes place above the crests and to the windward side of the ridges.

            b)    This zone usually extends about 5,000 ft. above the tops of the mountains, but when clouds are cumuliform, the zone may extend much higher.

3.    **Seasons.**  Icing may occur during any season of the year, but in temperate climates such as those that cover most of the contiguous United States, icing is more frequent in winter.

    a.    The freezing level is nearer the ground in winter than in summer leaving a smaller low-level layer of airspace free of icing conditions.

    b.    Cyclonic storms also are more frequent in winter, and the resulting cloud systems are more extensive.

    c.    Polar regions have the most dangerous icing conditions in spring and fall.

        1)    During the winter the air is normally too cold in the polar regions to contain the heavy concentrations of moisture necessary for icing, and most cloud systems are stratiform and are composed of ice crystals.

## G.  **Ground Icing**

1.    Frost, ice pellets, frozen rain, or snow may accumulate on parked aircraft.

    a.    You should remove all ice prior to takeoff.

2.    Water blown by propellers or splashed by wheels of an airplane as it taxis or runs through pools of water or mud may result in serious aircraft icing.

    a.    Ice may form in wheel wells, brake mechanisms, flap hinges, etc., and prevent proper operation of these parts.

3.    Ice on runways and taxiways can create traction and braking problems.

## H.  **Frost**

1.  Frost is a hazard to flying.

    a.  Pilots must remove all frost from airfoils prior to takeoff.

2.  Frost forms near the surface primarily in clear, stable air and with light winds -- conditions which in all other respects make weather ideal for flying.

    a.  Because of this, the real hazard is often minimized.

3.  Thin metal airfoils are especially vulnerable surfaces on which frost will form.

4.  Test data have shown that frost, ice, or snow formations having thickness and surface roughness similar to medium or course sandpaper on the leading edge and upper surfaces of a wing can reduce lift by as much as 30% and increase drag by 40%.

    a.  Changes in lift and drag can significantly increase stall speed, reduce controllability, and alter the airplane's flight characteristics.

    b.  Thicker or rougher frost, ice, or snow formations will have increasingly adverse effects, with the primary influence being surface roughness location on critical portions of an aerodynamic surface, e.g., the leading edge of a wing.

    c.  These adverse effects on the aerodynamic properties of an airfoil may result in sudden departure from the desired flight path without any prior indications or aerodynamic warning to the pilot.

        1)  Even a small amount of frost on airfoils may prevent an aircraft from becoming airborne at normal takeoff speed.

        2)  Also possible is that, once airborne, an aircraft could have insufficient margin of airspeed above stall so that moderate gusts or turning flight could produce incipient or complete stalling.

5.  Frost formation in flight offers a more complicated problem.  The extent to which it will form is still a matter of conjecture.

    a.  At most, it is comparatively rare.

## I.   **In Closing**

1.  Icing reduces aircraft efficiency by increasing weight, reducing lift, decreasing thrust, and increasing drag.

    a.  Icing also seriously impairs aircraft engine performance.

2.  Structural icing can occur as clear ice, rime ice, or a mixture of the two.

3.  Induction system icing, e.g., carburetor icing, lowers engine performance by reducing the intake of air necessary to support combustion.

4.  Instrument icing affects the airspeed and vertical speed indicators and the altimeter.  Also, ice forming on the radio antenna may result in failure of the communication and/or radio navigation systems.

5.  Icing is where you find it.  As with turbulence, icing may be local in extent and transient in character.

6.  Forecasters can identify regions in which icing is possible.

    a.  However, they cannot define the precise small pockets in which it occurs.

7.  You should plan your flight to avoid those areas where icing probably will be heavier than your aircraft can handle.

    a.  Also, you must be prepared to avoid or to escape the hazard when it is encountered en route.

8. Here are a few specific points to remember:

   a. Before takeoff, check weather for possible icing areas along your planned route.

     1) Check for pilot reports, and if possible talk to other pilots who have flown along your proposed route.

   b. If your aircraft is not equipped with deicing or anti-icing equipment, avoid areas of icing.

     1) Water (clouds or precipitation) must be visible and outside air temperature must be near 0°C or colder for structural ice to form.

   c. Always remove ice or frost from airfoils before attempting takeoff.

   d. In cold weather, avoid, when possible, taxiing or taking off through mud, water, or slush.

     1) If you have taxied through any of these, make a preflight check to ensure freedom of controls.

   e. When climbing out through an ice layer, climb at an airspeed a little faster than normal to avoid a stall.

   f. Use deicing or anti-icing equipment before accumulations of ice become too great.

     1) When such equipment becomes less than totally effective, change course or altitude to get out of the icing as rapidly as possible.

   g. If your aircraft is not equipped with a pitot-static system deicer, be alert for erroneous readings from your airspeed indicator, vertical speed indicator, and altimeter.

   h. In stratiform clouds, you can probably alleviate icing by changing to an altitude with above-freezing temperatures or to one colder than −10°C.

     1) An altitude change also may take you out of clouds.
     2) Rime icing in stratiform clouds can be very extensive horizontally.

   i. In frontal freezing rain, you may be able to climb or descend to a layer warmer than freezing.

     1) Temperature is always warmer than freezing at some higher altitude.

     2) If you are going to climb, move quickly; procrastination may leave you with too much ice.

       a) If you are going to descend, you must know the temperature and terrain below.

   j. Avoid cumuliform clouds if at all possible.  Clear ice may be encountered anywhere above the freezing level.

     1) Most rapid accumulations are usually at temperatures from 0°C to −15°C.

   k. Avoid abrupt maneuvers when your aircraft is heavily coated with ice because the aircraft has lost some of its aerodynamic efficiency.

   l. When "iced up," fly your landing approach with power.

9. Help your fellow pilots and the weather service by sending pilot reports when you encounter icing or when icing is forecast but none is encountered.

# END OF CHAPTER

# CHAPTER ELEVEN
# THUNDERSTORMS

Please take a few minutes to study each of the concepts listed above and anticipate/imagine what they are and how they relate to the other listed concepts.

## A.  Introduction

1.  This chapter looks at where and when thunderstorms occur, explains what creates a storm, and looks inside the storm at what goes on and what it can do to an aircraft.

2.  In some tropical regions, thunderstorms occur year-round.

   a.  In midlatitudes, they develop most frequently in spring, summer, and fall.
   b.  Arctic regions occasionally experience thunderstorms during summer.

3.  Figures 100 through 104 depict various thunderstorm activity.  Note the frequency shown by location (south-central and southeastern states) and by season (summer).

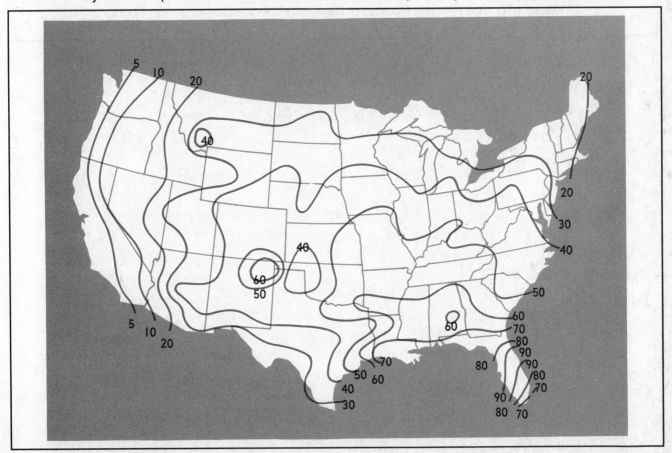

Figure 100.  The Average Number of Thunderstorms Each Year

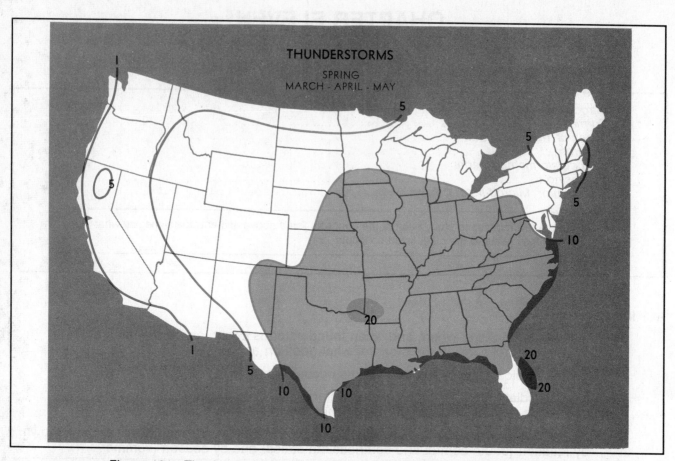

Figure 101.  The Average Number of Days with Thunderstorms during Spring

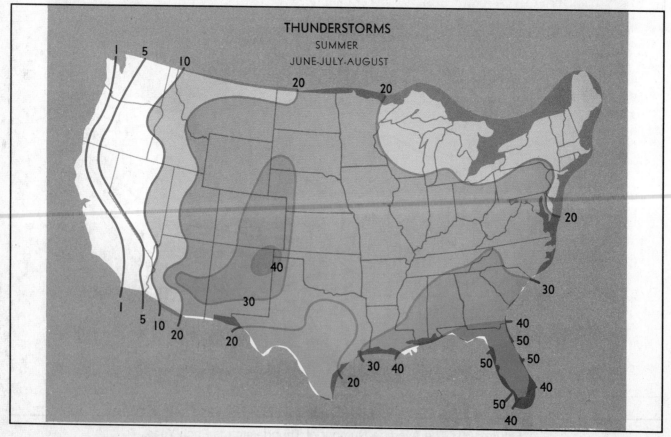

Figure 102.  The Average Number of Days with Thunderstorms during Summer

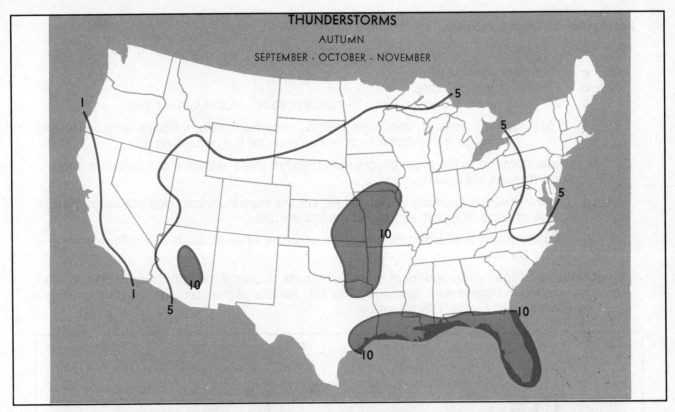

Figure 103.  The Average Number of Days with Thunderstorms during Fall

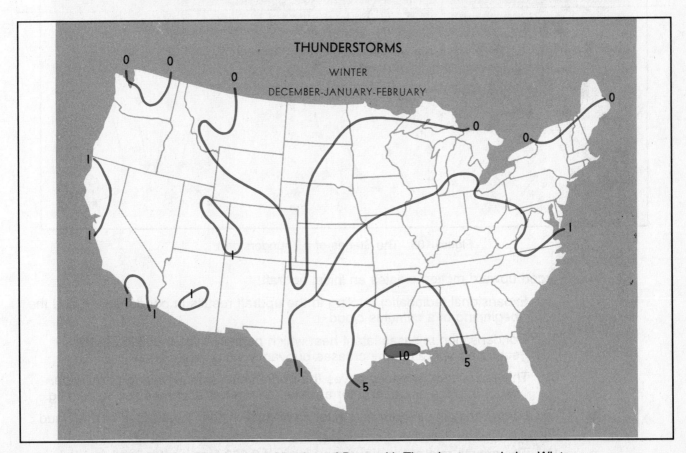

Figure 104.  The Average Number of Days with Thunderstorms during Winter

B.  **Formation of Thunderstorms**

1.  For a thunderstorm to form, the air must have

   a.  Sufficient water vapor.
   b.  An unstable lapse rate.
   c.  An initial upward boost (lifting) to start the storm process in motion.

   1)  Surface heating, converging winds, sloping terrain, a frontal surface, or any combination of these factors can provide the necessary lift.

2.  A thunderstorm cell's life cycle progresses through three stages -- the cumulus, the mature, and the dissipating.

   a.  It is virtually impossible to visually detect the transition from one stage to another; the change is subtle and by no means abrupt.

   b.  Furthermore, a thunderstorm may be a cluster of cells, each in a different stage of the life cycle.

3.  **Cumulus Stage.**  Although most cumulus clouds do not grow into thunderstorms, every thunderstorm begins as a cumulus.  The key feature of the cumulus stage is an updraft, as illustrated in Figure 105 (A) below.

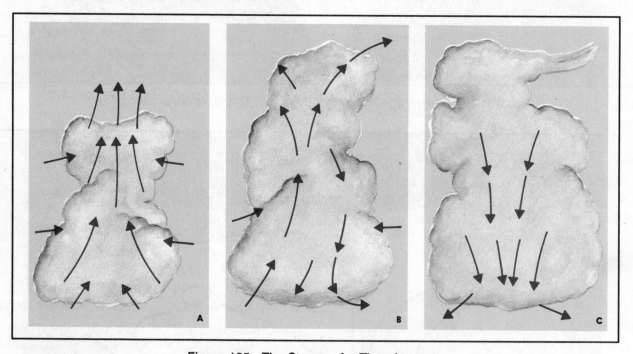

Figure 105.  The Stages of a Thunderstorm

   a.  Forced upward motion creates an initial updraft.

   1)  Expansional (adiabatic) cooling in the updraft results in condensation and the beginning of a cumulus cloud.

   2)  Condensation releases latent heat which partially offsets cooling in the saturated updraft and increases buoyancy within the cloud.

   3)  This increased buoyancy drives the updraft still faster, drawing more water vapor into the cloud, and for a while, the updraft becomes self-sustaining.

   b.  The updraft varies in strength and extends from very near the surface to the cloud top.

   1)  The growth rate of the cloud may exceed 3,000 fpm, so it is inadvisable to attempt to climb over rapidly building cumulus clouds.

      c.    Early during the cumulus stage, water droplets are quite small but grow to raindrop size as the cloud grows.

           1)    The upwelling air carries the liquid water above the freezing level, creating an icing hazard.

           2)    As the raindrops grow still heavier, they fall.

      d.    The cold rain drags air with it, creating a cold downdraft coexisting with the updraft; the cell has reached the mature stage.

    4.   **Mature Stage**. Precipitation beginning to fall from the cloud base is a sign that a downdraft has developed and that the cell has entered the mature stage.

      a.    Cold rain in the downdraft retards compressional (adiabatic) heating, and the downdraft remains cooler than surrounding air.

           1)    Thus, its downward speed is accelerated and can exceed 2,500 fpm.

           2)    The downrushing air spreads outward at the surface, as shown in Figure 105 (B) on page 118, producing strong, gusty surface winds, a sharp temperature drop, and a rapid rise in pressure.

                a)    The surface wind surge is a "plow wind" and its leading edge is the "first gust."

      b.    Meanwhile, updrafts reach a maximum with speeds sometimes exceeding 6,000 fpm.

           1)    Updrafts and downdrafts in close proximity create strong vertical shear and a very turbulent environment.

      c.    All thunderstorm hazards (discussed later in this chapter) reach their greatest intensity during the mature stage.

           1)    Duration of the mature stage is closely related to severity of the thunderstorm.

    5.   **Dissipating Stage**. Downdrafts characterize the dissipating stage of the thunderstorm cell, as shown in Figure 105 (C) on page 118, and the storm dies rapidly.

      a.    When rain has ended and downdrafts have abated, the dissipating stage is complete.

      b.    When all cells of the thunderstorm have completed this stage, only harmless cloud remnants remain.

    6.   Individual thunderstorms can measure from less than 5 mi. to more than 30 mi. in diameter.

      a.    Cloud bases range from a few hundred feet in very moist climates to 10,000 ft. or higher in drier regions.

           1)    Tops generally range from 25,000 to 45,000 ft. but occasionally extend above 65,000 ft.

## C.  Types of Thunderstorms

    1.   **Air mass thunderstorms** most often result from surface heating.

      a.    When the storm reaches the mature stage, rain falls through or immediately beside the updraft.

           1)    Falling precipitation induces frictional drag, retards the updraft and reverses it to a downdraft. Thus, the storm is self-destructive.

           2)    The downdraft and cool precipitation cool the lower portion of the storm and the underlying surface.

                a)    This cooling cuts off the inflow of water vapor, causing the storm to run out of energy and die.

           3)    A self-destructive cell usually has a life cycle of 20 min. to 1½ hr.

b.   Since air mass thunderstorms generally result from surface heating, they reach maximum intensity and frequency over land during middle and late afternoon.

   1)   Offshore, they reach a maximum during late hours of darkness when land temperature is coolest and cool air off the land flows over the relatively warm water.

2.   **Steady state thunderstorms** are usually associated with weather systems.

a.   Fronts, converging winds, and troughs aloft induce upward motion, spawning thunderstorms which often form into squall lines (see page 122).

   1)   Afternoon heating intensifies these storms.

b.   In a steady state storm, precipitation falls outside the updraft, as shown in Figure 106 below, allowing the updraft to continue unabated.

   1)   Thus, the mature stage updrafts become stronger and last much longer than in air mass storms -- hence, the name, "steady state."

   2)   A steady state cell may persist for several hours.

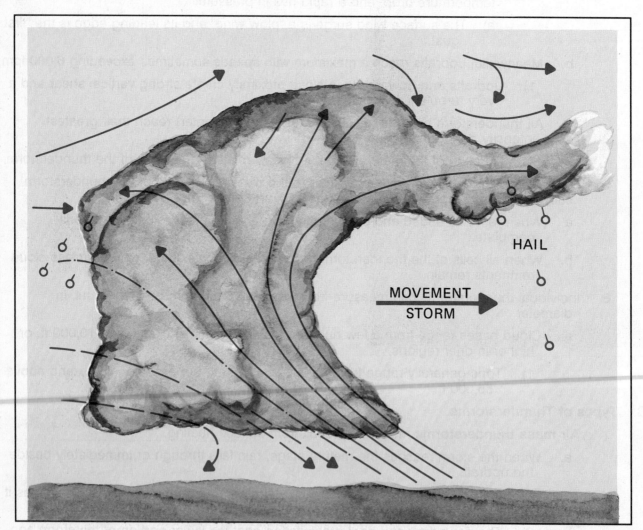

HAIL

MOVEMENT
STORM

Figure 106.  Schematic of the Mature Stage of a Steady State Thunderstorm Cell

D.  **Hazards**

   1.  **Tornadoes**

      a.  The most violent thunderstorms draw air into their cloud bases with great vigor.

         1)  If the incoming air has any initial rotating motion, it often forms an extremely concentrated vortex from the surface well into the cloud.

            a)  Meteorologists have estimated that wind in such a vortex can exceed 200 kt.; pressure inside the vortex is quite low.

         2)  The strong winds gather dust and debris, and the low pressure generates a funnel-shaped cloud extending downward from the cumulonimbus base.

            a)  If the cloud does not reach the surface, it is a *funnel cloud*.
            b)  If it touches a land surface, it is a *tornado*, Figure 107 below.
            c)  If it touches water, it is a *water spout*, Figure 108 below.

Figure 107.  A Tornado

Figure 108.  A Waterspout

      b.  Tornadoes may occur with isolated thunderstorms, but form more frequently with steady state thunderstorms associated with cold fronts or squall lines.

         1)  Reports or forecasts of tornadoes indicate that atmospheric conditions are favorable for violent turbulence.

      c.  Families of tornadoes have been observed as appendages of the main cloud extending several miles outward from the area of lightning and precipitation.

         1)  Thus, any cloud connected to a severe thunderstorm carries a threat of violence.

d.   Frequently, cumulonimbus mamma clouds (see Figure 110) occur in connection with violent thunderstorms and tornadoes.

1)   The cloud displays rounded, irregular pockets or festoons from its base and is a sign of violent turbulence and extreme instability.

Figure 110.  Cumulonimbus Mamma Clouds

2)   Surface aviation reports specifically mention this and other especially hazardous clouds.

e.   Tornadoes occur most frequently in the Great Plains states east of the Rocky Mountains.

1)   As shown in Figure 111 on page 123, however, they have occurred in every state.

f.   An aircraft entering a tornado vortex is almost certain to suffer structural damage.

1)   Since the vortex extends well into the cloud, any pilot inadvertently caught on instruments in a severe thunderstorm could encounter a hidden vortex.

2.   **Squall Lines**.  A squall line is a non-frontal, narrow band of active thunderstorms.

a.   Often it develops ahead of a cold front in moist, unstable air, but it may also develop in unstable air far removed from any front.

b.   The line may be too long to easily detour and too wide and severe to penetrate.

c.   It often contains severe steady-state thunderstorms and presents the single most intense weather hazard to aircraft.

d.   A squall line usually forms rapidly, generally reaching maximum intensity during the late afternoon and the first few hours of darkness.

e.   Figure 112 on page 123 is a photograph of an advancing squall line.

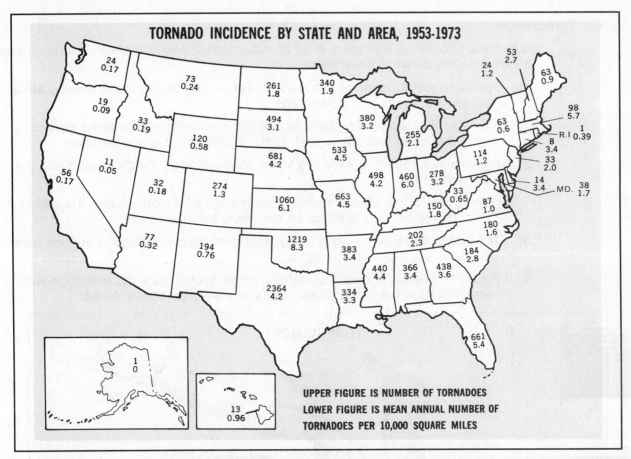

Figure 111.  Tornado Incidence by State and Area (1953-1973)

Figure 112.  Squall Line Thunderstorms

3. **Turbulence**

a.    Hazardous turbulence is present in *all* thunderstorms, and in a severe thunderstorm it can seriously damage an airframe.

    1)    The strongest turbulence within the cloud occurs as the result of wind shear between updrafts and downdrafts.

        a)    Outside the cloud, shear turbulence has been encountered several thousand feet above and 20 mi. laterally from a severe storm.

    2)    A low-level turbulent area is the shear zone between the plow wind and the surrounding air.

        a)    Often, a "roll cloud" on the leading edge of a storm marks the eddies in this shear and signifies an extremely turbulent zone.

    3)    The first gust causes a rapid and sometimes drastic change in surface wind ahead of an approaching storm.

    4)    Figure 113 below shows a schematic cross section of a thunderstorm with areas outside the cloud where turbulence may be encountered.

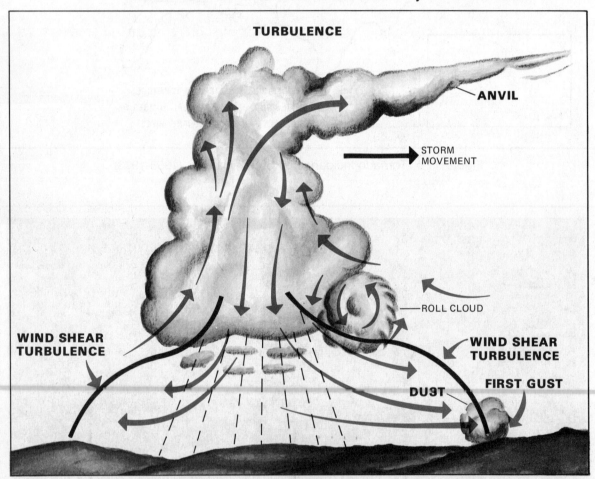

Figure 113.  Schematic Cross Section of a Thunderstorm

b.    It is almost impossible to hold a constant altitude in a thunderstorm, and maneuvering in an attempt to do so greatly increases stresses on the aircraft.

    1)    Stresses will be least if the aircraft is held in a constant *attitude* and allowed to "ride the waves."

4. **Hail**

    a. Hail competes with turbulence as the greatest thunderstorm hazard to aircraft.

       1) Supercooled drops above the freezing level begin to freeze.

         a) Once a drop has frozen, other drops attach and freeze to it, so the hailstone grows, sometimes into a huge iceball.

       2) Large hail occurs with severe thunderstorms usually built to great heights.

         a) Eventually the hailstones fall, possibly some distance from the storm core.

         b) Hail has been observed in clear air several miles from the parent thunderstorm.

    b. As hailstones fall below the freezing level, they begin to melt, and precipitation may reach the ground as either hail or rain.

       1) Rain at the surface does not mean the absence of hail aloft.

       2) You should anticipate possible hail with *any* thunderstorm, especially beneath the anvil of a large cumulonimbus.

    c. Hailstones larger than ½ in. in diameter can significantly damage an aircraft in a few seconds.

       1) Figure 114 below is a photograph of an aircraft flown through hail.

Figure 114. Hail Damage to an Aircraft

5.  **Icing**

    a.    Updrafts in a thunderstorm support abundant liquid water.

        1)    When carried above the freezing level, the water becomes supercooled.

        2)    When temperature in the upward current cools to about −15°C, much of the remaining water vapor sublimates as ice crystals.

            a)    Above this level, the amount of supercooled water decreases.

    b.    Supercooled water freezes on impact with an aircraft (see Part I, Chapter 10, Icing, beginning on page 105).

        1)    Clear icing can occur at any altitude above the freezing level, but at high levels, icing may be rime or mixed rime and clear.

        2)    The abundance of supercooled water makes clear icing very rapid between 0°C and −15°C, and encounters can be frequent in a cluster of cells.

6.  **Low Ceiling and Visibility**

    a.    Visibility generally is near zero within a thunderstorm cloud.

    b.    Ceiling and visibility also can become restricted in precipitation and dust between the cloud base and the ground.

        1)    The restrictions create the same problem as all ceiling and visibility restrictions.

    c.    The hazards are increased many times when associated with the other thunderstorm hazards of turbulence, hail, and lightning, which make precision instrument flying virtually impossible.

7.  **Effect on Altimeters**

    a.    Pressure usually falls rapidly with the approach of a thunderstorm, then rises sharply with the onset of the first gust and arrival of the cold downdraft and heavy rain showers, falling back to normal as the storm moves on.

        1)    This cycle of pressure change may occur in as little as 15 min.

    b.    If the altimeter setting is not corrected, the indicated altitude may be in error by over 100 ft.

8.  **Thunderstorm Electricity**

    a.    Electricity generated by thunderstorms is rarely a great hazard to aircraft, but it may cause damage and is annoying to flight crews.

        1)    Lightning is the most spectacular of the electrical discharges.

    b.    **Lightning**

        1)    A lightning strike can puncture the skin of an aircraft and can damage communication and electronic navigational equipment.

            a)    Lightning has been suspected of igniting fuel vapors causing explosion; however, serious accidents due to lightning strikes are extremely rare.

            b)    Nearby lightning can blind the pilot rendering him momentarily unable to navigate either by instrument or by visual reference.

            c)    Nearby lightning can also induce permanent errors in the magnetic compass.

            d)    Lightning discharges, even distant ones, can disrupt radio communications on low and medium frequencies.

      2) A few pointers on lightning:

        a) The more frequent the lightning, the more severe the thunderstorm.

        b) Increasing frequency of lightning indicates a growing thunderstorm.

        c) Decreasing lightning indicates a storm nearing the dissipating stage.

        d) At night, frequent distant flashes playing along a large sector of the horizon suggest a probable squall line.

  c. **Precipitation Static**

      1) A steady, high level of noise in radio receivers is called precipitation static and is caused by intense corona discharges from sharp metallic points and edges of flying aircraft.

        a) It is encountered often in the vicinity of thunderstorms.

        b) When an aircraft flies through clouds, precipitation, or a concentration of solid particles (ice, sand, dust, etc.), it accumulates a charge of static electricity.

        c) The electricity discharges onto a nearby surface or into the air, causing a noisy disturbance at lower frequencies.

      2) The corona discharge is weakly luminous and may be seen at night. Although it has a rather eerie appearance, it is harmless.

        a) It was named "St. Elmo's Fire" by Mediterranean sailors, who saw the brushy discharge at the top of ships' masts.

E. **Thunderstorms and Radar**

  1. Weather radar detects water droplets of precipitation size.

    a. The greater the number of drops, the stronger the echo; similarly, the larger the drops, the stronger the echo

    b. Drop size determines echo intensity to a much greater extent than does drop number.

  2. Meteorologists have shown that drop size is almost directly proportional to rainfall rate and the greatest rainfall rate is in thunderstorms.

    a. Thus, thunderstorms yield the strongest echoes.

    b. Hailstones usually are covered with a film of water and, therefore, act as huge water droplets giving the strongest echo of all types of precipitation.

    c. Showers show less intense echoes, and gentle rain and snow return the weakest of all echoes.

  3. Since the strongest echoes identify thunderstorms, they also mark the areas of greatest hazards.

    a. Realize, however, that severe turbulence associated with a thunderstorm can also occur outside and below the cell, and thus not appear on the radar.

    b. Radar information can be valuable both from ground-based radar for preflight planning and from airborne radar for severe weather avoidance.

  4. Thunderstorms build and dissipate rapidly, and they also may move rapidly.

    a. **Do not attempt to preflight plan a course between echoes.**

    b. The best use of ground radar information is to isolate general areas and coverage of echoes.

      1) You must avoid individual storms from in-flight observations either by visual sighting or by airborne radar.

5.    Airborne weather avoidance radar is, as its name implies, for avoiding severe weather, not for penetrating it.

    a.    Whether to fly into an area of radar echoes depends on echo intensity, spacing between the echoes, and your capabilities and those of your aircraft.

    b.    Remember that weather radar detects only precipitation drops; it does not detect minute cloud droplets.

    c.    **The radar scope provides no assurance of avoiding instrument weather due to clouds and fog.**

        1)    Your scope may be clear between intense echoes; this clear area does not necessarily mean you can fly between the storms and maintain visual sighting of them.

6.    The most intense echoes are severe thunderstorms.

    a.    Remember that hail may fall several miles from the cloud, and hazardous turbulence may extend as much as 20 mi. from the cloud.

        1)    Avoid the most intense echoes by at least 20 mi.; that is, echoes should be separated by at least 40 mi. before you fly between them.

        2)    As echoes diminish in intensity, you can reduce the distance by which you avoid them.

    b.    Figure 116 below illustrates use of airborne radar in avoiding thunderstorms.

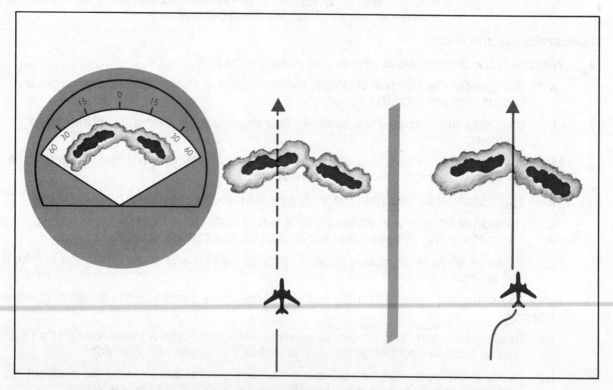

Figure 116.  Use of Airborne Radar to Avoid Heavy Precipitation and Turbulence

7. A **stormscope** senses and displays electrical discharges, as opposed to precipitation.

    a. By definition, a thunderstorm has lightning, which is a discharge of static electricity.

        1) Thus, an indication of discharge on the stormscope implies turbulence.

    b. However, a clear display only indicates a lack of electrical discharge.

        1) Convective activity, hazardous precipitation, and other thunderstorm hazards may still be present.

    c. Experts agree that a combination of stormscope and radar is the best thunderstorm detection system.

## F. In Closing

1. Thunderstorms progress through three stages: cumulus, mature, and dissipating.

2. Air mass thunderstorms generally result from surface heating, reaching maximum intensity and frequency over land during middle and late afternoon.

3. Steady state thunderstorms, usually associated with weather systems, may persist for several hours.

4. Hazards associated with thunderstorms are tornadoes, squall lines, turbulence, icing, hail, low ceiling and visibility, altimeter inaccuracy due to pressure changes, lightning, and precipitation static.

5. The strongest echoes on weather radar identify thunderstorms, and the most intense echoes are severe thunderstorms.

6. Weather radar detects precipitation, whereas a stormscope senses and displays electrical discharges.

7. Above all, remember this: **never regard any thunderstorm as "light"** even when radar observers report the echoes are of light intensity.

    a. **Avoiding thunderstorms is the best policy.**

8. The following are some DOs and DON'Ts of thunderstorm *avoidance*:

    a. Don't land or take off in the face of an approaching thunderstorm. A sudden wind shift or low level turbulence could cause loss of control.

    b. Don't attempt to fly under a thunderstorm even if you can see through to the other side. Turbulence under the storm could be disastrous.

    c. Do circumnavigate the entire area if the area has 6/10 thunderstorm coverage.

    d. Don't fly without airborne radar into a cloud mass containing scattered embedded thunderstorms. Scattered thunderstorms not embedded usually can be visually circumnavigated.

    e. Don't trust the visual appearance to be a reliable indicator of the turbulence associated with a thunderstorm.

    f. Do avoid by at least 20 NM any thunderstorm identified as severe or giving an intense radar echo. This is especially true under the anvil of a large cumulonimbus.

    g. Do clear the top of a known or suspected severe thunderstorm by at least 1,000 ft. altitude for each 10 kt. of wind speed at the cloud top. This rule would exceed the altitude capability of most aircraft.

    h. Do remember that vivid and frequent lightning indicates the probability of a severe thunderstorm.

    i. Do regard as severe any thunderstorm with tops 35,000 ft. or higher, whether the top is visually sighted or determined by radar.

9.  If you **cannot** avoid penetrating a thunderstorm, following are some DOs **before** entering the storm:

    a.  Tighten your safety belt, put on your shoulder harness if you have one, and secure all loose objects.

    b.  Plan your course to take you through the storm in a minimum time and *hold* it.

    c.  To avoid the most critical icing, establish a penetration altitude below the freezing level or above the level of −15°C.

    d.  Turn on pitot heat and carburetor or jet inlet heat. Icing can be rapid at any altitude and cause almost instantaneous power failure or loss of airspeed indication.

    e.  Establish power settings for reduced turbulence penetration airspeed recommended in your aircraft manual. Reduced airspeed lessens the structural stresses on the aircraft.

    f.  Turn up cockpit lights to highest intensity to lessen danger of temporary blindness from lightning.

    g.  If using automatic pilot, disengage altitude hold mode and speed hold mode. The automatic altitude and speed controls will increase maneuvers of the aircraft, thus increasing structural stresses.

    h.  If using airborne radar, tilt your antenna up and down occasionally.

        1)  Tilting it up may detect a hail shaft that will reach a point on your course by the time you do.

        2)  Tilting it down may detect a growing thunderstorm cell that may reach your altitude.

10. The following are some DOs and DON'Ts **during** thunderstorm penetration:

    a.  Do keep your eyes on your instruments. Looking outside the cockpit can increase danger of temporary blindness from lightning.

    b.  Don't change power settings; maintain settings for reduced airspeed.

    c.  Do maintain a constant *attitude*; let the aircraft "ride the waves." Maneuvers in trying to maintain constant altitude increase stresses on the aircraft.

    d.  Don't turn back once you are in a thunderstorm. A straight course through the storm most likely will get you out of the hazards most quickly. In addition, turning maneuvers increase stresses on the aircraft.

## G. Addendum: Microbursts

1.  Microbursts are small-scale intense downdrafts which, on reaching the surface, spread outward in all directions from the downdraft center. This causes the presence of both vertical and horizontal wind shears that can be extremely hazardous to all types and categories of aircraft, especially at low altitudes.

2.  Parent clouds producing microburst activity can be any of the low or middle layer convective cloud types.

    a.  Microbursts commonly occur within the heavy rain portion of thunderstorms, but also in much weaker, benign-appearing convective cells that have little or no precipitation reaching the ground.

3.   The life cycle of a microburst as it descends in a convective rain shaft is illustrated below.

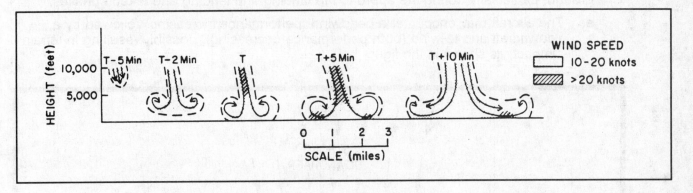

a.   T is the time the microburst strikes the ground.

4.   Characteristics of microbursts include:

a.   Size -- The microburst downdraft is typically less than 1 mi. in diameter as it descends from the cloud base to about 1,000-3,000 ft. above the ground.

1)   In the transition zone near the ground, the downdraft changes to a horizontal outflow that can extend to approximately 2½ mi. in diameter.

b.   Intensity -- The downdrafts can be as strong as 6,000 fpm.

1)   Horizontal winds near the surface can be as strong as 45 kt. resulting in a 90-kt. shear (headwind to tailwind change for a traversing aircraft) across the microburst.

2)   These strong horizontal winds occur within a few hundred feet of the ground.

c.   Visual signs -- Microbursts can be found almost anywhere there is convective activity.

1)   They may be embedded in heavy rain associated with a thunderstorm or in light rain in benign-appearing virga.

2)   When there is little or no precipitation at the surface accompanying the microburst, a ring of blowing dust may be the only visual clue of its existence.

d.   Duration -- An individual microburst will seldom last longer than 15 min. from the time it strikes the ground until dissipation.

1)   An important consideration for pilots is that the microburst intensifies for about 5 min. after it strikes the ground, with the maximum intensity winds lasting approximately 2 to 4 min.

2)   Once microburst activity starts, multiple microbursts in the same general area are not uncommon and should be expected.

3)   Sometimes microbursts are concentrated into a line structure, and under these conditions, activity may continue for as long as an hour.

5.   Microburst wind shear may create a severe hazard for aircraft within 1,000 ft. of the ground, particularly during the approach to landing and landing and takeoff phases.

a.   The aircraft may encounter a headwind (performance increasing) followed by a downdraft and tailwind (both performance decreasing), possibly resulting in terrain impact, as shown in the figure below.

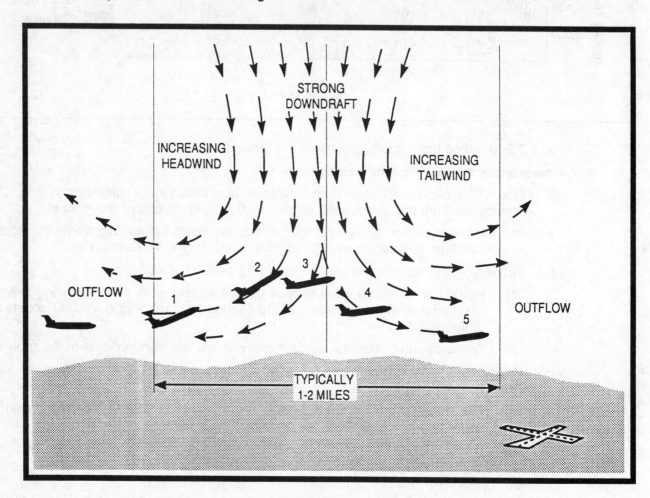

6.   Flight in the vicinity of suspected or reported microburst activity should always be avoided.

# END OF CHAPTER

# CHAPTER TWELVE
# COMMON IFR PRODUCERS

Please take a few minutes to study each of the concepts listed above and anticipate/imagine what they are and how they relate to the other listed concepts.

## A. Fog

1. Fog is a surface-based cloud composed of either water droplets or ice crystals.

   a. Fog is the most frequent cause of reducing surface visibility to below 3 SM, and is one of the most common and persistent weather hazards encountered in aviation.

      1) The rapidity with which fog can form makes it especially hazardous.

         a) It is not unusual for visibility to drop from VFR to less than 1 SM in a few minutes.

      2) It is primarily a hazard during takeoff and landing, but it is also important to VFR pilots who must maintain visual reference to the ground.

   b. A small temperature-dew point spread is essential for fog to form.

      1) Therefore, fog is prevalent in coastal areas where moisture is abundant.

         a) However, fog can occur anywhere.

      2) Abundant condensation nuclei enhance the formation of fog.

         a) Thus, fog is prevalent in industrial areas where by-products of combustion provide a high concentration of these nuclei.

      3) Fog occurs most frequently in the colder months, but the season and frequency of occurrence vary from one area to another.

   c. Fog is classified by the way it forms. Formation may involve more than one process. Fog may form

      1) By cooling air to its dew point.
      2) By adding moisture to air near the ground.

2.  **Radiation fog** is relatively shallow fog.  It may be dense enough to hide the entire sky or may conceal only part of the sky.

   a.   "Ground fog" is a form of radiation fog.

   1)   As viewed in flight, dense radiation fog may obliterate the entire surface below you.

   a)   A less dense fog may permit your observation of a small portion of the surface directly below you.

   b)   Tall objects such as buildings, hills, and towers may protrude upward through ground fog giving you fixed references for VFR flight.

   c)   Figure 117 below illustrates ground fog as seen from the air.

Figure 117.  Ground Fog as Seen from the Air

   b.   Conditions favorable for radiation fog are clear sky, little or no wind, and a small temperature-dew point spread (high relative humidity).

   1)   The fog forms almost exclusively at night or near daybreak.

   2)   Terrestrial radiation cools the ground; in turn, the cool ground cools the air in contact with it.

   a)   When the air is cooled to its dew point, fog forms.

   3)   When rain soaks the ground, followed by clearing skies, radiation fog is not uncommon the following morning.

   c.   Radiation fog is restricted to land because water surfaces cool little from nighttime radiation.

   1)   It is shallow when wind is calm.

   2)   Winds up to about 5 kt. mix the air slightly and tend to deepen the fog by spreading the cooling through a deeper layer.

   a)   Stronger winds disperse the fog or mix the air through a still deeper layer with stratus clouds forming at the tip of the mixing layer.

   d.   Ground fog usually "burns off" rather rapidly after sunrise.

   1)   Other radiation fog generally clears before noon unless clouds move in over the fog.

3. **Advection fog** forms when moist air moves over colder ground or water.

    a. It is most common along coastal areas but often develops deep in continental areas.

        1) At sea it is called "sea fog."

    b. Advection fog deepens as wind speed increases up to about 15 kt.

        1) Wind much stronger than 15 kt. lifts the fog into a layer of low stratus or stratocumulus.

    c. The west coast of the United States is quite vulnerable to advection fog.

        1) This fog frequently forms offshore as a result of cold water as shown in Figure 118 below and then is carried inland by the wind.

Figure 118. Advection Fog off the Coast of California

d.   During the winter, advection fog over the central and eastern United States results when moist air from the Gulf of Mexico spreads northward over cold ground as shown in Figure 119 below.

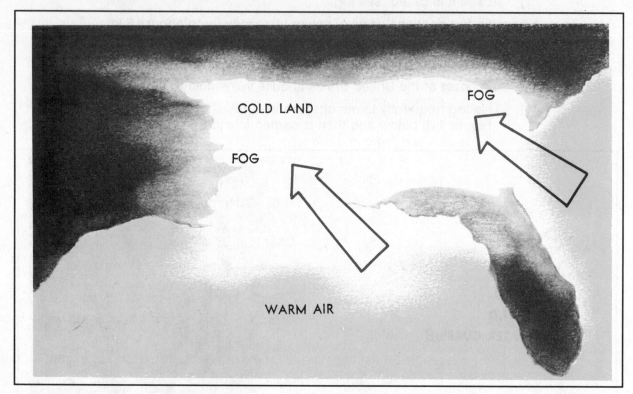

Figure 119.  Advection Fog over Southeastern U.S. and Gulf Coast

1)   The fog may extend as far north as the Great Lakes.

e.   Water areas in northern latitudes have frequent dense sea fog in summer as a result of warm, moist, tropical air flowing northward over colder Arctic waters.

f.   A pilot will notice little difference between flying over advection fog and over radiation fog except that skies may be cloudy above the advection fog.

1)   Also, advection fog is usually more extensive and much more persistent than radiation fog.

2)   Advection fog can move in rapidly regardless of the time of day or night.

4.  **Upslope fog** forms as a result of moist, stable air being cooled adiabatically as it moves up sloping terrain.

    a.  Once the upslope wind ceases, the fog dissipates.

    b.  Unlike radiation fog, it can form under cloudy skies.

    c.  Upslope fog is common along the eastern slopes of the Rockies and somewhat less frequent east of the Appalachians.

    d.  Upslope fog often is quite dense and extends to high altitudes.

5.  **Precipitation-induced fog** forms when relatively warm rain or drizzle falls through cool air, and evaporation from the precipitation saturates the cool air.

    a.  Precipitation-induced fog can become quite dense and continue for an extended period of time.

        1)  This fog may extend over large areas, completely suspending air operations.

        2)  It is most commonly associated with warm fronts, but can occur with slow moving cold fronts and with stationary fronts.

    b.  Fog induced by precipitation is in itself hazardous, as is any fog.

        1)  It is especially critical, however, because it occurs in the proximity of precipitation and other possible hazards such as icing, turbulence, and thunderstorms.

6.  **Ice fog** occurs in cold weather when the temperature is well below freezing and water vapor sublimates directly as ice crystals.

    a.  Conditions favorable for its formation are the same as for radiation fog except for cold temperature, usually –32°C (–25°F) or colder.

        1)  It occurs mostly in the Arctic regions, but is not unknown in middle latitudes during the cold season.

    b.  Ice fog can be blinding to someone flying in the direction of the Sun.

B.  **Low Stratus Clouds**

    1.  Stratus clouds, like fog, are composed of extremely small water droplets or ice crystals suspended in air.

    2.  Stratus and fog frequently exist together.

        a.  In many cases there is no real line of distinction between fog and stratus; rather, one gradually merges into the other.

    3.  Flight visibility may approach zero in stratus clouds.

    4.  Stratus tends to be lowest during night and early morning, lifting or dissipating due to solar heating during the late morning or afternoon.

        a.  Low stratus clouds often occur when moist air mixes with a colder air mass or in any situation where temperature-dew point spread is small.

C.  **Haze and Smoke**

1.   Haze is a concentration of salt particles or other dry particles not readily classified as dust or other phenomena.

   a.   It occurs in stable air, is usually only a few thousand feet thick, but sometimes may extend as high as 15,000 ft.

   b.   Haze layers often have definite tops above which horizontal visibility is good.

      1)   However, downward visibility from above a haze layer is poor, especially on a slant.

   c.   Visibility in haze varies greatly depending upon whether the pilot is facing the Sun.

      1)   Landing an aircraft into the Sun is often hazardous when haze is present.

2.   Smoke concentrations form primarily in industrial areas when air is stable.

   a.   It is most prevalent at night or early morning under a temperature inversion but it can persist throughout the day.

      1)   Figure 120 below illustrates smoke trapped under a temperature inversion.

Figure 120.  Smoke Trapped in Stagnant Air under an Inversion

3.   When skies are clear above haze or smoke, visibility generally improves during the day; however, the improvement is slower than the clearing of fog.

   a.   Fog evaporates, but haze or smoke must be dispersed by movement of air.

      1)   Haze or smoke may be blown away; or heating during the day may cause convective mixing that spreads the smoke or haze to a higher altitude, decreasing the concentration near the surface.

   b.   At night or in the early morning, radiation fog or stratus clouds often combine with haze or smoke.

      1)   The fog and stratus may clear rather rapidly during the day but the haze and smoke will linger.

      2)   A heavy cloud cover above haze or smoke may block sunlight, preventing dissipation; visibility will improve little, if any, during the day.

D.  **Blowing Restrictions to Visibility**

    1.    Strong wind lifts blowing dust in both stable and unstable air.

        a.    When air is unstable, dust is lifted to great heights (as much as 15,000 ft.) and may be spread over wide areas by upper winds.

            1)    Visibility is restricted both at the surface and aloft.

        b.    When air is stable, dust does not extend to as great a height as in unstable air and usually is not as widespread.

        c.    Dust, once airborne, may remain suspended and restrict visibility for several hours after the wind subsides.

    2.    Blowing sand is more local than blowing dust; the sand is seldom lifted above 50 ft. However, visibilities within it may be near zero.

        a.    Blowing sand may occur in any dry area where loose sand is exposed to strong wind.

    3.    Blowing snow can cause visibility at ground level to be near zero and the sky may become obscured when the particles are raised to great heights.

E.  **Precipitation**

    1.    Rain, drizzle, and snow are the forms of precipitation which most commonly present ceiling and/or visibility problems.

    2.    Drizzle or snow restricts visibility to a greater degree than rain.

        a.    Drizzle falls in stable air and, therefore, often accompanies fog, haze, or smoke, frequently resulting in extremely poor visibility.

        b.    Visibility may be reduced to zero in heavy snow.

    3.    Rain seldom reduces surface visibility below 1 SM except in brief, heavy showers, but it does limit cockpit visibility.

        a.    When rain streams over the aircraft windshield, freezes on it, or fogs over the inside surface, the pilot's visibility to the outside is greatly reduced.

F.  **Obscured or Partially Obscured Sky**

    1.    To be classified as obscuring phenomena, smoke, haze, fog, precipitation, or other visibility restricting phenomena must extend upward from the surface.

        a.    When the sky is totally hidden by the surface-based phenomena, the ceiling is the vertical visibility from the ground upward into the obscuration.

        b.    If clouds or part of the sky can be seen above the obscuring phenomena, the condition is defined as a partial obscuration.

            1)    A partial obscuration does not define a ceiling.

            2)    However, a cloud layer above a partial obscuration may constitute a ceiling.

    2.    An obscured ceiling differs from a cloud ceiling.

        a.    With a cloud ceiling you normally can see the ground and runway once you descend below the cloud base.

        b.    With an obscured ceiling, the obscuring phenomena restrict visibility between your altitude and the ground, and you have restricted slant visibility.

1)    Thus, you cannot always clearly see the runway or approach lights even after penetrating the level of the obscuration ceiling as shown in Figure 122 below.

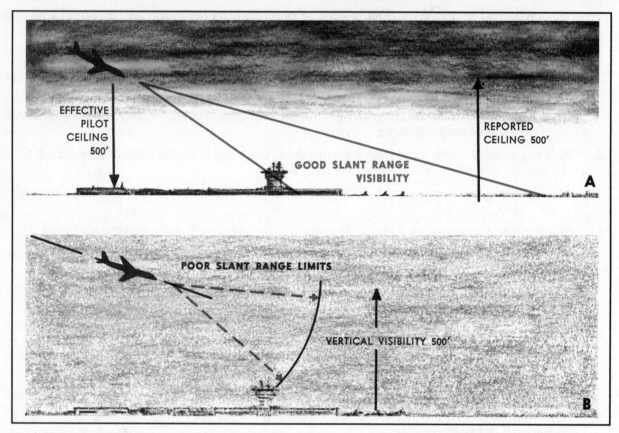

EFFECTIVE
PILOT
CEILING
500'

GOOD SLANT RANGE
VISIBILITY

REPORTED
CEILING 500'

**A**

POOR SLANT RANGE LIMITS

VERTICAL VISIBILITY 500'

**B**

Figure 122.  Difference in Visibility When a Ceiling Is Caused by a Layer Aloft (A) and by a Surface-Based Obscuration (B)

3.    Partial obscurations also present a visibility problem for the pilot on approach to land but usually to a lesser degree than the total obscuration.

a.    However, be especially aware of erratic visibility reduction in the partial obscuration.

1)    Visibility along the runway or on the approach can instantaneously become zero.

2)    This abrupt and unexpected reduction in visibility can be extremely hazardous, especially on touchdown.

## G.  How to Avoid IFR Weather

1.    In your preflight preparation, be aware of phenomena that may produce IFR or marginal VFR flight conditions.

a.    Current charts and special analyses along with forecast and prognostic charts are your best sources of information.

b.    No weather observation is more current or more accurate than the one you make through your cockpit window.

c.    Your understanding of IFR producers will help you make better preflight and in-flight decisions.

2.  Do not fly VFR in weather suitable only for IFR.

    a.  If you do, you endanger not only your own life but the lives of others both in the air and on the ground.

    b.  Remember, the single cause of the greatest number of general aviation fatal accidents is "continued VFR into adverse weather."

3.  Be especially alert for development of:

    a.  Fog the following morning when at dusk temperature-dew point spread is 10°C (15°F) or less, skies are clear, and winds are light.

    b.  Fog when moist air is flowing from a relatively warm surface to a colder surface.

    c.  Fog when temperature-dew point spread is 3°C (5°F) or less and decreasing.

    d.  Fog or low stratus when a moderate or stronger moist wind is blowing over an extended upslope.

        1)  Temperature and dew point converge at about 2°C (4°F) for every 1,000 ft. the air is lifted.

    e.  Steam fog when air is blowing from a cold surface (either land or water) over warmer water.

    f.  Fog when rain or drizzle falls through cool air.

        1)  This is especially prevalent during winter ahead of a warm front and behind a stationary front or stagnating cold front.

    g.  Low stratus clouds whenever there is an influx of low level moisture overriding a shallow cold air mass.

    h.  Low visibilities from haze and smoke when a high pressure area stagnates over an industrial area.

    i.  Low visibilities due to blowing dust or sand over semiarid or arid regions when winds are strong and the atmosphere is unstable.

        1)  This is especially prevalent in spring.

        2)  If the dust extends upward to moderate or greater heights, it can be carried many miles beyond its source.

    j.  Low visibility due to snow or drizzle.

    k.  An undercast when you must make a VFR descent.

4.  Expect little if any improvement in visibility when:

    a.  Fog exists below heavily overcast skies.
    b.  Fog occurs with rain or drizzle and precipitation is forecast to continue.
    c.  Dust extends to high levels and no frontal passage or precipitation is forecast.
    d.  Smoke or haze exists under heavily overcast skies.
    e.  A stationary high persists over industrial areas.

H.  **In Closing**

1.  Fog is classified by the way it forms.

    a.  Radiation fog is relatively shallow, forming at night or near daybreak when the air is cooled to its dew point.

    b.  Advection fog forms when moist air moves over colder ground or water.

    c.  Upslope fog forms as moist, stable air is cooled adiabatically while moving up sloping terrain.

    d.  Precipitation-induced fog forms when evaporation from relatively warm rain or drizzle saturates cooler air.

    e.  Ice fog occurs when the temperature is well below freezing, and water vapor sublimates directly as ice crystals.

2.  Low stratus clouds frequently exist with fog, and may cause flight visibility to approach zero.

3.  Haze and smoke must be dispersed by movement of air and will clear more slowly than fog or stratus clouds.

4.  Strong wind lifts blowing dust as high as 15,000 ft., restricting visibility at the surface and aloft.

    a.  Blowing sand and blowing snow can cause visibility to be near zero.

5.  Precipitation presents ceiling and/or visibility problems.

    a.  Drizzle or snow restricts visibility to a greater degree than rain.

6.  Surface-based phenomena can partially or completely obscure the sky.

    a.  The obscuring phenomena restrict visibility between your altitude and the ground, whereas with a cloud ceiling, you can see the ground and runway once you descend below the cloud base.

7.  Refer to topic G., How to Avoid IFR Weather, on page 140.  Your recognition of phenomena that may produce IFR conditions will help you make better preflight and in-flight decisions.

# END OF CHAPTER
# END OF PART I

# PART II
# AVIATION WEATHER -- OVER AND BEYOND

The following four chapters are separated into Part II to facilitate study of Part I by pilots without interest in these four special topics.  Our Part I outline is 142 pages long, and our Part II outline 67 pages long.

13.  High Altitude Weather
14.  Arctic Weather
15.  Tropical Weather
16.  Soaring Weather

It is less intimidating to group the first 12 chapters as Part I and Chapters 13 through 16 as Part II.  You can then study only the Part II topics that are relevant to you.

# CHAPTER THIRTEEN
# HIGH ALTITUDE WEATHER

Please take a few minutes to study each of the concepts listed above and anticipate/imagine what they are and how they relate to the other listed concepts.

## A. The Tropopause

1. The tropopause is a thin layer forming the boundary between the troposphere and stratosphere.

   a. The height of the tropopause varies from about 65,000 ft. over the Equator to 20,000 ft. or lower over the poles.

   b. The tropopause is not continuous but generally descends step-wise from the Equator to the poles.

      1) These steps occur as "breaks."

      2) Figure 123 below is a cross section of the troposphere and lower stratosphere showing the tropopause and associated features.

         a) Note the break between the tropical and the polar tropopauses.

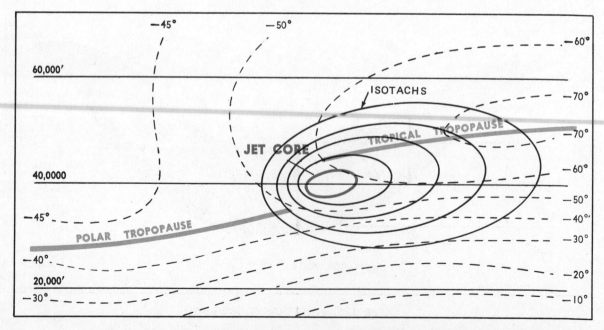

Figure 123. A Cross Section of the Upper Troposphere and Lower Stratosphere

2.   An abrupt change in temperature lapse rate characterizes the tropopause.

   a.   Note in Figure 123 how temperature above the tropical tropopause increases with height and how, over the polar tropopause, temperature remains almost constant with height.

3.   Temperature and wind vary greatly in the vicinity of the tropopause, affecting efficiency, comfort, and safety of flight.

   a.   Maximum winds generally occur at levels near the tropopause.

      1)   These strong winds create narrow zones of wind shear which often generate hazardous turbulence.

   b.   Preflight knowledge of temperature, wind, and wind shear is important to flight planning.

B.   **The Jet Stream**

1.   The jet stream, or jet, is a narrow, shallow, meandering river of maximum winds extending around the globe in a wavelike pattern.

   a.   A second jet stream is not uncommon, and three at one time are not unknown.

      1)   A jet may be as far south as the northern Tropics.
      2)   A jet in midlatitudes generally is stronger than one in or near the Tropics.

   b.   The jet stream typically occurs in a break in the tropopause as shown in Figure 123 on page 144.

      1)   Thus, a jet stream occurs in an area of intensified temperature gradients characteristic of the break.

2.   The concentrated winds, by arbitrary definition, must be 50 kt. or greater to classify as a jet stream.

   a.   The jet maximum (concentration of greatest wind) is not constant; rather, it is broken into segments, shaped somewhat like a boomerang as diagramed in Figure 125 below.

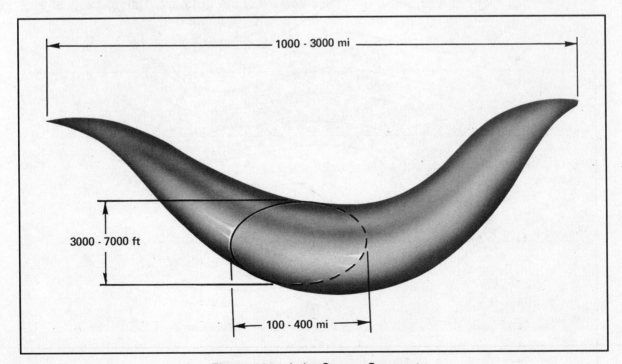

Figure 125.  A Jet Stream Segment

3.   Jet stream segments move with pressure ridges and troughs in the upper atmosphere.

   a.   In general they travel faster than pressure systems, and maximum wind speed varies as the segments progress through the systems.

   b.   In midlatitude, wind speed in the jet stream averages considerably stronger in winter than in summer.

      1)   Also, the jet shifts farther south in winter than in summer.

4.   In Figure 123 on page 144, the isotachs (lines of equal wind speed) illustrate that the maximum wind speed is at the jet core and that each isotach away from the jet core the wind speed decreases.

   a.   The rate of decrease of wind speed is considerably greater on the polar side than on the equatorial side; thus, the magnitude of wind shear is greater on the polar side than on the equatorial side.

5.   Figure 126 below shows a map with two jet streams.  The paths of the jets approximately conform to the shape of the pressure contours.

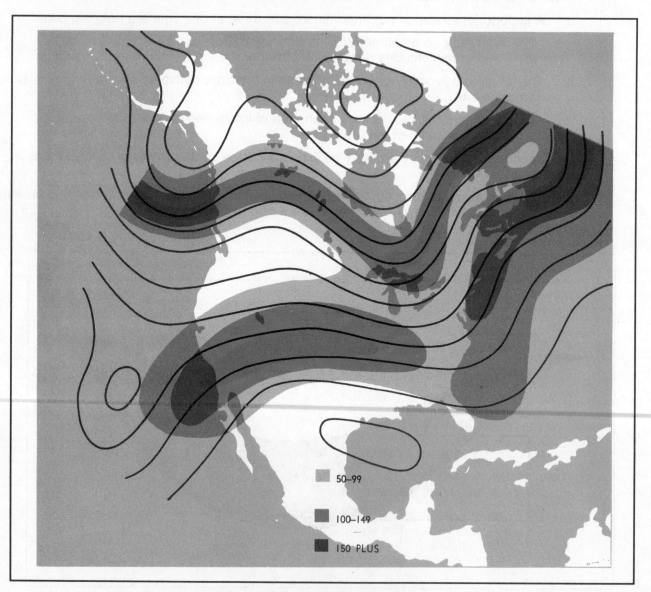

Figure 126.  Multiple Jet Streams

a.   Note how spacing of the pressure contours is closer and wind speeds higher in the vicinity of the jets than outward on either side.

1)   Thus, horizontal wind shear is evident on both sides of the jet and is greatest near the maximum wind segments.

6.   Strong, long-trajectory jet streams usually are associated with well-developed surface lows and frontal systems beneath deep upper troughs or lows.

a.   Cyclogenesis (i.e., development of a surface low) is usually south of the jet stream and moves nearer as the low deepens.

b.   The occluding low moves north of the jet, and the jet crosses the frontal system near the point of occlusion (right-hand side of Figure 127).

c.   Figure 127 below diagrams mean jet positions relative to surface systems.

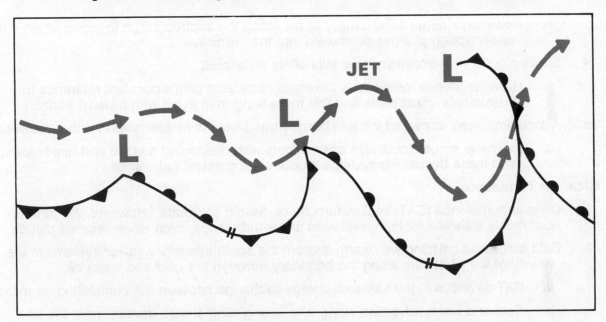

Figure 127.  Mean Jet Positions Relative to Surface Systems

d.   These long jets mark high level boundaries between warm and cold air and are favored places for cirriform cloudiness.

## C.  Cirrus Clouds

1.   Air travels in a corkscrew path around the jet core with upward motion on the equatorial side.

a.   Thus, when high level moisture is available, cirriform clouds form on the equatorial side of the jet.

b.   Jet stream cloudiness can form independently of well-defined pressure systems.

1)   Such cloudiness ranges primarily from scattered to broken coverage in shallow layers or streaks.

2)   The occasional fish hook and streamlined, wind-swept appearance of jet stream cloudiness always indicates very strong upper wind, usually quite far from developing or intense weather systems.

2.   The densest cirriform clouds occur with well-defined systems.  They appear in broad bands.

   a.   Cloudiness is rather dense in an upper trough, thickens downstream, and becomes most dense at the crest of the downwind ridge.

      1)   The clouds taper off after passing the ridge crest into the area of descending air.

   b.   The poleward boundary of the cirrus band often is quite abrupt and frequently casts a shadow on lower clouds, especially in an occluded frontal system.

3.   The upper limit of dense, banded cirrus is near the tropopause; a band may be either a single layer or multiple layers 10,000 ft. to 12,000 ft. thick.

   a.   Dense, jet stream cirriform cloudiness is most prevalent along midlatitude and polar jets.

   b.   However, a cirrus band usually forms along the subtropical jet in winter when a deep upper trough plunges southward into the Tropics.

4.   Cirrus clouds, in themselves, have little effect on aircraft.

   a.   However, dense, continuous coverage requires a pilot's constant reference to instruments; most pilots find this more tiring than flying with a visual horizon.

5.   A more important aspect of the jet stream cirrus shield is its association with turbulence.

   a.   Extensive cirrus cloudiness often occurs with deepening surface and upper lows; and these deepening systems produce the greatest turbulence.

## D.   Clear Air Turbulence

1.   Clear air turbulence (CAT) implies turbulence devoid of clouds.  However, the term is commonly reserved for high-level wind shear turbulence, even when in cirrus clouds.

2.   Cold outbreaks colliding with warm air from the south intensify weather systems in the vicinity of the jet stream along the boundary between the cold and warm air.

   a.   CAT develops in the turbulent energy exchange between the contrasting air masses.

   b.   Cold and warm advection along with strong wind shears develop near the jet stream, especially where curvature of the jet stream sharply increases in deepening upper troughs.

   c.   CAT is most pronounced in winter when temperature contrast is greatest between cold and warm air.

3.   CAT is found most frequently in an upper trough on the cold (polar) side of the jet stream.

   a.   Another frequent CAT location, shown in Figure 129 on page 149, is along the jet stream north and northeast of a rapidly deepening surface low.

4.   Even in the absence of a well-defined jet stream, CAT is often experienced in wind shears associated with sharply curved contours of strong lows, troughs, and ridges aloft, and in areas of strong, cold or warm air advection.

   a.   Also, mountain waves can create CAT.

      1)   Mountain wave CAT may extend from the mountain crests to as high as 5,000 ft. above the tropopause, and can range 100 mi. or more downstream from the mountains.

5.   CAT can be encountered where there seems to be no reason for its occurrence.

   a.   Strong winds may carry a turbulent volume of air away from its source region.

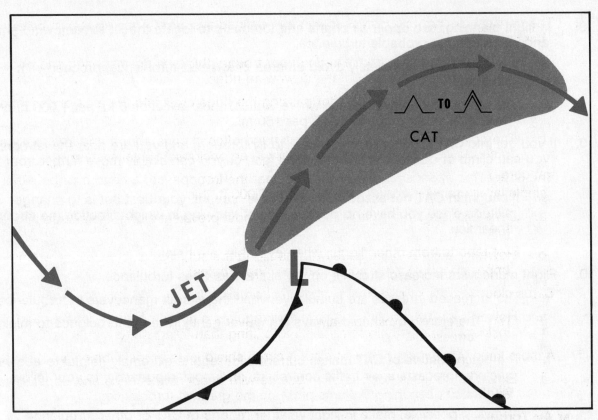

Figure 129.  A Frequent CAT Location

   b.   Turbulence intensity diminishes downstream, but some turbulence still may be
        encountered where it normally would not be expected.

   c.   CAT forecast areas are sometimes elongated to indicate probable turbulence drifting
        downwind from the main source region.

6.   A forecast of turbulence specifies a volume of airspace which is quite small when
     compared to the total volume of airspace used by aviation, but is relatively large
     compared to the localized extent of the hazard.

   a.   Since turbulence in the forecast volume is patchy, you can expect to encounter it
        only intermittently and possibly not at all.

   b.   A flight through forecast turbulence, on the average, encounters only light and
        annoying turbulence 10% to 15% of the time.

        1)   About 2% to 3% of the time there is a need to have all objects secured.
        2)   The odds that a pilot will experience control problems are about 1 in 500.

7.   Turbulence is greatest near the jet stream windspeed maxima, usually on the polar sides,
     where there is a combination of strong wind shear, curvature in the flow, and cold air
     advection.

   a.   In Figure 126 on page 146, these areas would be to the northwest of Vancouver
        Island, from north of the Great Lakes to east of James Bay and over the Atlantic
        east of Newfoundland.

        1)   Also, turbulence in the form of mountain waves is probable in the vicinity of
             the jet stream from southern California across the Rockies into the Central
             Plains.

8.    In flight planning, use upper air charts and forecasts to locate the jet stream, wind shears, and areas of most probable turbulence.

    a.    If impractical to completely avoid an area of forecast turbulence, proceed with caution.

    b.    You will do well to avoid areas where vertical shear exceeds 6 kt. per 1,000 ft. or horizontal shear exceeds 40 kt. per 150 mi.

9.    If you get into CAT rougher than you care to fly through, and you are near the jet core, you can climb or descend a few thousand feet, or you can simply move farther from the jet core.

    a.    If caught in CAT not associated with the jet stream, your best bet is to change altitude since you have no positive way of knowing in which direction the strongest shear lies.

    b.    Pilot reports from other flights, when available, are helpful.

10.    Flight maneuvers increase stresses on the aircraft, as does turbulence.

    a.    The increased stresses are cumulative when the aircraft maneuvers in turbulence.

        1)    Therefore, you should always maneuver gently when in turbulence to minimize stress.

    b.    The patchy nature of CAT makes current pilot reports extremely helpful to observers, briefers, forecasters, air traffic controllers, and most importantly, to your fellow pilots.

    c.    Always, if possible, make in-flight weather reports of CAT or other turbulence encounters.

        1)    Negative reports also help when no CAT is experienced where it normally might be expected.

## E.    Condensation Trails

1.    A condensation trail, popularly contracted to "contrail," is generally defined as a cloud-like streamer which is frequently generated in the wake of aircraft flying in clear, cold, humid air.

    a.    Two distinct types are observed -- exhaust trails and aerodynamic trails.

2.    **Exhaust Contrails.**  The exhaust contrail is formed by the addition to the atmosphere of sufficient water vapor from aircraft exhaust gases to cause saturation or supersaturation of the air.

    a.    Since heat is also added to the atmosphere in the wake of an aircraft, the addition of water vapor must be of such magnitude that it saturates or supersaturates the atmosphere despite the added heat.

3.    **Aerodynamic Contrails.**  In air that is almost saturated, aerodynamic pressure reduction around airfoils, engine nacelles, and propellers cools the air to saturation leaving condensation trails from these components.

    a.    This type of trail usually is neither as dense nor as persistent as exhaust trails.

        1)    However, under critical atmospheric conditions, an aerodynamic contrail may trigger the formation and spreading of a deck of cirrus clouds.

        2)    The induced layer may make necessary the strict use of instruments by a subsequent flight at that altitude.

   4.   **Dissipation Trails (Distrails).**  The term dissipation trail applies to a rift in clouds caused by the heat of exhaust gases from an aircraft flying in a thin cloud layer.

        a.   The exhaust gases sometimes warm the air to the extent that it is no longer saturated, and the affected part of the cloud evaporates.

        b.   The cloud must be both thin and relatively warm for a distrail to exist; therefore, they are not common.

F.   **Haze Layers**

   1.   Haze layers not visible from the ground are, at times, of concern at high altitude.

        a.   These layers are really cirrus clouds with a very low density of ice crystals.
        b.   The tops of these layers generally are very definite and are at the tropopause.
        c.   High-level haze occurs in stagnant air; it is rare in fresh outbreaks of cold polar air.
             1)   Cirrus haze is common in Arctic winter.
             2)   Sometimes ice crystals restrict visibility from the surface to the tropopause.

   2.   Visibility in the haze sometimes may be near zero, especially when one is facing the sun.

        a.   To avoid the poor visibility, climb into the lower stratosphere or descend below the haze.

        b.   This change may need to be several thousand feet.

G.   **Canopy Static**

   1.   Canopy static, similar to the precipitation static sometimes encountered at lower levels, is produced by particles brushing against plastic-covered aircraft surfaces.

        a.   The discharge of static electricity results in a noisy disturbance that interferes with radio reception.

        b.   Discharges can occur in such rapid succession that interference seems to be continuous.

   2.   Since dust and ice crystals in cirrus clouds are the primary producers of canopy static, usually you may eliminate it by changing altitude.

H.   **Icing**

   1.   Although icing at high altitudes is not as common or extreme as at low altitudes, it can occur.

        a.   It can form quickly on airfoils and exposed parts of jet engines.
        b.   Structural icing at high altitudes usually is rime, although clear ice is possible.

   2.   High-altitude icing generally forms in tops of tall cumulus buildups, anvils, and even in detached cirrus.

        a.   Clouds over mountains are more likely to contain liquid water than those over more gently sloping terrain because of the added lift of the mountains.

             1)   Thus, icing is more likely to occur and to be more hazardous over mountainous areas.

   3.   Because ice generally accumulates slowly at high altitudes, anti-icing equipment usually eliminates any serious problems.

        a.   However, anti-icing systems currently in use are not always adequate.

        b.   If such is the case, avoid the icing problem by changing altitude or by varying course to remain clear of the clouds.

I.  **Thunderstorms**

1.  A well-developed thunderstorm may extend upward through the troposphere and penetrate the lower stratosphere.

    a.  Sometimes the main updraft in a thunderstorm may toss hail out the top or the upper portions of the storm.

        1)  An aircraft may thus encounter hail in clear air at a considerable distance from the thunderstorm, especially under the anvil cloud.

    b.  Turbulence may be encountered in clear air for a considerable distance both above and around a growing thunderstorm.

2.  Thunderstorm avoidance rules apply equally at high altitudes.

    a.  When flying in the clear, visually avoid all thunderstorm tops.

        1)  In a severe thunderstorm situation, avoid tops by at least 20 mi.

    b.  When you are on instruments, weather avoidance radar helps·you in avoiding thunderstorm hazards.

        1)  If in an area of severe thunderstorms, avoid the most intense echoes by at least 20 mi.

J.  **In Closing**

1.  High-altitude weather phenomena include

    a.  The tropopause -- the thin layer forming the boundary between the troposphere and the stratosphere

        1)  Characterized by an abrupt change in temperature lapse rate.

    b.  The jet stream -- a river of maximum winds (50 kt. or greater) extending around the globe in a wavelike pattern.

    c.  Cirrus clouds -- when dense, require you to refer constantly to your instruments.

        1)  The jet stream cirrus shield is associated with turbulence.

    d.  Clear air turbulence (CAT) -- a term commonly reserved for high-level wind shear turbulence.

    e.  Condensation trails -- cloud-like streamers that are frequently generated in the wake of aircraft flying in clear, cold, humid air.

    f.  Haze layers -- really cirrus clouds with a low density of ice crystals.

    g.  Canopy static -- discharge of static electricity causing interference with radio reception.

    h.  Icing  can occur at high altitudes, although it is not as common or extreme as at low altitudes.

    i.  Thunderstorms -- may extend upward through the troposphere and penetrate the lower stratosphere.  Thunderstorm avoidance rules apply equally at high altitudes.

2.  Use upper air charts and forecasts (see Part III of this book) to locate areas to avoid.

# END OF CHAPTER

# CHAPTER FOURTEEN
# ARCTIC WEATHER

Please take a few minutes to study each of the concepts listed above and anticipate/imagine what they are and how they relate to the other listed concepts.

## A.    Definition

1.    The Arctic is the region shown in Figure 131 below which lies north of the Arctic Circle (66½° latitude).

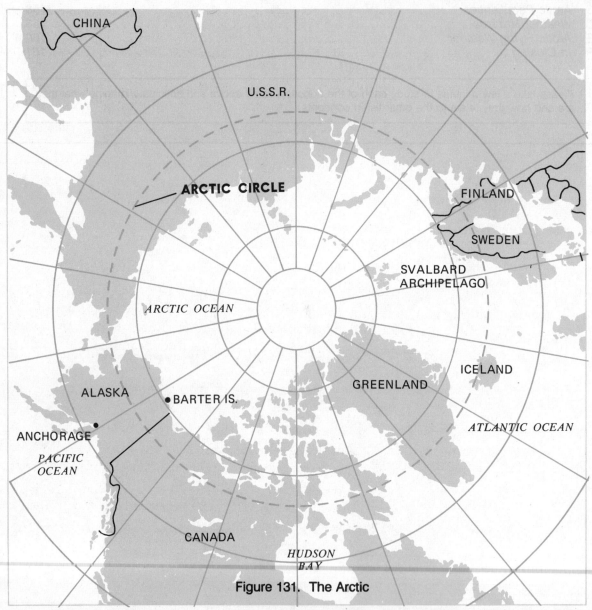

Figure 131.  The Arctic

2.    However, this discussion includes Alaskan weather even though much of Alaska lies south of the Arctic Circle.

B.  **Climate, Air Masses, and Fronts**

    1.   The climate of any region is largely determined by the amount of energy received from the sun.

        a.   Local characteristics of the area also influence climate.

    2.   **Long Days and Nights**

        a.   A profound seasonal change in length of day and night occurs in the Arctic because of the Earth's tilt and its revolution around the sun.

            1)   Figure 132 below shows that any point north of the Arctic Circle has autumn and winter days when the sun stays below the horizon all day and days in spring and summer with 24 hr. of sunshine.

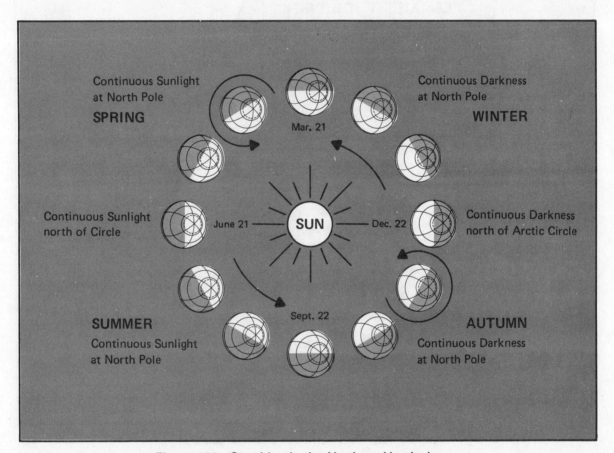

Figure 132.  Sunshine in the Northern Hemisphere

            2)   The number of these days increases toward the North Pole; there the sun stays below the horizon for 6 months and shines continuously during the other 6 months.

        b.   Twilight in the Arctic is prolonged because of the shallow angle of the sun below the horizon.

            1)   In more northern latitudes, it persists for days when the sun remains just below the horizon.

            2)   This abundance of twilight often makes visual reference possible at night.

### 3.   Land and Water

a.   Figure 133 below shows the water and land distribution in the Arctic.

Figure 133.  The Permanent Arctic Ice Pack

1)   Arctic mountain ranges are effective barriers to air movement.
2)   Large masses of air stagnate over the inland continental areas.

    a)   Thus, the Arctic continental areas are air mass source regions.

b.   A large portion of the Arctic Ocean is covered throughout the year by a deep layer of ice -- the permanent ice pack as shown in Figure 133 above.

1)   Even though the ocean is ice-covered through much of the year, the ice and water below contain more heat than the surrounding cold land, thus moderating the climate to some extent.

2)   Oceanic and coastal areas have a milder climate during winter than would be expected and a cool climate in summer.

    a)   As opposed to large water bodies, large land areas show a more significant seasonal temperature variation.

4.  **Temperature**

   a.  The Arctic is very cold in winter, but due to local terrain and the movement of pressure systems, occasionally some areas are surprisingly warm.

      1)  During winter, coastal areas average about 20° warmer than the interior.

      2)  During summer, interior areas are pleasantly warm with many hours of sunshine.

   b.  Coastal areas have relatively cool, short summers due to their proximity to water.

5.  **Clouds and Precipitation**

   a.  Cloudiness over the Arctic is at a minimum during winter, reaching a maximum in summer and fall.

      1)  Spring also brings many cloudy days.

      2)  During summer afternoons, scattered cumulus clouds forming over the interior occasionally grow into thundershowers.

         a)  These thundershowers, usually circumnavigable, move generally from northeast to southwest in the polar easterlies, which is opposite the general movement in midlatitudes.

   b.  Precipitation in the Arctic is generally light.

      1)  Annual amounts over the ice pack and along the coastal areas are only 3 to 7 in.

         a)  The interior is somewhat wetter, with annual amounts of 5 to 15 in.

      2)  Precipitation falls mostly in the form of snow over ice caps and oceanic areas and mostly as summer rain over interior areas.

6.  **Wind**

   a.  Strong winds occur more often along the coasts than elsewhere.

   b.  The frequency of high winds in coastal areas is greatest in fall and winter.

   c.  Wind speeds are generally light in the continental interior during the entire year, but are normally at their strongest during summer and fall.

7.  **In winter, air masses** form over the expanded ice pack and adjoining snow-covered land areas.

   a.  These air masses are characterized by very cold surface air, very low humidity, and strong low-level temperature inversions.

   b.  Occasionally, air from unfrozen ocean areas flows northward over the Arctic.

      1)  These intrusions of moist, cold air account for most of the infrequent wintertime cloudiness and precipitation in the Arctic.

8.  **During the summer**, the top layer of the Arctic permafrost melts leaving very moist ground, and the open water areas of the Polar Basin expand markedly.

   a.  Thus, the entire area becomes more humid, relatively mild, and semimaritime in character.

   b.  The largest amount of cloudiness and precipitation occurs inland during the summer months.

9.  **Fronts.** Occluded fronts are the rule.

   a.  Weather conditions with occluded fronts are much the same in the Arctic as elsewhere -- low clouds, precipitation, poor visibility, and sudden fog formation.

   b.  Fronts are much more frequent over coastal areas than over the interior.

C.  **Arctic Peculiarities**

1.  **Effects of Temperature Inversion**

    a.  The intense low-level inversion over the Arctic during much of the winter causes sound (including people's voices) to carry over extremely long distances.

    b.  Light rays are bent as they pass at low angles through the inversion.

        1)  This bending creates an effect known as looming -- a form of mirage that causes objects beyond the horizon to appear above the horizon.

        2)  Mirages distorting the shape of the sun, moon, and other objects are common with the low-level inversions.

2.  **Aurora Borealis**

    a.  Certain energy particles from the sun strike the Earth's magnetic field and are carried along the lines of force where they tend to lower and converge near the geomagnetic poles.

        1)  The energy particles then pass through rarefied gases of the outer atmosphere, illuminating them in much the same way as an electrical charge illuminates neon gas in neon signs.

    b.  The Aurora Borealis takes place at high altitudes above the Earth's surface.

        1)  The highest frequency of observations is over the northern United States and northward.

        2)  Displays of aurora vary from a faint glow to an illumination of the Earth's surface equal to a full moon.

        3)  They frequently change shape and form and are also called dancing lights or northern lights.

3.  **Light Reflection by Snow-Covered Surfaces**

    a.  Much more light is reflected by snow-covered surfaces than by darker surfaces.

    b.  Snow often reflects Arctic sunlight sufficiently to blot out shadows, thus markedly decreasing the contrast between objects.

        1)  Dark distant mountains may be easily recognized, but a crevasse normally directly in view may be undetected due to lack of contrasts.

4.  **Light from Celestial Bodies**

    a.  Illumination from the moon and stars is much more intense in the Arctic than in lower latitudes.

        1)  Pilots have found that light from a half-moon over a snow-covered field may be sufficient for landing.

        2)  Even illumination from the stars creates visibility far beyond that found elsewhere.

    b.  Only under heavy overcast skies does the night darkness in the Arctic begin to approach the degree of darkness in lower latitudes.

D.  **Weather Hazards**

1.  **Fog** limits landing and takeoff in the Arctic more than any other visibility restriction.

    a.  Water-droplet fog is the main hazard to aircraft operations in coastal areas during the summer.

    b.  Ice fog is the major restriction in winter.

2. **Ice fog** is common in the Arctic.

   a. It forms in moist air during extremely cold, calm conditions in winter, occurring often and tending to persist.

   b. Effective visibility is reduced much more in ice fog when one is looking toward the sun.

   c. Ice fog may be produced both naturally and artificially.

   d. Ice fog affecting aviation operations most frequently is produced by the combustion of aircraft fuel in cold air.

      1) When the wind is very light and the temperature is about –30°F or colder, ice fog often forms instantaneously in the exhaust gases of automobiles and aircraft.

      2) It lasts from as little as a few minutes to days.

3. **Steam fog,** often called "sea smoke," forms in winter when cold, dry air passes from land areas over comparatively warm ocean waters.

   a. Moisture evaporates rapidly from the water surface; but since the cold air can hold only a small amount of water vapor, condensation takes place just above the surface of the water and appears as "steam" rising from the ocean.

   b. This fog is composed entirely of water droplets that often freeze quickly and fall back into the water as ice particles.

   c. Low-level turbulence can occur and icing can become hazardous.

4. **Advection fog,** which may be composed either of water droplets or of ice crystals, is most common in winter and is often persistent.

   a. Advection fog forms along coastal areas when comparatively warm, moist, oceanic air moves over cold land.

      1) If the land areas are hilly or mountainous, lifting of the air results in a combination of low stratus and fog.

         a) The stratus and fog quickly diminish inland.

   b. Lee sides of islands and mountains usually are free of advection fog because of drying due to compressional heating as the air descends downslope.

   c. Icing in advection fog is in the form of rime and may become quite severe.

5. **Blowing Snow**

   a. Over the frozen Arctic Ocean and along the coastal areas, blowing snow and strong winds are common hazards during autumn and winter.

   b. Blowing snow is a greater hazard to flying operations in the Arctic than in midlatitudes because the snow is "dry" and fine and can be picked up easily by light winds.

      1) Winds in excess of 8 kt. may raise the snow several feet off the ground, obliterating objects such as runway markers.

      2) A sudden increase in surface wind may cause an unlimited visibility to drop to near zero in a few minutes.

         a) This sudden loss of visibility occurs frequently without warning in the Arctic.

      3) Stronger winds sometimes lift blowing snow to heights above 1,000 ft. and produce drifts over 30 ft. deep.

6. **Icing** is most likely in spring and fall, but is also encountered in winter.

    a. During spring and fall, icing may extend to upper levels along frontal zones.

    b. While icing is mostly a problem over water and coastal areas, it does exist inland.

        1) It occurs typically as rime, but a combination of clear and rime icing is not unusual in coastal mountains.

7. **Frost**

    a. In coastal areas during spring, fall, and winter, heavy frost and rime may form on aircraft parked outside, especially when fog or ice fog is present.

    b. This frost should be removed; it reduces lift and is especially hazardous if surrounding terrain requires a rapid rate of climb.

8. **Whiteout** is a visibility restriction phenomenon that occurs in the Arctic when a layer of cloudiness of uniform thickness overlies a snow- or ice-covered surface.

    a. Parallel rays of the sun are broken up and diffused when passing through the cloud layer so that they strike the snow surface from many angles.

        1) The diffused light then reflects back and forth countless times between the snow and the cloud eliminating all shadows.

        2) The result is a loss of depth perception.

    b. Buildings, people, and dark-colored objects appear to float in the air, and the horizon disappears.

        1) Low-level flight over icecap terrain or landing on snow surfaces becomes dangerous.

E. **Arctic Flying Weather**

1. A great number of pilots who fly Alaska and the Arctic are well-seasoned. They are eager to be of help and are your best sources of information.

    a. Before flying the Arctic, be sure to learn all you can about your proposed route.

2. Generally, flying conditions in the Arctic are good when averaged over the entire year.

    a. However, areas of Greenland compete with the Aleutians for the world's worst weather.

        1) These areas are exceptions.

3. Whiteouts, in conjunction with overcast skies, often present a serious hazard, especially for visual flight.

    a. Many mountain peaks are treeless and rounded rather than ragged, making them unusually difficult to distinguish under poor visibility conditions.

4. **Oceanic and Coastal Areas**

    a. In oceanic and coastal areas, predominant hazards change with the seasons.

        1) In summer, the main hazard is fog in coastal areas.

    b. In winter, ice fog is the major restriction to aircraft operation.

        1) Blowing and drifting snow often restrict visibility as well.

        2) Storms and well-defined frontal passages frequent the coastal areas accompanied by turbulence, especially in the coastal mountains.

    c. Icing is most frequent in spring and fall and may extend to high levels in active, turbulent frontal zones.

        1) Fog is also a source of icing when the temperature is colder than freezing.

5. **Continental Areas**

    a. Over the continental interior, good flying weather prevails much of the year, although during winter, ice fog often restricts aircraft operations.

        1) In terms of ceiling and visibility, the summer months provide the best flying weather.

        2) Thunderstorms develop on occasion during the summer, but they usually can be circumnavigated without much interference with flight plans.

## F. In Closing

1. The Arctic climate is influenced by local characteristics:

    a. Long days and nights
    b. Land and water distribution
    c. Temperature
    d. Clouds and precipitation
    e. Wind
    f. Air masses forming over the expanded ice pack in winter
    g. Melting of the top layer of the Arctic permafrost
    h. Fronts

2. Other phenomena peculiar to the Arctic are

    a. Mirages caused by the effects of low-level temperature inversions.

        1) Also, sound carries over extremely long distances.

    b. Aurora Borealis.
    c. Light reflection by snow-covered surfaces.
    d. Light from celestial bodies.

3. Weather hazards include

    a. Fog,
    b. Blowing snow,
    c. Icing,
    d. Frost,
    e. Whiteout.

4. Interior areas generally have good flying weather, but coastal areas and Arctic slopes often are plagued by low ceiling, poor visibility, and icing.

5. Whiteout conditions over ice and snow covered areas often cause pilot disorientation.

6. Flying conditions are usually worse in mountain passes than at reporting stations along the route.

7. Routes through the mountains are subject to strong turbulence, especially in and near passes.

8. Beware of a false mountain pass that may lead to a dead end.

9. Thundershowers sometimes occur in the interior during May through August.  They are usually circumnavigable and generally move from northeast to southwest.

10. Always file a flight plan.  Stay on regularly traversed routes, and if downed, stay with your plane.

11. If lost during summer, fly down-drainage, that is, downstream.  Most airports are located near rivers, and chances are you can reach a landing strip by flying downstream.

    a. If forced down, you will be close to water on which a rescue plane can land.  In summer, the tundra is usually too soggy for landing.

12.  Weather stations are few and far between.

   a.  Adverse weather between stations may go undetected unless reported by a pilot in flight.

   b.  A report confirming good weather between stations is also just as important.

   c.  Help yourself and your fellow pilots by reporting weather en route.

# END OF CHAPTER

# CHAPTER FIFTEEN
# TROPICAL WEATHER

## A. Definition

1. Technically, the Tropics lie between latitudes 23½°N and 23½°S.

   a. However, weather typical of this region sometimes extends as much as 45° from the Equator.

2. One generally thinks of the Tropics as uniformly rainy, warm, and humid.

   a. However, the Tropics contain both the wettest and the driest regions of the world.

## B. Circulation

1. In Part I, Chapter 4, Wind, beginning on page 29, we learned that wind blowing out of the subtropical high pressure belts toward the Equator form the northeast and southeast trade winds of the two hemispheres.

   a. These trade winds converge in the vicinity of the Equator where air rises.

      1) This convergence zone is the "intertropical convergence zone" (ITCZ).

   b. In some areas of the world, seasonal temperature differences between land and water areas generate rather large circulation patterns that overpower the trade wind circulation.

      1) These areas are monsoon regions.

2. **Subtropical High Pressure Belts**

   a. If the surface under the subtropical high pressure belts were all water of uniform temperature, the high pressure belts would be continuous highs around the globe.

      1) The belts would be areas of descending or subsiding air and would be characterized by strong temperature inversions and very little precipitation.

      2) However, land surfaces at the latitudes of the high pressure belts are generally warmer throughout the year than are water surfaces.

         a) Thus, the high pressure belts are broken into semipermanent high pressure anticyclones over oceans with troughs or lows over continents.

   b. The subtropical highs shift southward during the Northern Hemisphere winter and northward during summer.

      1) The seasonal shift, the height and strength of the inversion, and terrain features determine weather in the subtropical high pressure belts.

c.  **Continental Weather**

1)  Along the west coasts of continents under a subtropical high, the air is stable.

a)  The inversion is strongest and lowest where the east side of an anticyclone overlies the west side of a continent.

b)  Moisture is trapped under the inversion; fog and low stratus occur frequently.

i)  However, precipitation is rare since the moist layer is shallow and the air is stable.

c)  Heavily populated areas also add contaminants to the air which, when trapped under the inversion, create an air pollution problem.

2)  The situation on eastern continental coasts is just the opposite.

a)  The inversion is weakest and highest where the west side of an anticyclone overlies the eastern coast of a continent.

b)  Convection can penetrate the inversion, and showers and thunderstorms often develop.

3)  Low ceiling and fog often prevent landing at a west coast destination, but a suitable alternate generally is available a few miles inland.

a)  An alternate selection may be more critical for an eastern coast destination because of widespread instability and associated hazards.

d.  **Weather over Open Sea**

1)  Under a subtropical high over the open sea, cloudiness is scant.

2)  The few clouds that do develop have tops from 3,000 to 6,000 ft., depending on the height of the inversion.

3)  Ceiling and visibility are generally quite ample for VFR flight.

e.  **Island Weather**

1)  An island under a subtropical high receives very little rainfall because of the persistent temperature inversion.

2)  Surface heating over some larger islands causes light convective showers.

3)  Cloud tops are only slightly higher than over open water.

4)  Temperatures are mild, showing small seasonal and diurnal changes.

3.  **Trade Wind Belts**

a.  Figures 138 and 139 on page 165 show prevailing winds throughout the Tropics for July and January.

1)  Note that trade winds blowing out of the subtropical highs over ocean areas are predominantly northeasterly in the Northern Hemisphere and southeasterly in the Southern Hemisphere.

b.  The inversion from the subtropical highs is carried into the trade winds and is known as the "trade wind inversion."

1)  As in a subtropical high, the inversion is strongest where the trades blow away from the west coast of a continent and weakest where they blow onto an eastern continental shore.

2)  Daily variations from these prevailing directions are small except during tropical storms.

a)  As a result, weather at any specific location in a trade wind belt varies little from day to day.

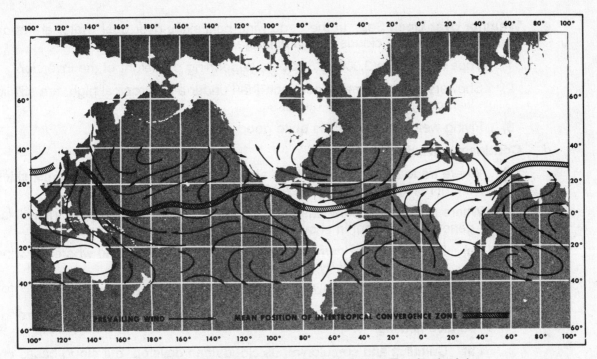

Figure 138.  Prevailing Winds throughout the Tropics in July

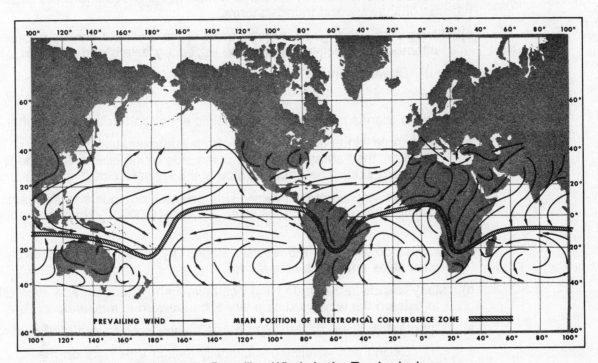

Figure 139.  Prevailing Winds in the Tropics in January

c.  **Weather over Open Sea.**  In the trade wind belt, skies over open water are about one-half covered by clouds on the average.

   1)  Tops range from 3,000 to 8,000 ft., depending on height of the inversion.

   2)  Showers, although more common than under a subtropical high, are still light with comparatively little rainfall.

   3)  Flying weather generally is quite good.

d.  **Continental Weather**

   1)  Where trade winds blow offshore along the west coasts of continents, skies are generally clear and the area is quite arid.

   2)  Where trade winds blow onshore on the east sides of continents, rainfall is generally abundant in showers and occasional thunderstorms.

      a)  Rainfall may be carried a considerable distance inland where the winds are not blocked by a mountain barrier.

      b)  Inland areas blocked by a mountain barrier are deserts.

   3)  Afternoon convective currents are common over arid regions due to strong surface heating.

      a)  Cumulus and cumulonimbus clouds can develop, but cloud bases are high and rainfall is scant because of the low moisture content.

   4)  Flying weather along eastern coasts and mountains is subject to the usual hazards of showers and thunderstorms.

      a)  Flying over arid regions is good most of the time, but can be turbulent in afternoon convective currents; be especially aware of dust devils.

         i)  Blowing sand or dust sometimes restricts visibility.

e.  **Island Weather**

   1)  Mountainous islands have the most dramatic effect on trade wind weather.

      a)  Since trade winds are consistently from approximately the same direction, they always strike the same side of the island; this side is the windward side.  The opposite side is the leeward side.

      b)  Winds blowing up the windward side produce copious and frequent rainfall, although cloud tops rarely exceed 10,000 ft.  Thunderstorms are rare.

      c)  Downslope winds on the leeward slopes dry the air leaving relatively clear skies and much less rainfall.

      d)  Many islands in the trade wind belt have lush vegetation and even rain forests on the windward side while the leeward is semiarid.

      e)  The greatest flying hazard near these islands is obscured mountain tops.

         i)  Ceiling and visibility occasionally restrict VFR flight on the windward side in showers.

         ii)  IFR weather is virtually nonexistent on leeward slopes.

   2)  Islands without mountains have little effect on cloudiness and rainfall.

      a)  Afternoon surface heating increases convective cloudiness slightly, but shower activity is light.

b)  Any island in either the subtropical high pressure belt or trade wind belt enhances cumulus development even though tops do not reach great heights.

    i)  Thus, a cumulus top higher than the average tops of surrounding cumulus usually marks the approximate location of an island.

    ii)  If it becomes necessary to ditch in the ocean, look for a tall cumulus.

4.  **The Intertropical Convergence Zone (ITCZ).**  Converging winds in the intertropical convergence zone (ITCZ) force air upward.

a.  The inversion typical of the subtropical high and trade wind belts disappears.

    1)  Figures 138 and 139 show the ITCZ and its seasonal shift.

    2)  The ITCZ is well marked over tropical oceans but is weak and ill-defined over large continental areas.

b.  **Weather over Islands and Open Water.**  Convection in the ITCZ carries huge quantities of moisture to great heights.

    1)  Showers and thunderstorms frequent the ITCZ and tops to 30,000 ft. or higher are common, as shown in Figure 137 below.

        a)  Precipitation is copious.

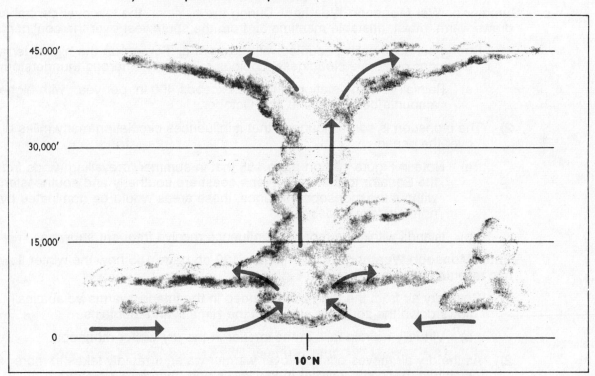

Figure 137.  Vertical Cross Section Illustrating Convection in the ITCZ

    2)  Since convection dominates the ITCZ, there is little difference in weather over islands and open sea under the ITCZ.

    3)  Flying through the ITCZ usually presents no great problem if one follows the usual practice of avoiding thunderstorms.

      4)   Since the ITCZ is ill-defined over continents, we will not attempt to describe ITCZ continental weather as such.

         a)   Continental weather ranges from arid to rain forests and is more closely related to the monsoon than to the ITCZ.

## 5. Monsoon

a.   Refer to Figures 23 and 24 in Part I, Chapter 4, Wind, on page 35. You can see that over the large land mass of Asia, the subtropical high pressure breaks down completely.

    1)   Asia is covered by an intense high during the winter and a well-developed low during the summer.

    2)   You can also see the same over Australia and central Africa, although the seasons are reversed in the Southern Hemisphere.

b.   The cold, high pressures in winter cause wind to blow from the deep interior outward and offshore.

    1)   In summer, wind direction reverses, and warm, moist air is carried far inland into the low pressure area.

    2)   This large-scale seasonal wind shift is the monsoon.

       a)   The most notable monsoon is that of southern and southeastern Asia.

c.   **Summer or Wet Monsoon Weather.** During the summer, the low over central Asia draws warm, moist, unstable maritime air from the southwest over the continent.

    1)   Strong surface heating coupled with rising of air flowing up the higher terrain produces extensive cloudiness, copious rain, and numerous thunderstorms.

       a)   Rainfall at some stations in India exceeds 400 in. per year, with highest amounts between June and October.

    2)   The monsoon is so pronounced that it influences circulation many miles out over the ocean.

       a)   Note in Figure 138 on page 165 that in summer, prevailing winds from the Equator to the south Asian coast are southerly and southeasterly; without the monsoon influence, these areas would be dominated by northeasterly trades.

       b)   Islands within the monsoon influence receive frequent showers.

d.   **Winter Monsoon Weather.** Note in Figure 139 on page 165 how the winter flow has reversed from that shown in Figure 138.

    1)   Cold, dry air from the high plateau deep in the interior warms adiabatically as it flows down the southern slopes of the Himalayan Mountains.

       a)   Virtually no rain falls in the interior in the dry winter monsoon.

    2)   As the dry air moves offshore over warmer water, it rapidly takes in more moisture, becomes warmer in low levels and, therefore, unstable.

       a)   Rain is frequent over offshore islands and even along coastal areas after the air has had a significant over-water trajectory.

    3)   The Philippine Islands are in an area of special interest.

       a)   During the summer, they are definitely in southerly monsoon flow and are subjected to abundant rainfall.

       b)   In winter, wind over the Philippines is northeasterly -- in the transition zone between the northeasterly trades and the monsoon flow.

       c)    It is academic whether we call the phenomenon the trade winds or monsoon; in either case, it produces abundant rainfall.

           i)    The Philippines have a year-round humid, tropical climate.

  e.   **Other Monsoon Areas.**  Australia in July (Southern Hemisphere winter) is an area of high pressure with predominantly offshore winds as shown in Figure 138 on page 165.

    1)   Most of the continent is dry during the winter.

    2)   In January (Figure 139 on page 165), winds are onshore into the continental low pressure.

       a)    However, most of Australia is rimmed by mountains, and coastal regions are wet where the onshore winds blow up the mountain slopes.

       b)    The interior is arid where downslope winds are warmed and dried.

    3)   Central Africa is known for its humid climate and jungles.

       a)    Note in Figures 138 and 139 on page 165 that prevailing wind is onshore much of the year over these regions.

       b)    Some regions are wet the year round; others have the seasonal monsoon shift and have a summer wet season and a winter dry season.

    4)   In the Amazon Valley of South America during the Southern Hemisphere winter (July), southeast trades, as shown in Figure 138 on page 165, penetrate deep into the valley bringing abundant rainfall which contributes to the jungle climate.

       a)    In January, the ITCZ moves south of the valley as shown in Figure 139 on page 165.

       b)    The northeast trades are caught up in the monsoon, cross the Equator, and also penetrate the Amazon Valley.

       c)    The jungles of the Amazon result largely from monsoon winds.

  f.   **Flying Weather in Monsoons**

    1)   During the winter monsoon, excellent flying weather prevails over dry interior regions.

       a)    Over water, you must pick your way around showers and thunderstorms.

    2)   In the summer monsoon, VFR flight over land is often restricted by low ceilings and heavy rain.

       a)    IFR flight must cope with the hazards of thunderstorms.

    3)   Freezing level in the Tropics is quite high -- 14,000 ft. or higher -- so icing is restricted to high levels.

## C.  Transitory Systems

  1.   A **wind shear line** found in the Tropics mainly results from midlatitude influences.

    a.   In Part I, Chapter 8, Air Masses and Fronts, beginning on page 73, we stated that an air mass becomes modified when it flows from its source region.

      1)   By the time a cold air mass originating in high latitudes reaches the Tropics, temperature and moisture are virtually the same on both sides of the front.

      2)   A shear line, or wind shift, is all that remains.

b.   A shear line also results when a semi-permanent high splits into two cells, inducing a trough, as shown in Figure 140 below.

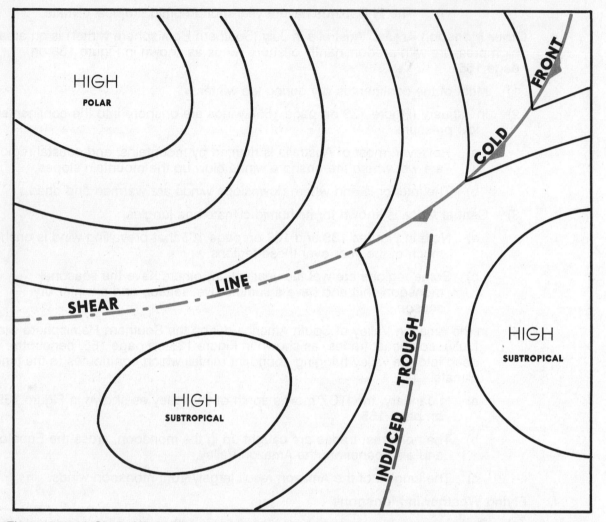

Figure 140.  A Shear Line and an Induced Trough Caused by a Polar High Pushing into the Tropics

c.   These shear lines are zones of convergence creating forced upward motion.

1)   Consequently, considerable thunderstorm and rain shower activity occurs along a shear line.

2.  **Troughs** in the atmosphere, generally at or above 10,000 ft. move through the Tropics, especially along the poleward fringes.

    a.  As a trough moves to the southeast or east, it spreads middle and high cloudiness over extensive areas to the east of the trough line.

    b.  Occasionally, a well-developed trough will extend deep into the Tropics, and a closed low forms at the equatorial end of the trough.

        1)  The low then may separate from the trough and move westward, producing a large amount of cloudiness and precipitation.

        2)  If this occurs in the vicinity of a strong subtropical jet stream, extensive and sometimes dense cirrus and some convective and clear air turbulence often develop.

    c.  Troughs and lows aloft produce considerable amounts of rainfall in the Tropics, especially over land areas where mountains and surface heating lift air to saturation.

        1)  Low pressure systems aloft contribute significantly to the record 460 in. average annual rainfall on Mt. Waialeale on Kauai, Hawaii.

        2)  Other mountainous areas of the Tropics are also among the wettest spots on earth.

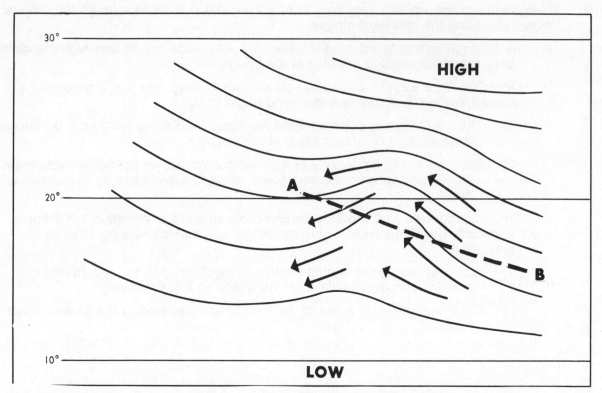

Figure 142.  A Northern Hemisphere Easterly Wave Moving from (B) to (A)

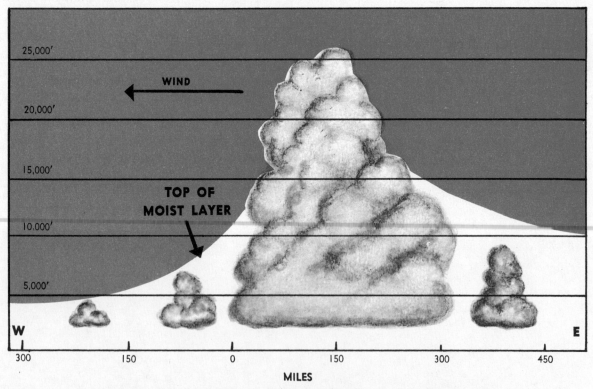

Figure 143.  Vertical Cross Section along Line A–B in Figure 142

3.   **Tropical waves** (also called easterly waves) are common tropical weather disturbances, normally occurring in the trade wind belt.

   a.   In the Northern Hemisphere, they usually develop in the southeastern perimeter of the subtropical high pressure systems.

   b.   They travel from east to west around the southern fringes of these highs in the prevailing easterly circulation of the Tropics.

   c.   Surface winds in advance of a wave are somewhat more northerly than the usual trade wind direction.

   d.   As the wave approaches, as shown in Figure 142 on page 172, pressure falls; as it passes, surface wind shifts to the east-southeast or southeast.

   1)   The typical wave is preceded by very good weather but followed by extensive cloudiness, as shown in Figure 143 on page 172, and often by rain and thunderstorms.

   2)   The weather activity is roughly in a north-south line.

   e.   Tropical waves occur in all seasons, but are more frequent and stronger during summer and early fall.

   1)   Pacific waves frequently affect Hawaii.

   2)   Atlantic waves occasionally move into the Gulf of Mexico, reaching the U.S. coast.

4.   **Tropical cyclone** is a general term for any low that originates over tropical oceans.

   a.   Tropical cyclones are classified according to their intensity based on average 1-min. wind speeds.

   1)   Wind gusts in these storms may be as much as 50% higher than the average 1-min. wind speeds.

   b.   Tropical cyclone international classifications are:

   1)   Tropical Depression -- highest sustained winds up to 34 kt.,
   2)   Tropical Storm -- highest sustained winds of 35 through 64 kt., and
   3)   Hurricane or Typhoon -- highest sustained winds 65 kt. or more.

   c.   Strong tropical cyclones are known by different names in different regions of the world.

   1)   A tropical cyclone in the Atlantic and eastern Pacific is a "hurricane."
   2)   In the western Pacific, "typhoon."
   3)   Near Australia and in the Indian Ocean, simply "cyclone."

d.  **Development**.  Prerequisite to tropical cyclone development are optimal sea surface temperature under weather systems that produce low-level convergence and cyclonic wind shear.

1)  Favored breeding grounds are tropical (easterly) waves, troughs aloft, and areas of converging northeast and southeast trade winds along the intertropical convergence zone.

2)  The low-level convergence associated with these systems, by itself, will not support development of a tropical cyclone.

  a)  The system must also have horizontal outflow -- divergence -- at high tropospheric levels.

  b)  This combination creates a "chimney," in which air is forced upward causing clouds and precipitation.

  c)  Condensation releases large quantities of latent heat which raises the temperature of the system and accelerates the upward motion.

   i)  The rise in temperature lowers the surface pressure which increases low-level convergence.  This draws more moisture-laden air into the system.

  d)  When these chain-reaction events continue, a huge vortex is generated which may culminate in hurricane force winds.

3)  Figure 144 below shows regions of the world where tropical cyclones frequently develop.  Notice that they usually originate between latitudes 5° and 20°.

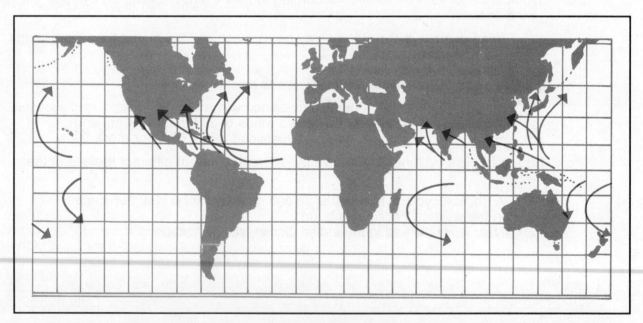

Figure 144.  Principal Regions Where Tropical Cyclones Form and Their Favored Directions of Movement

  a)  Tropical cyclones are unlikely within 5° of the Equator because the Coriolis force is so small near the Equator that it will not turn the winds enough for them to flow around a low-pressure area.

  b)  Winds flow directly into an equatorial low and rapidly fill it.

e.   **Movement.** Tropical cyclones in the Northern Hemisphere usually move in a direction between west and northwest while in low latitudes.

   1)   As these storms move toward the midlatitudes, they come under the influence of the prevailing westerlies.

      a)   At this time, the storms are under the influence of two wind systems, i.e., the trade winds at low levels and prevailing westerlies aloft.

      b)   Thus, a storm may move very erratically and may even reverse course, or circle.

   2)   Finally, the prevailing westerlies gain control and the storm recurves toward the north, then to the northeast, and finally to the east-northeast.

      a)   By this time, the storm is well into midlatitudes.

f.   **Decay.** As the storm curves toward the north or east, it usually begins to lose its tropical characteristics and acquires characteristics of lows in middle latitudes.

   1)   Cooler air flowing into the storm gradually weakens it.

      a)   If the storm tracks along a coastline or over the open sea, it gives up slowly, carrying its fury to areas far removed from the Tropics.

      b)   If the storm moves well inland, it loses its moisture source and weakens from starvation and increased surface friction, usually after leaving a trail of destruction and flooding.

   2)   When a storm takes on middle latitude characteristics, it is said to be "extratropical," meaning "outside the Tropics."

      a)   Tropical cyclones produce weather conditions that differ somewhat from those produced by their higher latitude cousins.

g.   **Weather in a Tropical Depression**

   1)   While in its initial developing stage, the cyclone is characterized by a circular area of broken to overcast clouds in multiple layers.

      a)   Embedded in these clouds are numerous showers and thunderstorms.

      b)   Rain shower and thunderstorm coverage varies from scattered to almost solid.

   2)   The diameter of the cloud pattern varies from less than 100 mi. in small systems to well over 200 mi. in large ones.

h.   **Weather in Tropical Storms and Hurricanes**

   1)   As cyclonic flow increases, the thunderstorms and rain showers form broken or solid lines paralleling the wind flow that is spiraling into the center of the storm.

      a)   These lines are the spiral rain bands frequently seen on radar.

      b)   These rain bands continually change as they rotate around the storm.

      c)   Rainfall in the rain bands is very heavy, reducing ceiling and visibility to near zero.

   2)   Winds are usually very strong and gusty and, consequently, generate violent turbulence.

   3)   Between the rain bands, ceilings and visibilities are somewhat better, and turbulence generally is less intense.

4)    The "eye" usually forms in the tropical storm stage and continues through the hurricane stage.

    a)    In the eye, skies are free of turbulent cloudiness, and wind is comparatively light.

    b)    The average diameter of the eye is between 15 and 20 mi., but sometimes it is as small as 7 mi. and rarely more than 30 mi. in diameter.

    c)    Surrounding the eye is a wall of cloud that may extend above 50,000 ft.

        i)    This "wall cloud" contains deluging rain and the strongest winds of the storm.

        ii)    Maximum wind speeds of 175 kt. have been recorded in some storms.

5)    Figure 145 below is a radar display of a mature hurricane.

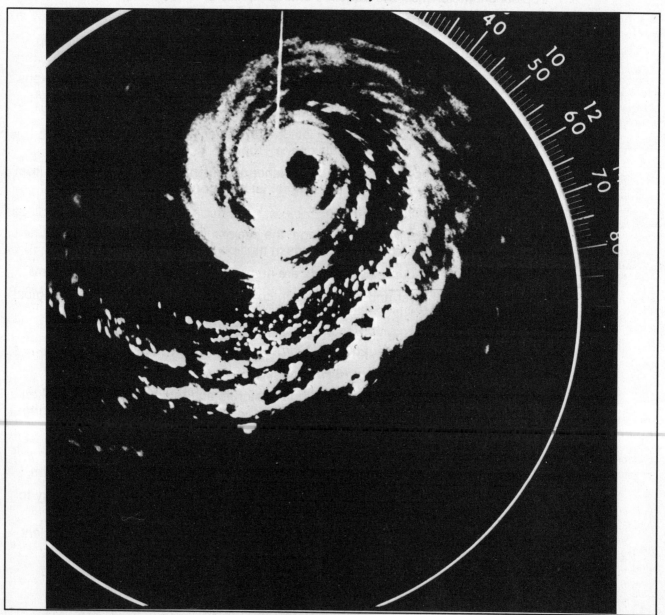

Figure 145.  Radar Display of a Hurricane

i.  **Detection and Warning.** The National Weather Service has a specialized hurricane forecast and warning service center at Miami, Florida, which maintains a constant watch for the formation and development of tropical cyclones.

   1) Weather information from land stations, ships at sea, reconnaissance aircraft, long-range radars, and weather satellites is fed into the center.

      a) The center forecasts the development, movement, and intensity of tropical cyclones.

      b) Forecasts and warnings are issued to the public and aviation interests by field offices of the National Weather Service.

j.  **Flying.** All pilots except those especially trained to explore tropical storms and hurricanes should AVOID THESE DANGEROUS STORMS.

   1) Occasionally, jet aircraft have been able to fly over small and less intense storms, but the experience of weather research aircraft shows hazards at all levels within them.

   2) Tops of thunderstorms associated with tropical cyclones frequently exceed 50,000 ft.

      a) Winds in a typical hurricane are strongest at low levels, decreasing with altitude.

         i) However, research aircraft have frequently encountered winds in excess of 100 kt. at 18,000 ft.

      b) Aircraft at low levels are exposed to sustained, pounding turbulence due to the surface friction of the fast-moving air.

      c) Turbulence increases in intensity in spiral rain bands and becomes the most violent in the wall cloud surrounding the eye.

   3) An additional hazard encountered in hurricanes is erroneous altitude readings from pressure altimeters.

      a) These errors are caused by the large pressure difference between the periphery of the storm and its center.

      b) One research aircraft lost almost 2,000 ft. true altitude traversing a storm while the pressure altimeter indicated a constant altitude of 5,000 ft.

   4) In short, tropical cyclones are very hazardous, so avoid them!

      a) To bypass the storm in a minimum of time, fly to the right of the storm to take advantage of the tailwind.

      b) If you fly to the left of the storm, you will encounter strong headwinds which may exhaust your fuel supply before you reach a safe landing area.

D.  **In Closing**

1.   The circulation basic to weather in the Tropics includes

   a.   Subtropical high-pressure belts which shift southward during the Northern Hemisphere winter and northward during summer.

   b.   Trade wind belts -- Trade winds blowing out of the subtropical highs over ocean areas are predominantly northeasterly in the Northern Hemisphere.

   c.   Intertropical Convergence Zone (ITCZ) -- In the ITCZ, converging winds force air upward.

      1)   Showers and thunderstorms are common.

   d.   Monsoon regions -- Cold, high pressures over the land in winter cause wind to blow outward and offshore.

      1)   In summer, wind direction reverses, and warm, moist air is carried far inland into the low-pressure area.

2.   Transitory (or migrating) tropical weather producers include

   a.   Wind shear lines, which are zones of convergence creating forced upward motion.

      1)   Considerable thunderstorm and rain shower activity occurs along a shear line.

   b.   Troughs aloft, which produce large amounts of rainfall in the Tropics, especially in mountainous areas.

   c.   Tropical waves, which normally occur in the trade wind belt.

      1)   Typically, they are preceded by very good weather but followed by extensive cloudiness, and often by rain and thunderstorms.

   d.   Tropical cyclone -- any low that originates over tropical oceans.  International classifications based on average wind speeds are

      1)   Tropical depression,
      2)   Tropical storm,
      3)   Hurricane or typhoon.

         a)   Tropical storms and hurricanes are extremely hazardous, and you should avoid them.

# END OF CHAPTER

# CHAPTER SIXTEEN
# SOARING WEATHER

> Please take a few minutes to study each of the concepts listed above and anticipate/imagine what they are and how they relate to the other listed concepts.

A. **Introduction** -- Soaring bears the relationship to flying that sailing bears to power boating.

   1.   Soaring has made notable contributions to meteorology.

      a.   For example, soaring pilots have probed thunderstorms and mountain waves with findings that have made flying safer for all pilots.

   2.   A sailplane must have auxiliary power to become airborne, such as a winch, a ground tow, or a tow by a powered aircraft.

      a.   Once the sailcraft is airborne and the tow cable released, performance depends on the weather and the skill of the pilot.

      b.   Forward thrust comes from gliding downward relative to the air just as thrust is developed in a power-off glide by a conventional aircraft.

          1)   Therefore, to gain or maintain altitude, the soaring pilot must rely on upward motion of the air.

   3.   To a sailplane pilot, "lift" means the rate of climb (s)he can achieve in an up-current, while "sink" denotes his/her rate of descent in a downdraft or in neutral air.

      a.   "Zero sink" means that upward currents are just strong enough to enable the pilot to hold altitude but not to climb.

      b.   Sailplanes are highly efficient machines; a sink rate of a mere 2 ft. per second (fps) provides an airspeed of about 40 kt., and a sink rate of 6 fps gives an airspeed of about 70 kt.

   4.   In lift, a sailplane usually flies 35 to 40 kt. with a sink rate of about 2 fps.

      a.   Therefore, to remain airborne, the pilot must have an upward air current of at least 2 fps.

      b.   There is no point in trying to soar until weather conditions favor vertical speeds greater than the minimum sink rate of the aircraft.

      c.   These vertical currents develop from several sources, which categorize soaring into five classes:

          1)   Thermal soaring
          2)   Frontal soaring
          3)   Sea breeze soaring
          4)   Ridge or hill soaring
          5)   Mountain wave soaring

B. **Thermal Soaring**

1. A thermal is simply the updraft in a small-scale convective current.

2. A soaring aircraft is always sinking relative to the air.

   a. To maintain or gain altitude, therefore, the soaring pilot must spend sufficient time in thermals to overcome the normal sink of the aircraft as well as to regain altitude lost in downdrafts.

      1) The pilot usually circles at a slow airspeed in a thermal and then darts on a beeline to the next thermal, as shown in Figure 147 below.

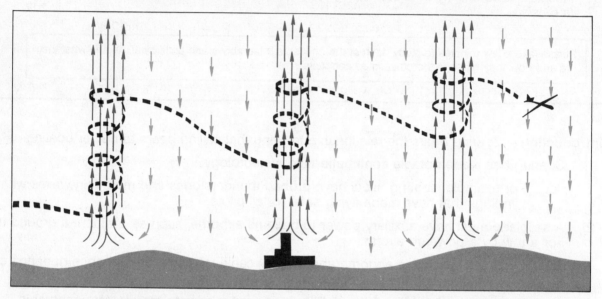

Figure 147. Thermals Generally Occur over a Small Portion of an Area While Downdrafts Predominate

3. Low-level heating is prerequisite to thermals; and this heating is mostly from the sun, although it may be augmented by man-made heat sources such as chimneys, factories, and cities.

   a. Cool air must sink to force warm air upward in thermals.

      1) Therefore, in small-scale convection, thermals and downdrafts coexist side by side.

   b. The net upward displacement of air must equal the net downward displacement.

      1) Fast rising thermals generally cover a small percentage of a convective area, while slower downdrafts predominate over the remaining greater portion, as diagramed in Figure 147 above.

   c. Since thermals depend on solar heating, thermal soaring is virtually restricted to daylight hours with considerable sunshine.

      1) Air tends to become stable at night due to low-level cooling by terrestrial radiation, often resulting in an inversion at or near the surface.

      2) Stable air suppresses convection, and thermals do not form until the inversion "burns off" or lifts sufficiently to allow soaring beneath the inversion.

         a) The earliest that soaring may begin varies from early forenoon to early afternoon, depending on the strength of the inversion and the amount of solar heating.

4. **Locating Thermals.** Since convective thermals develop from uneven heating at the surface, the most likely place for a thermal is above a surface that heats readily.

  a. When the sky is cloudless, the soaring pilot must look for those surfaces that heat most rapidly and seek thermals above those areas.

   1) Barren sandy or rocky surfaces, plowed fields, stubble fields surrounded by green vegetation, cities, factories, chimneys, etc., are good thermal sources.

  b. The angle of the sun profoundly affects location of thermals over hilly landscapes.

   1) During early forenoon, the sun strikes eastern slopes more directly than others; thus, the most favorable areas for thermals are eastern slopes.

   2) The favorable areas move to southern slopes during midday.

   3) In the afternoon, they move to western slopes before they begin to weaken as the evening sun sinks toward the western horizon.

  c. Surface winds must converge to feed a rising thermal; so when you sight a likely spot for a thermal, look for dust or smoke movement near the surface.

   1) If you can see dust or smoke "streamers" from two or more sources converging on the spot as shown in Figure 148 (A), you have chosen wisely.

    a) If, however, the streamers diverge as shown in Figure 148 (B), a downdraft most likely hovers over the spot and it's time to move on.

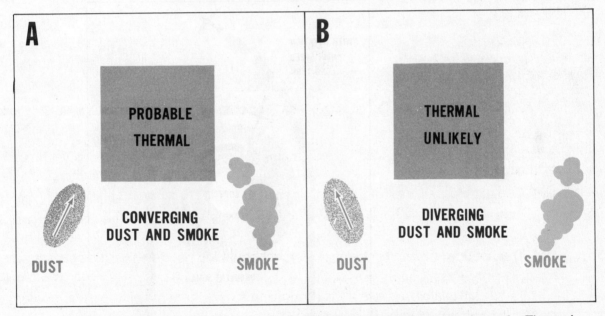

Figure 148. Using Surface Dust and Smoke Movement as Indications of a Thermal

   2) Rising columns of smoke from chimneys and factories mark thermals augmented by man-made sources.

    a) They are good sources of lift if upward speed is great enough to support the aircraft and if they are broad enough to permit circling.

   3) Towns or cities may provide thermals; but to use a thermal over a populated area, the pilot must have sufficient altitude to glide clear of the area in the event the thermal subsides.

d.   Dust devils occur under sunny skies over sandy or dusty, dry surfaces and are sure signs of strong thermals with lots of lift.

   1)   To tackle this excellent source of lift, you must use caution.

      a)   The thermals are strong and turbulent and are surrounded by areas of little lift or possibly areas of sink.

   2)   If approaching the dust devil at too low an altitude, an aircraft may sink to an altitude too low for recovery.

      a)   A recommended procedure is to always approach the whirling vortex at an altitude 500 ft. or more above the ground.

      b)   At this altitude, you have enough airspace for maneuvering in the event you get into a downdraft or turbulence too great for comfort.

   3)   A dust devil may rotate either clockwise or counterclockwise.

      a)   Before approaching the dusty column, determine its direction of rotation by observing dust and debris near the surface.

   4)   You should enter against the direction of rotation.

      a)   Figure 149 below diagrams a horizontal cross section of a clockwise rotating dust devil and ways of entering it.

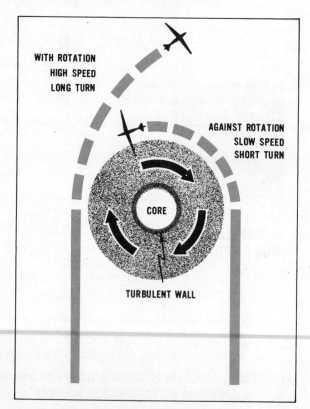

Figure 149.  Horizontal Cross Section of a Dust
Devil Rotating Clockwise

      b)   If you enter with the direction of rotation as on the left, the wind speed is added to your airspeed, giving you a fast circling speed -- probably too great to remain in the thermal.

         i)   Against the rotation as on the right, wind speed is subtracted from airspeed, giving you a slow circling speed.

5) Stay out of the eye of the vortex.

    a) Centrifugal force in the center throws air outward, greatly reducing pressure within the hollow center.

    b) The rarified air in the center provides very little lift, and the wall of the hollow center is very turbulent.

6) If you are 500 ft. or more above the ground but having trouble finding lift, the dust devil is well worth a try.

    a) If the thermal is sufficiently broad to permit circling within it, you have it made.

    b) The dust column may be quite narrow, but this fact does not necessarily mean the thermal is narrow; the thermal may extend beyond the outer limits of visible dust.

    c) Approach the dusty column against the direction of rotation at minimum airspeed.

    d) Enter the column near the outer edge of the dust and stay away from the hollow vortex core.

    e) Remain alert; you are circling little more than a wing span away from violent turbulence.

e. Wind causes a thermal to lean with altitude.  When seeking the thermal supporting soaring birds or aircraft, you must make allowance for the wind.

1) The thermal at lower levels usually is upwind from your high-level visual cue.

2) A thermal may not be continuous from the surface upward to the soaring birds or sailplane; rather it may be in segments or bubbles.

    a) If you are unsuccessful in finding the thermal where you expect it, seek elsewhere.

f. When convective clouds develop, thermal soaring is usually at its best and the problem of locating thermals is greatly simplified.

1) Cumulus clouds are positive signs of thermals, but thermals grow and die.

    a) A cloud grows with a rising thermal, but when the thermal dies, the cloud slowly evaporates.

    b) Because the cloud dissipates *after* the thermal ceases, the pilot who can spot the difference between a growing and dying cumulus has enhanced his/her soaring skill.

2) The warmest and most rapidly rising air is at the center of the thermal.

    a) Therefore, the cloud base will be highest in the center, giving a concave shape to the cloud base as shown in the left and center of Figure 150 on page 184.

    b) When the thermal ceases, the base assumes a convex shape as shown on the right.

    c) Another cue to look for is the outline of the cloud sides and top.

        i) The outline of the growing cumulus is firm and sharp.

        ii) The dying cumulus has fragmentary sides and lacks the definite outline as shown on the right of Figure 150.

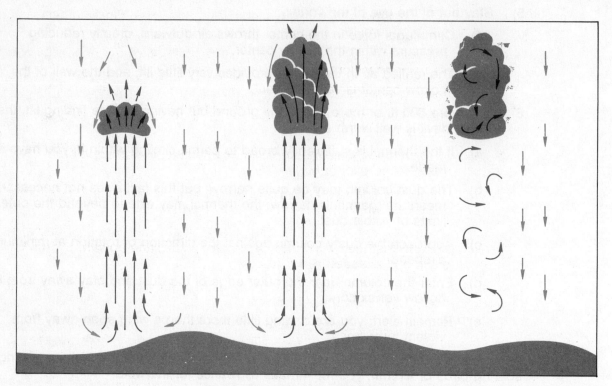

Figure 150.  Cumulus Clouds and Thermals

    d)   You can expect to find a thermal beneath either of the growing cumuli in Figure 150.

        i)   On the average, the infant cumulus on the left would be the better choice because of its longer life expectancy.

        ii)  This, of course, is a gamble, since all cumuli do not grow to the same size.

3)   As a cumulus cloud grows, it may shade the surface that generated it.

    a)   The surface cools, temporarily arresting the thermal.

    b)   As the cloud dissipates or drifts away with the wind, the surface again warms and regenerates the thermal.

    c)   This intermittent heating is one way in which thermals occur as segments or bubbles.

4)   Although abundant convective cloud cover reduces thermal activity, we cannot quote a definite amount that renders thermals too weak for soaring.

    a)   About 5/10 cover seems to be a good average approximation.

    b)   Restriction of thermals by cumulus cloudiness first becomes noticeable at low levels.

    c)   A sailplane may be unable to climb more than a few hundred feet at a low altitude while pilots at higher levels are maintaining height in or just beneath 6/10 to 8/10 convective cloud cover.

g. When air is highly unstable, the cumulus cloud can grow into a more ambitious towering cumulus or cumulonimbus.

   1) The energy released by copious condensation can increase buoyancy until the thermals become violent.

      a) Towering cumulus can produce showers.

   2) The cumulonimbus is the thunderstorm cloud producing heavy rain, hail, and icing.

      a) Well-developed *towering cumulus and cumulonimbus are for the experienced pilot only*.

      b) Some pilots find strong lift in or near convective precipitation, but they avoid hail, which can seriously batter the aircraft.

   3) Violent thermals just beneath and within these highly developed clouds often are so strong that they will continue to carry a sailplane upward even with nose down and airspeed at the redline.

h. Dense, broken, or overcast middle and high cloudiness shade the surface, cutting off surface heating and convective thermals.

   1) On a generally warm, bright day but with thin or patchy middle or high cloudiness, cumulus may develop, but the thermals are few and weak.

   2) The high-level cloudiness may drift by in patches.

      a) Thermals may surge and wane as the cloudiness decreases and increases.

   3) Never anticipate optimal thermal soaring when plagued by these mid- and high-level clouds.

   4) Altocumulus castellanus clouds, which are middle-level convective clouds, develop in updrafts at and just below the cloud levels.

      a) They do not extend upward from the surface.

      b) If a sailplane can reach levels near the cloud bases, the updrafts with altocumulus castellanus can be used in the same fashion as thermals formed by surface convection.

      c) The problem is reaching the convective level.

i. Wet ground favors thermals less than dry ground since wet ground heats more slowly.

   1) Some flat areas with wet soil such as swamps and tidewater areas have reputations for being poor thermal soaring areas.

      a) Convective clouds may be abundant but thermals generally are weak.

   2) Showery precipitation from scattered cumulus or cumulonimbus is a sure sign of unstable air favorable for thermals.

      a) When showers have soaked the ground in localized areas, however, downdrafts are almost certain over these wet surfaces. Avoid shower-soaked areas when looking for lift.

5.   **Thermal Structure.**  Thermals are as varied as trees in a forest.  No two are exactly alike.

   a.   When surface heating is intense and continuous, a thermal, once begun, continues for a prolonged period in a steady column as in Figure 153 below.

   1)   Sometimes called the "chimney thermal," this type seems from experience to be most prevalent.

   2)   In the chimney thermal, lift is available at any altitude below a climbing sailplane or soaring birds.

   b.   When heating is slow or intermittent, a "bubble" may be pinched off and forced upward; after an interval ranging from a few minutes to an hour or more, another bubble forms and rises as in Figure 154 below.

   1)   As explained earlier, intermittent shading by cumulus clouds forming atop a thermal is one reason for the bubble thermal.

   2)   A sailplane or birds may be climbing in a bubble, but an aircraft attempting to enter the thermal at a lower altitude may find no lift.

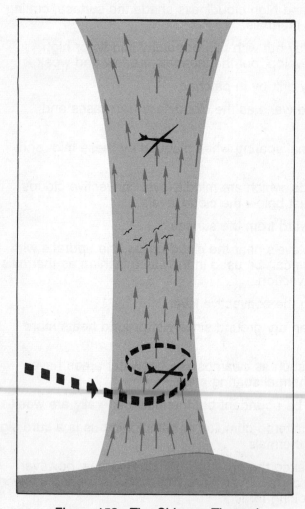

Figure 153.  The Chimney Thermal

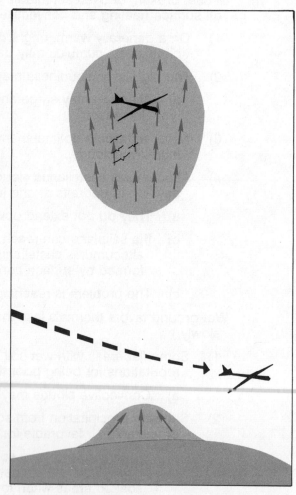

Figure 154.  Thermal Bubble

c.  A favored theoretical structure of some bubble thermals is the vortex shell which is much like a smoke ring blown upward as diagrammed in Figure 155 below.

1)  Lift is strongest in the center of the ring; downdrafts may occur in the edges of the ring or shell; and outside the shell, one would expect weak downdrafts.

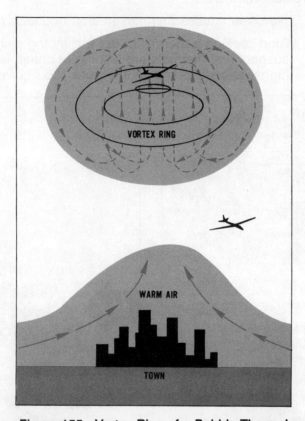

**Figure 155.  Vortex Ring of a Bubble Thermal**

d.   Wind and wind shear.  Thermals develop with a calm condition or with light, variable wind.  However, it seems that a surface wind of 5 to 10 kt. favors more organized thermals.

1)   A surface wind in excess of 10 kt. usually means stronger winds aloft, resulting in vertical wind shear.

   a)   This shear causes thermals to lean noticeably.

   b)   When seeking a thermal under a climbing sailplane and you know or suspect that thermals are leaning in shear, look for lift upwind from the higher aircraft as shown in Figure 156 below.

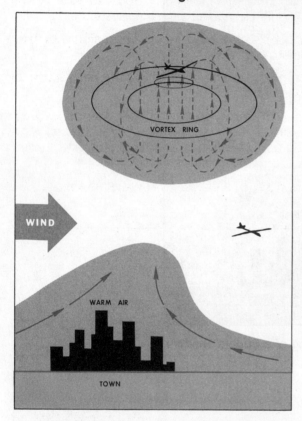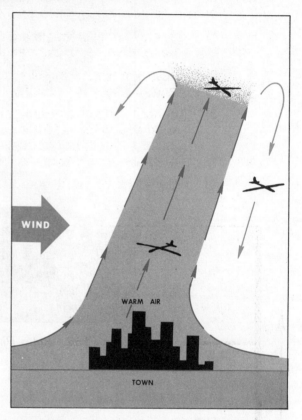

Figure 156.  Wind Causes Thermals to Lean

2)   Effect of shear on thermals depends on the relative strength of the two.

   a)   Strong thermals can remain fairly well organized with strong vertical wind shear; surface wind may even be at the maximum that will allow a safe launch.

   b)   Weak thermals are disorganized and ripped to shreds by strong vertical wind shear; individual thermal elements become hard to find and often are too small to use for lift.

   c)   A shear in excess of 3 kt. per 1,000 ft. distorts thermals to the extent that they are difficult to use.

3)   No critical surface wind speed can tell us when to expect such a shear.  However, shearing action often is visible in cumulus clouds.

   a)   A cloud sometimes leans but shows a continuous chimney.

   b)   At other times, the clouds are completely segmented by the shear.

   c)   Remember, however, that this shearing action is at cloud level; thermals below the clouds may be well organized.

4)   We must not overlook one other vital effect of the low-level wind shear.

   a)   On final approach for landing, the aircraft is descending into decreasing headwind.

   b)   Inertia of the aircraft into the decreasing wind causes a drop in airspeed. The decrease in airspeed may result in loss of control and perhaps a stall.

   c)   A good rule is to add one knot airspeed to normal approach speed for each knot of surface wind.

e.   Thermal streets are bands of thermals which become organized into straight lines parallel to each other, often providing lift over a considerable distance.

   1)   Generally, these streets are parallel to the wind, but on occasion they have been observed at right angles to the wind.

   2)   They form when wind direction changes little throughout the convective layer and the layer is capped by very stable air.

   3)   The formation of a broad system of evenly spaced streets is enhanced when wind speed reaches a maximum within the convective layer; that is, wind increases with height from the surface upward to a maximum and then decreases with height to the top of the convective layer.

   4)   Figure 158 below diagrams conditions favorable for thermal streeting.

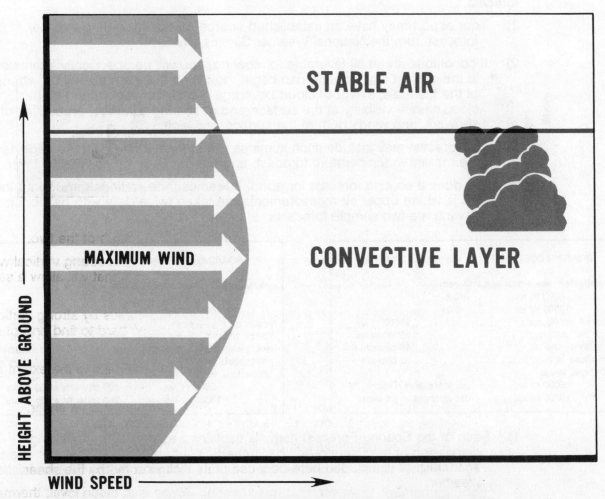

Figure 158.  Conditions Favorable for Thermal Streeting

5) Thermal streets may occur in clear air, or they may be indicated by cumulus clouds, which appear as long, narrow, parallel bands.

a) If cumulus clouds mark thermal streets, the top of the convective layer is approximately the height of the cloud tops.

6) The distance between streets in such a system is two to three times the general depth of the convective layer.

a) If convective clouds are present, this distance is two to three times the height of the cloud tops.

7) Downdrafts between these thermal streets are usually at least moderate and sometimes strong.

8) Cumulus cloud streets frequently form in the United States behind cold fronts in the cold air of polar outbreaks in which relatively flat cumuli develop.

9) Cloud streets are advantageous for sailplane pilots because, rather than circling in isolated thermals and losing height between them, the pilot soaring under a thermal street can maintain almost continuous, straight flight.

6. **Height and Strength of Thermals**

a. Since thermals are a product of instability, height of thermals depends on the depth of the unstable layer, and their strength depends on the degree of instability.

b. Most likely you will be soaring from an airport with considerable soaring activity -- possibly the home base of a soaring club -- and you are interested in a soaring forecast.

1) Your airport may have an established source of a daily soaring weather forecast from the National Weather Service (NWS).

2) If conditions are at all favorable for soaring, you will be specifically interested in the earliest time soaring can begin, how high the thermals will be, strength of the thermals, extent of cloud coverage -- both convective and higher cloudiness -- visibility at the surface and at soaring altitudes, probability of showers, and winds both at the surface and aloft.

3) The forecast may include such items as the thermal index (TI) (see page 194), the maximum temperature forecast, and the depth of the convective layer.

c. The NWS does a soaring forecast for about 60 radiosonde stations throughout the U.S. (this is where upper air measurements are taken twice daily with balloons). The following are two sample forecasts:

```
SOARING FORECAST          DATE...3/14/1986...122

THERMAL INDEX...MINUS SIGN INDICATES INSTABILITY
     5000 FT ASL ........ −10.5
    10000 FT ASL ........ −6.0
HEIGHT OF THE −3 INDEX ..................... 13600 FT ASL
TOP OF THE LIFT ............................ 16900 FT ASL
MAX TEMPERATURE ............................ 46 DEGREES F
FIRST USABLE LIFT .......................... 35 DEGREES F
UPPER LEVEL WINDS
     5000 FT ASL ........ /// DEGREES AT // KNOTS
    10000 FT ASL ........ 015 DEGREES AT 15 KNOTS
```

```
SOARING FORECAST          DATE...3/14/1986...122

THERMAL INDEX...MINUS SIGN INDICATES INSTABILITY
     5000 FT ASL ........ −6.0
    10000 FT ASL ........ 0.0
HEIGHT OF THE −3 INDEX ..................... 7400 FT ASL
TOP OF THE LIFT ............................ 9900 FT ASL
MAX TEMPERATURE ............................ 46 DEGREES F
FIRST USABLE LIFT .......................... 42 DEGREES F
UPPER LEVEL WINDS
     5000 FT ASL ........ 105 DEGREES AT 03 KNOTS
    10000 FT ASL ........ 260 DEGREES AT 05 KNOTS
```

1) Each of the Soaring Forecast items is explained as part of the following discussion of the Pseudo-Adiabatic chart. Your author feels that this chart is academic. It is included here because of its inclusion by the FAA in *Aviation Weather*.

2) The FAA is going to delete or modify "Top of the Lift" and "First Usable Lift."

## 7.    The Pseudo-Adiabatic Chart

a.    The pseudo-adiabatic chart is used to graphically compute adiabatic changes in vertically moving air and to determine stability.  It can be used to explain how the information in the Soaring Forecast is computed.

1)    It has five sets of lines shown in Figure 160 below.  These lines are:

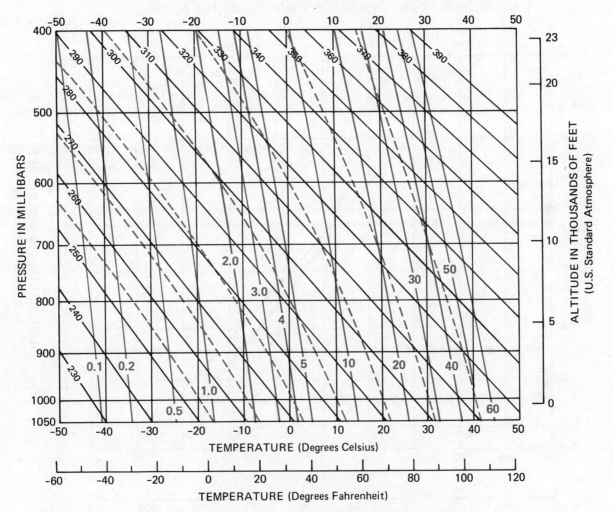

Figure 160.  The Pseudo-Adiabatic Chart

a)    Pressure in millibars (horizontal lines),

b)    Temperature in degrees Celsius (vertical lines),

c)    Dry adiabats (sloping black lines),

d)    Lines of constant water vapor or mixing ratio of water vapor to dry air (solid red lines), and

e)    Moist adiabats (dashed red lines).

2)    The chart also has an altitude scale in thousands of feet above sea level (ASL) in a standard atmosphere along the right margin and a Fahrenheit temperature scale across the bottom.

3)    The chart used in actual practice has a much finer grid than the one shown in Figure 160 above.

4)   The following examples deal with dry thermals; and since the red lines in Figure 160 on page 191 concern moist adiabatic changes, they are omitted from the examples.

a)   If you care to delve deeper into use of the chart, you will find moist adiabatic processes even more complicated than dry processes.

b.   An upper air observation, or sounding, is plotted on the pseudo-adiabatic chart as shown by the heavy, solid black line in Figure 161 below.

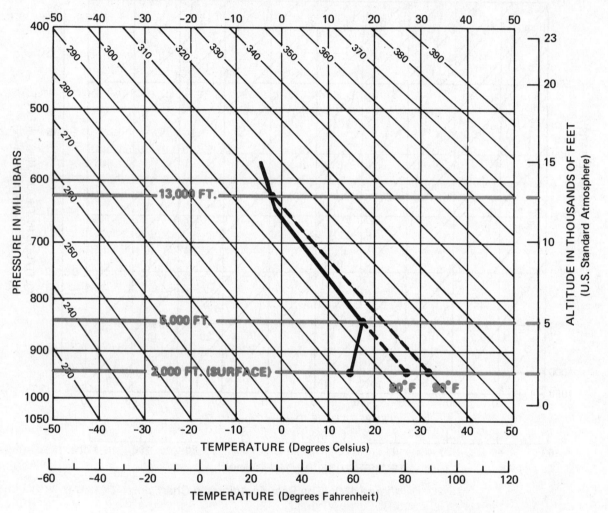

Figure 161.  An Upper Air Observation Plotted on the Pseudo-Adiabatic Chart

1)   This plotting is the vertical temperature profile at the time the radiosonde observation was taken.

2)   It is the actual or existing lapse rate.

3)   Blue lines are added to the illustration showing appropriate altitudes to help you interpret the chart.

c.   Depth of Convective Layer (Height of Thermals)

1)   We know that for air to be unstable, the existing lapse rate must be equal to or greater than the dry adiabatic rate of cooling.

a)   In other words, in Figure 161 on page 192, the solid black line representing the plotted existing lapse rate would slope parallel to or slope more than the dry adiabats. Obviously it does not from the surface to 5,000 ft. ASL.

b)   Therefore, at the time the sounding was taken, the air was stable; there was no convective or unstable layer, and thermals were nonexistent. Thermal soaring was impossible.

2)   As the surface temperature rises during the day, air is warmed and forced upward, (i.e., convection), cooling at the dry adiabatic rate.

a)   This movement continues until the temperature of the air moving upward is the same as the surrounding air.

3)   Assume that the sounding (as shown by the solid black line) in Figure 161 on page 192 was made at sunrise with a surface temperature of 59°F (15°C). The forecast temperature at noon is 80°F and the maximum temperature of the day is forecast at 90°F. By using the chart in Figure 161 on page 192, you can determine the height of the thermals at these times.

a)   Plot 80°F (about 27°C) at the surface elevation and draw a dashed line parallel to the dry adiabats (sloping lines) to the point at which it intersects the sounding (solid line), which is 5,000 ft. ASL.

i)   Convection lifts the warmer air to a level at which it cools adiabatically (represented by the dashed line) to the temperature of the surrounding air.

ii)   The thermal height is 5,000 ft. ASL (3,000 ft. AGL).

b)   Repeat the process using the maximum temperature of 90°F (about 30°C). The thermal height is 13,000 ft. ASL (11,000 ft. AGL).

i)   The Soaring Forecast provides the maximum forecast temperature for a specific station.

4)   Remember that we are talking about dry thermals.

a)   If convective clouds form below the indicated maximum thermal height, they will greatly distort the picture.

b)   However, if cumulus clouds do develop, thermals below the cloud base should be strengthened.

c)   If more higher clouds develop than were forecast, they will curtail surface heating, and the maximum temperature will most likely be cooler than forecast.

i)   Thermals will be weaker and will not reach as high an altitude.

5)   The Soaring Forecast provides you with the thermal height. It is listed as "top of the lift" (see the Soaring Forecast on page 190).

6)   The Soaring Forecast also provides you with the surface temperature that will provide usable lift, and it is listed as "first usable lift."

d.   Thermal Index (TI)

1)   Since thermals depend on sinking cold air forcing warm air upward, strength of thermals depends on the temperature difference between the sinking air and the rising air -- the greater the temperature difference the stronger the thermals.

a)   To arrive at an approximation of this difference, the forecaster computes a thermal index (TI).

2)   A thermal index may be computed for any level; but for the Soaring Forecast they are computed for 5,000 ft. ASL (850-mb) and 10,000 ft. ASL (700-mb) as shown on the Soaring Forecast samples on page 190.

a)   These levels are selected because they are in the altitude domain of routine soaring and because temperature data are routinely available for these two levels (i.e., Constant Pressure Analysis Chart).

3)   Three temperature values are needed -- the observed 850-mb and 700-mb temperatures and the forecast maximum temperature.

4)   Assume a sounding as in Figure 162 on page 195  with an 850-mb temperature of 15°C, a 700-mb temperature of 10°C, and forecast maximum of 86°F (30°C).

a)   Plot the three temperatures, using care to place the maximum temperature plot at field elevation (2,000 ft. in Figure 162 on page 195).

b)   Now draw a line (the black dashed line) through the maximum temperature parallel to the dry adiabats.

c)   Note that the dashed line intersects the 850-mb level at 20°C and the 700-mb level at 4°C.

d)   Subtract these temperatures from actual sounding temperatures at corresponding levels.

e)   Note the difference is −5°C at 850 mb (15 − 20 = −5) and +6°C at 700 mb (10 − 4 = +6).

f)   These values are the TI's at the two levels.

5)   Strength of thermals is proportional to the magnitude of the negative value of the TI.

a)   A TI of −8 or −10 predicts very good lift and a long soaring day.

b)   Thermals with this high a negative value will be strong enough to hold together even on a windy day.

c)   A TI of −3 indicates a very good chance of sailplanes reaching the altitude of this temperature difference.

i)   The height of the −3 index is given on the Soaring Forecast, as shown on page 190.

d)   A TI of −2 to zero leaves much doubt; and a positive TI offers even less hope of thermals reaching the altitude.

6) Remember that the TI is a forecast value.

    a) A miss in the forecast maximum or a change in temperature aloft can alter the picture considerably.

    b) The example in Figure 162 below should promise fairly strong thermals to above 5,000 ft. (TI of −5).

        i) The "top of the lift" is indicated at the intersection of the dashed line and solid line. This value is given in the Soaring Forecast.

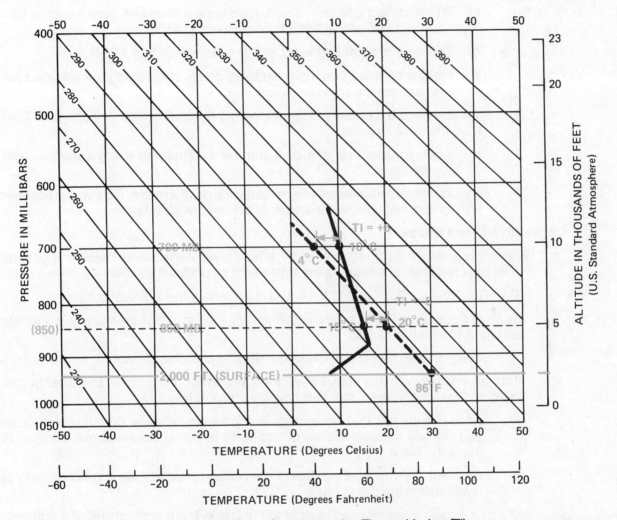

Figure 162. Computing the Thermal Index (TI)

7) The Soaring Forecast also provides the winds from the upper air sounding at the 5,000-ft. and 10,000-ft. levels. See the Soaring Forecast on page 190.

8) Often the National Weather Service will have no upper air sounding taken near a soaring base. Forecasts must be based on a simulated sounding derived from distant observations.

    a) At other times, for some reason a forecast may not be available.

    b) Furthermore, you can often augment the forecast with local observations.

e.    Do It Yourself.  The first step in determining height and strength of thermals is to obtain a local sounding.

1)    Send your tow aircraft aloft about sunrise and simply read outside air temperatures from the aircraft thermometer and altitudes from the altimeter.

a)    Read temperatures at 500-ft. intervals for about the first 2,000 ft. and at 1,000-ft. intervals at higher altitudes.

b)    The information may be radioed back to the ground, or may be recorded in flight and analyzed after landing.

c)    When using the latter method, read temperatures on both ascent and descent and average the temperatures at each level.

d)    This type of sounding is an airplane observation or APOB.

e)    Plot the sounding on the pseudo-adiabatic chart using the altitude scale rather than the pressure scale.

2)    Next you need a forecast maximum temperature.  Perhaps you can pick this up from the local forecast.

a)    If not, you can use your best judgment comparing today's weather with yesterday's.

3)    Although these procedures are primarily for dry thermals, they work reasonably well for thermals below the bases of convective clouds.

8.  **Convective Cloud Bases**

a.    Soaring experience suggests a shallow, stable layer immediately below the general level of convective cloud bases through which it is difficult to soar.

1)    This layer is 200 to 600 ft. thick and is known as the *sub-cloud layer*.

b.    The layer appears to act as a filter allowing only the strongest thermals to penetrate it and form convective clouds.

1)    Strongest thermals are beneath developing cumulus clouds.

c.    Thermals intensify within a convective cloud, but evaporation cools the outer edges of the cloud causing a downdraft immediately surrounding it.

1)    Add to this the fact that downdrafts predominate between cumulus clouds, and you can see the slim chance of finding lift between clouds above the level of the cloud base.

2)    In general, thermal soaring during convective cloud activity is practical only at levels below the cloud base.

d.    In Part I, Chapter 6, we learned to estimate height in thousands of feet of a convective cloud base by dividing the surface temperature-dew point spread by 2.5 (for Celslus) or 4.4 (for Fahrenheit).

1)    If the rising column were self-contained -- that is, if no air were drawn into the sides of the thermal -- the method would give a fairly accurate height of the base.  However, this is not the case.

2)    Air is entrained or drawn into the sides of the thermal, which lowers the water vapor content of the thermal, allowing it to reach a somewhat higher level before condensation occurs.

a)    Bases of the clouds are generally 10% to 15% higher than the computed height.

3)   Entrainment is a problem; observers and forecasters can only estimate its effect.

   a)   Until a positive technique is developed, heights of cumulus bases will tend to be reported and forecast too low.

9.   **Cross-Country Thermal Soaring**

   a.   A pilot can soar cross-country using either isolated thermals or thermal streets.

   1)   When using isolated thermals, (s)he gains altitude circling in thermals and then proceeds toward the next thermal in the general direction of his/her cross-country.

   2)   Under a thermal street, (s)he may be able to proceed with little if any circling if his/her chosen course parallels the thermal streets.

   3)   (S)he can obtain the greatest distance by flying in the direction of the wind.

   b.   In the central and eastern United States, the most favorable weather for cross-country soaring occurs behind a cold front.

   1)   Four factors contribute to making this pattern ideal.

   a)   The cold polar air is usually dry, and thermals can build to relatively high altitudes.

   b)   The polar air is colder than the ground, and thus, the warm ground aids solar radiation in heating the air.

   i)   Thermals begin earlier in the morning and last later in the evening.
   ii)   On occasions, soarable lift has been found at night.

   c)   Quite often, colder air at high altitudes moves over the cold, low-level outbreak, intensifying the instability and strengthening the thermals.

   d)   The wind profile frequently favors thermal streeting -- a real boon to speed and distance.

   c.   The same four factors may occur with cold frontal passages over mountainous regions in the western United States.

   1)   However, rugged mountains break up the circulation; and homogeneous conditions extend over smaller areas than over the eastern parts of the country.

   2)   The western mountain regions and particularly the desert southwest have one decided advantage.

   a)   Air is predominantly dry with more abundant daytime thermal activity favoring cross-country soaring, although it may be for shorter distances.

   d.   Among the world's most favorable tracks for long distance soaring is a high plains corridor along the east slope of the Rocky Mountains stretching from southwest Texas to Canada.

   1)   Terrain in the corridor is relatively flat and high with few trees; terrain surface ranges from barren to short grass.

   a)   These surface features favor strong thermal activity.

   2)   Prevailing wind is southerly and moderately strong, giving an added boost to northbound cross-country flights.

## C.  Frontal Soaring

1.  Warm air forced upward over cold air above a frontal surface can provide lift for soaring.

    a.  However, good frontal lift is transitory and accounts for a very small portion of powerless flight.

    b.  Seldom will you find a front parallel to your desired cross-country route, and seldom will it stay in position long enough to complete a flight.

        1)  A slowly moving front provides only weak lift.

        2)  A fast moving front often plagues the soaring pilot with cloudiness and turbulence.

2.  A front can occasionally provide excellent lift for a short period.

    a.  On a cross-country, you may be riding wave or ridge lift and need to move over a flat area to take advantage of thermals.

        1)  A front may offer lift during your transition.

3.  Fronts are often marked by a change in cloud type or amount.

    a.  However, the very presence of clouds may deter you from flying into the front.

        1)  Spotting a dry front is difficult.

    b.  Knowing that a front is in the vicinity and studying your aircraft reaction can tell you when you are in the frontal lift.

        1)  Staying in the lift is another problem.
        2)  Observing ground indicators of surface wind helps.

4.  An approaching front may enhance thermal or hill soaring.

    a.  An approaching front or a frontal passage most likely will disrupt a sea breeze or mountain wave.

## D.  Sea Breeze Soaring

1.  In many coastal areas during the warm seasons, a pleasant breeze from the sea occurs almost daily.

    a.  Caused by the heating of land on warm, sunny days, the sea breeze usually begins during early forenoon, reaches a maximum during the afternoon, and subsides around dusk after the land has cooled.

    b.  The leading edge of the cool sea breeze forces warmer air inland to rise as shown in Figure 165 on page 199.

        1)  Rising air from the land returns seaward at higher altitude to complete the convective cell.

    c.  A sailplane pilot operating in or near coastal areas often can find lift generated by this convective circulation.

        1)  The transition zone between the cool, moist air from the sea and the warm, drier air inland is often narrow and is a shallow, ephemeral kind of pseudo-cold front.

2.  Sometimes the wedge of cool air is called a **sea breeze front**.

    a.  If sufficient moisture is present, a line of cumuliform clouds just inland may mark the front.

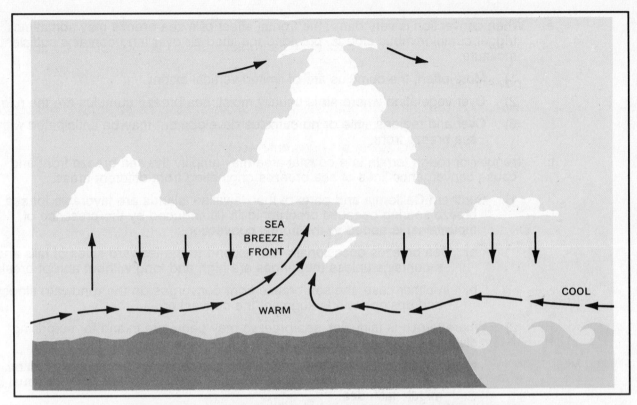

Figure 165.  Schematic Cross Section through a Sea Breeze Front

1) Whether marked by clouds or not, the upward moving air at the sea breeze front occasionally is strong enough to support soaring flight.

b. Within the sea breeze (i.e., between the sea breeze front and the ocean) the air is usually stable, and normally no lift may be expected at lower levels.

1) However, once airborne, pilots occasionally have found lift at higher levels in the return flow aloft.

2) A visual indication of this lift is cumulus extending seaward from the sea breeze front.

c. A large difference in land and sea water temperature intensifies the convective cell generating a sea breeze.

1) Where coastal waters are quite cool, such as along the California coast, and land temperatures warm rapidly in the daytime, the sea breeze becomes pronounced, penetrating perhaps 50 to 75 mi. inland at times.

2) Copious sunshine and cool sea waters favor a well-developed sea breeze front.

d. The sea breeze is a local effect.

1) Strong pressure gradients with a well-developed pressure system can overpower the sea breeze effect.

2) Winds will follow the direction and speed dictated by the strong pressure gradient.

a) Therefore, a sea breeze front is most likely when pressure gradient is weak and wind is light.

e.    When convection is very deep, the frontal effect of a sea breeze may sometimes trigger cumulonimbus clouds, provided the lifted air over land contains sufficient moisture.

1)    More often, the cumulus are of limited vertical extent.

2)    Over vegetation where air is usually moist, sea breeze cumulus are the rule.

3)    Over arid regions, little or no cumulus development may be anticipated with a sea breeze front.

f.    Irregular or rough terrain in a coastal area may amplify the sea breeze front and cause convergence lines of sea breezes originating from different areas.

1)    Southern California and parts of the Hawaiian Islands are favorable for sea breeze soaring because orographic lift (lift induced by the presence of mountains) is added to the frontal convection.

a)    Sea breezes occasionally may extend to the leeward sides of hills and mountains unless the ranges are high and long without abrupt breaks.

b)    In either case, the sea breeze front converges on the windward slopes, and upslope winds augment the convection.

2)    Where terrain is fairly flat, sea breezes may penetrate inland for surprising distances but with weaker lift along the sea breeze front.

a)    In the Tropics, sea breezes sometimes penetrate as much as 150 mi. inland, whereas an average of closer to 50 mi. inland is more usual in middle latitudes.

3)    Sea breezes reaching speeds of 15 to 25 kt. are not uncommon.

g.    When a sea breeze front develops, visual observations may provide clues to the extent of lift that you may anticipate.

1)    Expect little or no lift on the seaward side of the front when the sea air is markedly devoid of convective clouds or when the sea breeze spreads low stratus inland.

a)    However, some lift may be present along the leading edge of the sea breeze or just ahead of it.

2)    Expect little or no lift on the seaward side of the front when visibility decreases markedly in the sea breeze air.

a)    This is an indicator of stable air within the sea breeze.

3)    A favorable visual indication of lift along the sea breeze front is a line of cumulus clouds marking the front; cumuli between the sea breeze front and the ocean also indicate possible lift within the sea breeze air, especially at higher levels.

a)    Cumulus bases in the moist sea air are often lower than along the front.

4)    When a sea breeze front is devoid of cumulus but converging streamers of dust or smoke are observed, expect convection and lift along the sea breeze front.

5)    Probably the best combination to be sighted is cumuli and converging dust or smoke plumes along the sea breeze front as it moves upslope over hills or mountains.

a)    The upward motion is amplified by the upslope winds.

6)    A difference in visibility between the sea air and the inland air often is a visual clue to the leading edge of the sea breeze.

a)    Visibility in the sea air may be restricted by haze while visibility inland is unrestricted.

b) On the other hand, the sea air may be quite clear while visibility inland is restricted by dust or smoke.

3. **Local Sea Breeze Explorations**

a. Unfortunately, a sea breeze front is not always easy to find, and it is likely that many an opportunity for sea breeze soaring goes unnoticed.

1) As yet, little experience has been accrued in locating a belt of sea breeze lift without visual clues such as clouds, haze, or converging smoke or dust plumes.

b. The sea breeze front moving from the Los Angeles coastal plain into the Mojave Desert has been dubbed the "Smoke Front."

1) It has intense thermal activity and offers excellent lift along the leading edge of the front.

2) Associated with the sea breeze that moves inland over the Los Angeles coastal plain are two important zones of convergence, shown in Figure 166 below.

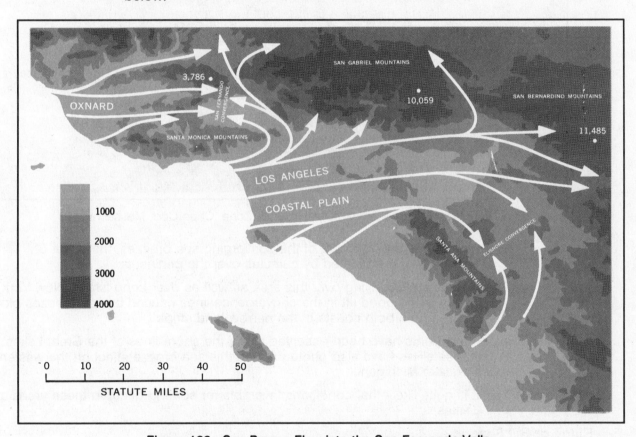

Figure 166. Sea Breeze Flow into the San Fernando Valley

a) One convergence line is the "San Fernando Convergence Zone."
b) A larger scale zone is in the Elsinore area, also shown in Figure 166.

3) This convergence zone apparently generates strong vertical currents because soaring pilots fly back and forth across the valley along the line separating smoky air to the north from relatively clear air to the south.

4) Altitudes reached depend upon the stability, but usually fall within the 6,000 ft. to 12,000 ft. MSL range for the usual dry thermal type lift.

a) Seaward, little or no lift is experienced in the sea breeze air marked by poor visibility.

c.   Figure 167 below shows converging air between sea breezes flowing inland from opposite coasts of the Cape Cod Peninsula.

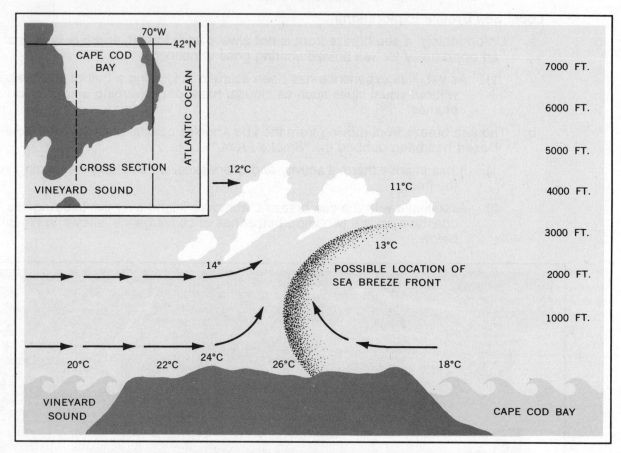

Figure 167.  Sea Breeze Convergence Zone, Cape Cod, Massachusetts

1)   Later in the development of the converging sea breezes, the onset of convection is indicated by cumulus over the peninsula.

2)   Sailplane pilots flying over this area as well as over Long Island, New York, have found good lift in the convergence lines caused by sea breezes blowing inland from both coasts of the narrow land strips.

d.   Sea breeze fronts have been observed along the shore lines of the Great Lakes. Weather satellites have also photographed this sea breeze effect on the western shore of Lake Michigan.

1)   It is quite likely that conditions favorable for soaring occur in those areas at times.

E.  **Ridge or Hill Soaring**

1.   Wind blowing toward hills or ridges flows upward, over, and around the abrupt rises in terrain.  The upward-moving air creates lift which is sometimes excellent for soaring.

a.   Figure 168 on page 203 is a schematic showing area of best lift.

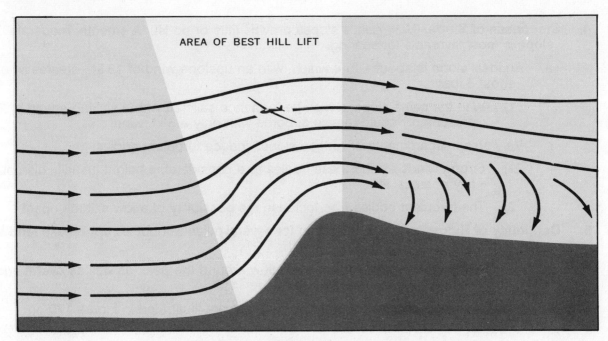

AREA OF BEST HILL LIFT

Figure 168.  Schematic Cross Section of Airflow over a Ridge

2.   **Wind.**  To create lift over hills or ridges, wind direction should be within about 30° to 40° of being perpendicular to the ridge line.

   a.   A sustained speed of 15 kt. or more usually generates enough lift to support a sailplane.

   b.   Height of the lift usually is two or three times the height of the rise from the valley floor to the ridge crest.

   c.   Strong winds tend to increase turbulence and low-level eddies without an appreciable increase in the height of the lift.

3.   **Stability** affects the continuity and extent of lift over hills or ridges.

   a.   Stable air allows relatively streamlined upslope flow.

      1)   A pilot experiences little or no turbulence in the steady, uniform area of best lift shown in Figure 168 above.

      2)   Since stable air tends to return to its original level, air spilling over the crest and downslope is churned into a snarl of leeside eddies, also shown in Figure 168 above.

      3)   Thus, stable air favors smooth lift but troublesome leeside low-altitude turbulence.

   b.   When the airstream is moist and unstable, upslope lift may release the instability, generating strong convective currents and cumulus clouds over windward slopes and hill crests.

      1)   The initially laminar flow is broken up into convective cells.

      2)   While the updrafts produce good lift, strong downdrafts may compromise low-altitude flight over rough terrain.

      3)   As with thermals, the lift will be transitory rather than smooth and uniform.

4.  **Steepness of Slope.**  Very gentle slopes provide little or no lift.  A smooth, moderate slope is most favorable for soaring.

    a.  An ideal slope is about 1 to 4 which, with an upslope wind of 15 kt., creates lift of about 6 fps.

        1)  With the same slope, a high-performance sailcraft with a sinking speed of 2 fps presumably could remain airborne with only a 5-kt. wind!

    b.  Very steep escarpments or rugged slopes induce turbulent eddies.

        1)  Strong winds extend these eddies to a considerable height, usually disrupting any potential lift.

        2)  The turbulent eddies also increase the possibility of a low-altitude upset.

5.  **Continuity of Ridges.**  Ridges extending for several miles without abrupt breaks tend to provide uniform lift throughout their length.

    a.  In contrast, a single peak diverts wind flow around the peak as well as over it and thus is less favorable for soaring.

    b.  Some wind flow patterns over ridges and hills are illustrated in Figure 170 on page 205.

        1)  Deviations from these patterns depend on wind direction and speed, on stability, on slope profile, and on general terrain roughness.

6.  **Soaring in Upslope Lift**

    a.  The soaring pilot, always alert, must remain especially so in seeking or riding hill lift.

    b.  When air is unstable, do not venture too near the slope.

        1)  You can identify unstable air either by the updrafts and downdrafts in dry thermals or by cumulus building over hills or ridges.

        2)  Approaching at too low an altitude may suddenly put you in a downdraft, forcing an inadvertent landing.

    c.  When winds are strong, surface friction may create low-level eddies even over relatively smooth slopes.

        1)  Also, friction may drastically reduce the effective wind speed near the surface.

        2)  When climbing at low altitude toward a slope under these conditions, be prepared to turn quickly toward the valley if you lose lift.

            a)  Renew your attempt to climb, farther from the hill.

    d.  If winds are weak, you may find lift only very near the sloping surface.  Then you must hug the slope to find needed lift.

        1)  However, avoid this procedure if there are indications of up  and downdrafts.

        2)  In general, for any given slope, keep your distance from the slope proportional to wind speed.

    e.  Leeward of hills and ridges is an area where wind is blocked by the obstruction.  In soaring circles this area is called the "wind shadow."

        1)  In the wind shadow, downdrafts predominate as shown in Figure 168 on page 203.

        2)  If you stray into the wind shadow at an altitude near or below the altitude of the ridge crest, you may be forced into an unscheduled and possibly rough landing.

        3)  Try to stay within the area of best lift shown in Figure 168 on page 203.

Figure 170.  Wind Flow over Various Types of Terrain

F.    **Mountain Wave Soaring**

    1.    The main attraction of soaring in mountain waves stems from the continuous lift to great heights.

        a.    Soaring flights to above 35,000 ft. have frequently been made in mountain waves.

        b.    Once a soaring pilot has reached the rising air of a mountain wave, (s)he has every prospect of maintaining flight for several hours.

        c.    While mountain wave soaring is related to ridge or hill soaring, the lift in a mountain wave is on a larger scale and is less transitory than lift over smaller rises in terrain.

        d.    Figure 171 below is a cross section of a typical mountain wave.

    2.    **Formation.**  When strong winds blow across a mountain range, large standing waves occur downwind from the mountains and upward to the tropopause.

        a.    The waves may develop singly, but more often they occur as a series of waves downstream from the mountains.

            1)    While the waves remain almost stationary, strong winds are blowing through them.

        b.    You may compare a mountain wave to a series of waves formed downstream from a submerged rocky ridge in a fast flowing creek or river.

            1)    Air dips sharply immediately to the lee of a ridge, then rises and falls in a wave motion downstream.

        c.    A strong mountain wave requires:

            1)    Marked stability in the airstream disturbed by the mountains.

                a)    Rapidly building cumulus over the mountains visually marks the air unstable; convection, evidenced by the cumulus, tends to deter wave formation.

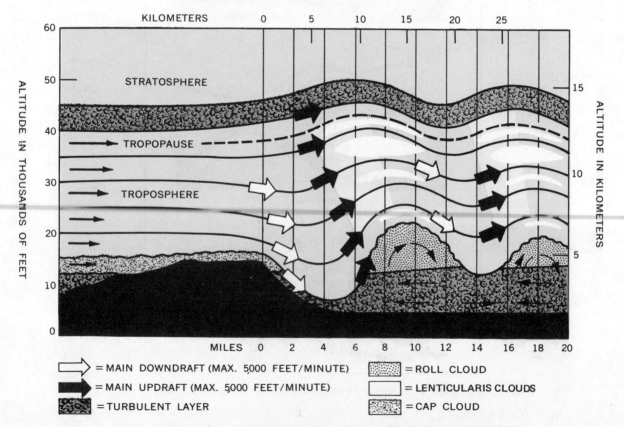

**Figure 171.  Schematic Cross Section of a Mountain Wave**

2)   Wind speed at the level of the summit should exceed a minimum which varies from 15 to 25 kt. depending on the height of the range.

   a)   Upper winds should increase or at least remain constant with height up to the tropopause.

3)   Wind direction should be within 30° normal to the range.  Lift diminishes as winds more nearly parallel the range.

### 3.   Wave Length and Amplitude

a.   Wave length is the horizontal distance between crests of successive waves and is usually between 2 and 25 mi.

   1)   In general, wave length is controlled by wind component perpendicular to the ridge and by stability of the upstream flow.

      a)   Wave length is directly proportional to wind speed and inversely proportional to stability.

   2)   Figure 172 below illustrates wave length and also amplitude.

b.   Amplitude of a wave is the vertical dimension and is half the altitude difference between the wave trough and crest.

   1)   In a typical wave, amplitude varies with height above the ground.

      a)   It is least near the surface and near the tropopause.

   2)   Greatest amplitude is roughly 3,000 to 6,000 ft. above the ridge crest.

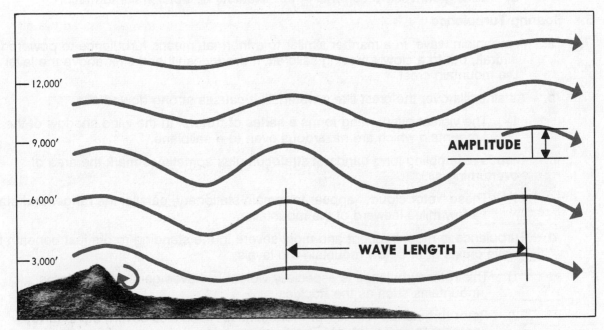

Figure 172.  Wave Length and Amplitude

   3)   Wave amplitude is controlled by size and shape of the ridge as well as wind and stability.

      a)   A shallow layer of great stability and moderate wind produces a greater wave amplitude than does a deep layer of moderate stability and strong winds.

   4)   Also, the greater the amplitude, the shorter is the wave length.

c.   Waves offering the strongest and most consistent lift are those with great amplitude and short wave length.

4.   **Visual Indicators**

a.   If the air has sufficient moisture, lenticular (lens-shaped) clouds mark wave crests.

   1)   Cooling of air ascending toward the wave crest saturates the air, forming clouds.

   2)   Warming of air descending beyond the wave crest evaporates the cloud.

   3)   Thus, by continuous condensation windward of the wave crest and evaporation leeward, the cloud appears stationary although wind may be blowing through the wave at 50 kt. or more.

   4)   Lenticular clouds in successive bands downstream from the mountain mark a series of wave crests.

b.   Spacing of lenticulars marks the wave length.

   1)   Clearly identifiable lenticulars also suggest larger wave amplitude than clouds which barely exhibit lenticular form.

   2)   These cloud types along with stratiform clouds on the windward slopes and along the mountain crest indicate the stability favorable to mountain wave soaring.

c.   Thunderstorms or rapidly building cumulus over mountains mark the air unstable.

   1)   As they reach maturity, the thunderstorms often drift downwind across leeward valleys and plains.

   2)   Strong convective currents in the unstable air deter wave formation.

5.   **Soaring Turbulence**

a.   A mountain wave, in a manner similar to a thermal, means turbulence to powered aircraft, but to a slowly moving sailcraft, it produces lift and sink above the level of the mountain crest.

b.   As air spills over the crest like a waterfall, it causes strong downdrafts.

   1)   The violent overturning forms a series of "rotors" in the wind shadow of the mountain which are hazardous even to a sailplane.

c.   Clouds resembling long bands of stratocumulus sometimes mark the area of overturning air.

   1)   These "rotor clouds" appear to remain stationary, parallel the range, and stand a few miles leeward of the mountains.

d.   Turbulence is most frequent and most severe in the standing rotors just beneath the wave crests at or below mountain-top levels.

   1)   This rotor turbulence is especially violent in waves generated by large mountains such as the Rockies.

   2)   Rotor turbulence with lesser mountains is much less severe but is always present to some extent.  The turbulence is greatest in well-developed waves.

6.   **Favored Areas**

a.   Mountain waves occur most frequently along the central and northern Rockies and the northern Appalachians.

   1)   Occasionally, waves form to the lee of mountains in Arkansas, Oklahoma, and southwestern Texas.

b.   Weather satellites have observed waves extending great distances downwind from the Rocky Mountains; one series extended for nearly 700 mi.

   1)   The more usual distance is 150 to 300 mi.

c.  While Appalachian waves are not as strong as those over the Rockies, they occur frequently, and satellites have observed them at an average of 115 mi. downwind.

    1)  Wave length of these waves averages about 10 NM.

7.  **Riding the Waves.**  You often can detect a wave by the uncanny smoothness of your climb.

a.  On first locating a wave, turn into the wind and attempt to climb directly over the spot where you first detected lift *provided* you can remain at an altitude above the level of the mountain crest.

    1)  After cautiously climbing well up into the wave, attempt to determine dimensions of the zone of lift.

        a)  If the wave is over rugged terrain, it may be impossible and unnecessary to determine the wave length.

            i)  Lift over such terrain is likely to be in patchy bands.

        b)  Over more even terrain, the wave length may be easy to determine and use in planning the next stage of flight.

b.  Wave clouds are a visual clue in your search for lift.

    1)  The wave-like shape of lenticulars is usually more obvious from above than from below.

    2)  Lift should prevail from the crest of the lenticulars upwind about one-third the wave length.

    3)  When your course takes you across the waves, climb on the windward side of the wave and fly as quickly as possible to the windward side of the next wave.

    4)  Wave lift of 300 to 1,200 fpm is not uncommon.

        a)  Soaring pilots have encountered vertical currents exceeding 3,000 fpm, the strongest ever reported being 8,000 fpm.

## G.  In Closing

1.  Thermal soaring -- The pilot circles in a thermal (updraft) to gain height and heads to the next thermal to regain altitude lost in the downdrafts between them.

a.  Thermals depend on solar heating, so thermal soaring is virtually restricted to daylight hours with considerable sunshine.

    1)  The most likely place for a thermal is above a surface that heats rapidly, e.g., barren sandy or rocky surfaces, plowed fields, stubble fields surrounded by green vegetation, cities, factories, and chimneys.

    2)  Dust devils are excellent sources of lift, but you must use caution because the thermals are strong and turbulent and are surrounded by areas of little lift or possibly of sink.

    3)  Thermal soaring is usually at its best when convective clouds develop.

        a)  Cumulus clouds are positive signs of thermals.

    4)  Wet ground is less favorable for thermals than dry ground because wet ground heats more slowly.

b.  No two thermals are exactly alike.

    1)  Some form as chimneys whereas others form bubbles of warm air that rise.

    2)  Wind shear can cause the thermal to lean or, at times, can completely break it up.

3)  Thermals may become organized into thermal streets.

    a)  A pilot can maintain generally continuous flight under a thermal street and seldom have to circle.

c.  Daily soaring weather forecasts are available from the National Weather Service.

    1)  You can determine height and strength of thermals by obtaining a local sounding and plotting it on a pseudo-adiabatic chart.

d.  You can soar cross-country using isolated thermals or thermal streets.

2.  Frontal soaring accounts for a very small portion of powerless flight.

    a.  A front can provide excellent lift for a short period.
    b.  An approaching front may enhance thermal or hill soaring.

3.  Southern California and parts of the Hawaiian Islands are favorable for sea breeze soaring because orographic lift augments convection from the sea breeze front.

    a.  Sea breeze fronts are not always easy to find, and visual clues may not be present.

        1)  A sea breeze front moves from the Los Angeles coastal plain into the Mojave Desert.

        2)  Sailplane pilots have found good lift in the convergence lines caused by sea breezes over Cape Cod, MA and Long Island, NY.

        3)  Sea breeze fronts have also been observed along the shorelines of the Great Lakes.

4.  Ridge or hill soaring -- Wind blowing toward hills or ridges flows upward, over, and around the abrupt rises in terrain.  Upward-moving air with a wind speed of 15 kt. usually generates enough lift to support a sailplane.

    a.  A smoother, moderate slope is most favorable for soaring.
    b.  When soaring in upslope lift, you must remain alert for unstable air.

5.  Mountain wave soaring is related to ridge or hill soaring, but the lift in a mountain wave is on a larger scale and is less transitory than lift over smaller rises in terrain.

    a.  Mountain waves occur most frequently along the central and northern Rockies and the northern Appalachians.

6.  Records are made to be broken.  Altitude and distance records are a prime target of many sailplane enthusiasts.

    a.  Distance records may be possible by flying a combination of lift sources such as thermal, frontal, ridge, or wave.

    b.  Altitude records are set in mountain waves.

        1)  Altitudes above 46,000 ft. have been attained over the Rocky Mountains.

        2)  Soaring flights to more than 24,000 ft. have been made in Appalachian waves.

        3)  Flights as high as 20,000 ft. have been recorded from New England to North Carolina.

7.  We hope that this chapter has given you an insight into the minute variations in weather that profoundly affect a soaring aircraft.

    a.  When you have remained airborne for hours without power, you have met a unique challenge and experienced a singular thrill of flying.

# END OF CHAPTER
# END OF PART II

# PART III
# AVIATION WEATHER SERVICES

*Aviation Weather Services* is published periodically by the FAA/NWS to keep pilots abreast of weather maps and other services available from their FSS.  AC 00-45E was published in December 1999.  We have prepared the following 26 chapters in Part III based on AC 00-45E.

| FAA **Table of Contents** AC 00-45E | Gleim **Table of Contents** This Part III |
|---|---|
| 1. The Aviation Weather Service Program<br>2. Aviation Routine Weather Report (METAR)<br>3. Pilot and Radar Reports, Satellite Pictures, and Radiosonde Additional Data (RADATs) | 1. The Aviation Weather Service Program<br>2. Aviation Routine Weather Report (METAR)<br>3. Pilot Weather Reports (PIREPs)<br>4. Radar Weather Report (SD)<br>5. Satellite Weather Pictures<br>6. Radiosonde Additional Data (RADAT) |
| 4. Aviation Weather Forecasts | 7. Terminal Aerodrome Forecast (TAF)<br>8. Aviation Area Forecast (FA)<br>9. Transcribed Weather Broadcasts (TWEB) Text Products<br>10. In-Flight Aviation Weather Advisories (WST, WS, WA)<br>11. Winds and Temperatures Aloft Forecast (FD)<br>12. Center Weather Service Unit (CWSU) Products<br>13. Hurricane Advisory (WH)<br>14. Convective Outlook (AC)<br>15. Severe Weather Watch Bulletin (WW) |
| 5. Surface Analysis Chart<br>6. Weather Depiction Chart<br>7. Radar Summary Chart<br>8. Constant Pressure Analysis Charts<br>9. Composite Moisture Stability Chart<br>10. Winds and Temperatures Aloft Charts<br>11. Significant Weather Prognostic Charts | 16. Surface Analysis Chart<br>17. Weather Depiction Chart<br>18. Radar Summary Chart<br>19. Constant Pressure Analysis Chart<br>20. Composite Moisture Stability Chart<br>21. Winds and Temperatures Aloft Charts<br>22. U.S. Low-Level Significant Weather Prog<br>23. High-Level Significant Weather Prog |
| 12. Convective Outlook Chart<br>13. Volcanic Ash Advisory Center Products<br>14. Turbulence Locations, Conversion and Density Altitude Tables, Contractions and Acronyms, Schedule of Products, National Weather Service Station Identifiers, WSR-88D Sites, and Internet Addresses | 24. Convective Outlook Chart<br>25. Volcanic Ash Advisory Center (VAAC) Products<br>26. Other Weather-Related Information |

We have divided the 14 sections (chapters) of AC 00-45E into 26 chapters in our book to facilitate study of various weather reports and weather forecasts.  The chapter titles are aligned so you can see which AC 00-45E chapters were broken up into additional chapters in this book.

# CHAPTER ONE
# THE AVIATION WEATHER SERVICE PROGRAM

Please take a few minutes to study each of the concepts listed above and anticipate/imagine what they are and how they relate to the other listed concepts.

A.   **Weather Service:  Aviation Effort**.  Providing weather service to aviation is a joint effort of the National Weather Service (NWS), the Federal Aviation Administration (FAA), the Department of Defense (DOD) Weather Service, and other aviation-oriented groups and individuals.

1.   Because of international flights and a need for worldwide weather forecasts, foreign weather services also have a vital input into our service.

2.   This chapter discusses the civilian agencies of the federal government and their observation and communication services to the aviation community.

B.   **National Oceanic and Atmospheric Administration (NOAA)**

1.   NOAA is an agency of the Department of Commerce and is one of the leading scientific agencies in the U.S. government.

2.   Among its six major divisions are the National Environmental Satellite Data and Information Service and the NWS.

C.   **National Environmental Satellite Data and Information Service (NESDIS)**

1.   NESDIS is located in Washington, D.C., and is responsible for directing the weather satellite program.

2.   Satellite images are available to NWS meteorologists and a wide range of other users for operational use.

3.   The Satellite Analysis Branch (SAB) coordinates the satellite and other known information for the NOAA Volcanic Hazards Alert program under an agreement with the FAA.

a.   The SAB works with the NWS as part of the Volcanic Ash Advisory Center (VAAC), located in Washington, D.C.

D.   **National Weather Service (NWS)** collects and analyzes meteorological and hydrological data and subsequently prepares forecasts on a national, hemispheric, and global scale.  The following is a description of the NWS facilities tasked with these duties.

1.   **National Centers for Environmental Prediction (NCEP)**

a.   Under the NCEP are nine separate national centers, each with its own specific mission.  They are the Climate Prediction Center, Space Environment Center, Marine Prediction Center, Hydrometeorological Prediction Center, Environmental Modeling Center, NCEP Center Operations, Storm Prediction Center, Aviation Weather Center, and Tropical Prediction Center.

b.   We will discuss the centers (items 2. through 6. on the next page) that provide services to the aviation community.

2.   The **National Center Operations (NCO)**, located in Washington, D.C., is the focal point of the NWS's weather processing system.

   a.   From worldwide weather reports, the NCO prepares automated weather analysis charts and guidance forecasts for use by NWS offices and other users.

   b.   The winds and temperatures aloft forecast is one example of an NCO product that is specifically prepared for aviation.

   c.   The NCO works with the SAB to fulfill the VAAC responsibilities to the aviation community regarding potential volcanic ash hazards to aviation.

3.   The **Storm Prediction Center (SPC)** is responsible for monitoring and forecasting severe weather over the 48 contiguous states.  The center also develops severe weather forecasting techniques and conducts research.

   a.   The SPC products include convective outlooks and forecasts, as well as severe weather watches.

   b.   The SPC is located in Norman, OK, near the heart of the area most frequently affected by severe thunderstorms.

4.   The **Aviation Weather Center (AWC)**, located in Kansas City, MO, issues warnings, forecasts, and analyses of hazardous weather for aviation interests.

   a.   The AWC identifies existing or imminent weather hazards to aircraft in flight and creates warnings for transmission to the aviation community.

      1)   The AWC also produces operational forecasts of weather conditions expected during the next two days that will affect domestic and international aviation interests.

   b.   As a Meteorological Watch Office (MWO) under the regulations of the International Civil Aviation Organization (ICAO), the AWC meteorologists prepare and issue aviation area forecasts (FA) and in-flight aviation weather advisories for the 48 contiguous states.

5.   The **Tropical Prediction Center (TPC)** is located in Miami, FL.  The National Hurricane Center (NHC), a branch of the TPC, issues hurricane advisories for the Atlantic, the Caribbean, the Gulf of Mexico, the eastern Pacific, and adjacent land areas.

   a.   The TPC also develops hurricane forecasting techniques and conducts hurricane research.

   b.   The Central Pacific Hurricane Center, located in Honolulu, HI, issues advisories for the central Pacific Ocean.

   c.   The Tropical Analysis and Forecast Branch (TAFB) of the TPC prepares and distributes tropical weather, aviation and marine forecasts, and warnings.

      1)   As an MWO, the TAFB meteorologists prepare and issue aviation forecasts, SIGMETs (significant meteorological information), and convective SIGMETs for their tropical Flight Information Region (FIR).

6.   The **Hydrometeorological Prediction Center (HPC)**, located in Camp Springs, MD, prepares the surface analysis chart.

7.   A **Weather Forecast Office (WFO)** prepares and issues various public, marine, and aviation forecasts and weather warnings for its area of responsibility.

   a.   In support of aviation, WFOs issue terminal aerodrome forecasts (TAF) and transcribed weather broadcasts (TWEB).

   b.   As MWOs, the Guam and Honolulu WFOs issue aviation area forecasts and in-flight advisories.

8. The **Alaskan Aviation Weather Unit (AAWU)**, located in Anchorage, AK, prepares and disseminates to the FAA and the Internet a suite of graphic products, including a graphic FA and a 24- and 36-hr. forecast of significant weather.

   a. As an MWO, the AAWU meteorologists prepare and issue International SIGMETs within the Anchorage Continental FIR, Anchorage Oceanic FIR, and the Anchorage Arctic FIR, as well as domestic FAs and AIRMETs (airman's meteorological information) for Alaska and the adjacent coastal waters.

   b. The AAWU is one of nine VAACs worldwide preparing Volcanic Ash Advisory Statements (VAAS) for the same FIRs listed in item a. above.

E. **Federal Aviation Administration (FAA)** is a part of the Department of Transportation and provides a wide range of services to the aviation community.  The following is a description of those FAA facilities involved with aviation weather and pilot services.

   1. **Automated Flight Service Station (AFSS)**

      a. The FAA supports 61 automated FSSs (AFSS) and 14 non-automated FSS facilities.

         1) The 14 non-automated FSSs are located in remote areas in Alaska and some are seasonal or part-time facilities.

      b. The AFSSs provide more aviation weather briefing service than any other government service outlet.

         1) An AFSS provides preflight and in-flight briefings, transcribed weather briefings, and scheduled and unscheduled weather broadcasts, and it furnishes weather support to flights in its area.

      c. As a starting point for a preflight weather briefing, you may wish to listen to one of the following two recorded weather briefings an AFSS can provide.  For a more detailed briefing, you can contact the AFSS directly.

         1) The **transcribed weather broadcast (TWEB)** provides continuous aeronautical and meteorological information on selected nondirectional beacon (NDB) and very high frequency (VHF) omnidirectional range (VOR) frequencies.

            a) At TWEB equipment locations controlling two or more VORs, the one used least for ground-to-air communications, preferably the nearest VOR, may be used as a TWEB outlet simultaneously with a NDB facility.

            b) The sequence and content of a TWEB follows:

               i)   Introduction

               ii)  Synopsis

               iii) Adverse conditions.  This information is extracted from in-flight aviation weather advisories, center weather advisories (CWA), and alert severe weather watch bulletins (AWW).

               iv)  TWEB route forecasts

               v)   Winds aloft forecast.  This forecast is broadcast for the location nearest to the TWEB site.  The forecast should include the levels for 3,000 ft. to 12,000 ft. but shall always include at least two forecast levels above the surface.

               vi)  Radar reports.  Local or pertinent radar weather reports (SD) are used.  If there is access to real-time weather radar equipment, the observed data are summarized using the SDs to determine precipitation type, intensity, movement, and height.

vii) Aviation routine weather reports (METAR). METARs are recorded beginning with the local reports, then the remainder of the reports beginning with the first station east of true north and continuing clockwise around the TWEB location.

viii) Density altitude. The temperature and the statement "check density altitude" are included in the TWEB for any station with a field elevation at or above 2,000 ft. MSL and meets a certain temperature criteria.

ix) Pilot weather reports (PIREP). PIREPs are summarized. If the weather conditions meet soliciting requirements, a request for PIREPs will be added.

x) Alert notices (ALNOT), if applicable

xi) Closing statement

c) At selected locations, telephone access to the TWEB has been provided (TEL-TWEB). Telephone numbers for this service are found in the FSS and National Weather Service Telephone Numbers section of the *Airport/ Facility Directory (A/FD)*.

2) **Telephone Information Briefing Service (TIBS)** is provided by AFSSs and provides continuous telephone recordings of meteorological and/or aeronautical information.

a) Specifically, TIBS provides area and/or route briefings, airspace procedures, and special announcements (if applicable) concerning aviation interests.

b) TIBS should also contain, but is not limited to,

i) Aviation routine weather reports (METARs)
ii) Terminal aerodrome forecasts (TAFs)
iii) Winds/temperatures aloft forecast

c) TIBS is available 24 hr. a day. The recorded information will be updated as conditions change or forecasts are updated.

i) At some AFSSs, TIBS service may not be reduced from 2200 to 0600 local time. During this time, an announcement must be recorded and will include a time when full service will resume.

d) TIBS can be accessed by use of 800-WX-BRIEF (800-992-7433) toll-free.

i) A touchtone telephone is necessary to access the TIBS program.

e) Each AFSS will provide at least four route and/or area briefings. An area briefing will encompass a 50-NM radius around a specified airport.

i) You will have access to Notices to Airmen (NOTAM) information either through TIBS or access to a briefer.

f) The order and content of the TIBS recording is as follows:

i) Introduction -- a statement that includes the time the information was recorded and the route and/or area of coverage

ii) Adverse conditions -- a summary of in-flight aviation weather advisories, CWAs, AWWs, and any other available information that may adversely affect flight in the route or area

iii) VFR not recommended (VNR) statement -- a statement issued when current or forecast conditions would make flight under VFR doubtful

iv)   Synopsis -- a brief statement describing the type, location, and movement of weather systems or air masses that might affect the route or the area

v)    Current conditions -- a summary of current weather conditions over the route or the area

vi)   Density altitude -- a "check density altitude" statement for any weather reporting station at an airport with a field elevation at or above 2,000 ft. MSL and meets certain temperature criteria.

vii)  En route forecast -- a summary of appropriate forecast data in logical order (climb out, en route, and descent)

viii) Winds aloft -- a summary of winds aloft forecast for the route or the area for levels through 12,000 ft.

ix)   Request for PIREPs, if appropriate

x)    NOTAM information that affects the route or the area -- NOTAM information that may be available either from the TIBS recording or by access to a FSS briefer

xi)   Military training activity

xii)  ALNOTs, if appropriate

xiii) Closing announcement

d.   If you are already in flight and you need weather information and assistance, the following services are provided by FSSs.  They can be accessed over the proper radio frequencies that are listed on aeronautical charts and the *Airport/Facility Directory*.

1)   **Hazardous Inflight Weather Advisory Service (HIWAS)** is a continuous broadcast service (24 hr. a day) over selected VORs of in-flight aviation weather advisories, i.e., AIRMETs, SIGMETs, convective SIGMETs, severe weather forecast alerts (AWW), center weather advisories (CWA), and urgent PIREPs.

a)   The HIWAS broadcast will state that there are no hazardous weather advisories if none exist.

b)   The HIWAS broadcast area is defined as the area within 150 NM of a HIWAS outlet.

c)   HIWAS broadcasts will not be interrupted or delayed except in emergency situations.

d)   Once the HIWAS broadcast is updated, an announcement will be made on all communications and navigational aid (NAVAID) frequencies (except emergency) and En Route Flight Advisory Service (EFAS).

e)   In the event that a HIWAS outlet is out of service, an announcement will be made on all communications and NAVAID frequencies except emergency and EFAS.

f)   The HIWAS broadcast will include the following:

i)    A statement of introduction, including the appropriate area(s) and a recording time

ii)   A summary of in-flight aviation advisories, CWAs, AWWs, and any other weather not included in a current hazardous weather advisory

iii)  A request for PIREPs, if applicable

         iv)   A recommendation to contact a FSS or Flight Watch for additional details concerning hazardous weather

      2)    The **En Route Flight Advisory Service (EFAS)**, or Flight Watch, is a weather service from selected FSSs on a common frequency (122.0 MHZ) below flight level (FL) 180 and on assigned discrete frequencies to aircraft at FL 180 and above.

         a)   The purposes of EFAS is to provide en route aircraft with timely and meaningful weather information tailored to the type of flight intended, route of flight, and altitude.

         b)   Additionally, Flight Watch is a focal point for rapid receipt and dissemination of pilot reports.

         c)   To use this service, call FLIGHT WATCH.

            i)    EXAMPLE: "(Oakland) Flight Watch, this is . . ."

         d)   EFAS is normally available throughout the conterminous U.S. and Puerto Rico from 6 a.m. to 10 p.m., local time.

         e)   Figure 1-1 on the next page indicates the sites where EFAS and associated outlets are located.

2.    The **Air Traffic Control System Command Center (ATCSCC)**, also known as "central flow control," is located in Herndon, VA. ATCSCC has the mission of balancing air traffic demand with system capacity. This balance ensures maximum safety and efficiency for the National Airspace System (NAS) while minimizing delays.

    a.   The ATCSCC utilizes the Traffic Management System, aircraft situation display, monitor alert, follow-on functions, and direct contact with the air route traffic control center (ARTCC) and terminal radar approach control facility (TRACON) traffic management units to manage flow on a national as well as local level.

    b.   Because weather is the most common reason for air traffic delays and reroutings, the ATCSCC is supported full-time by Air Traffic Control System Command Center Weather Unit Specialists (ATCSCCWUS).

      1)    These FAA specialists are responsible for the dissemination of meteorological information as it pertains to national air traffic flow management.

3.    An **Air Route Traffic Control Center (ARTCC)** is an en route radar facility established to provide air traffic control service to aircraft operating on IFR flight plans within controlled airspace and principally during the en route phase of flight.

    a.   When equipment capabilities and controller workload permit, certain advisory/assistant services may be provided to VFR aircraft.

    b.   En route controllers become familiar with pertinent weather information and remain aware of current weather information needed to perform air traffic control duties.

      1)    An en route controller will advise pilots of hazardous weather that may impact operations within 150 NM of the controller's assigned sector or area of jurisdiction.

4.    The **Center Weather Service Unit (CWSU)** is a joint agency aviation weather support team located at each ARTCC that provides weather consultation, forecasts, and advice to managers and staff within the ARTCC and other supported FAA facilities.

    a.   The CWSU is composed of NWS meteorologists and FAA traffic management personnel, the latter being assigned as Weather Coordinators.

      1)    The flow or exchange of weather information between the CWSU meteorologists and the FAA personnel in the ARTCC is the responsibility of the Weather Coordinator.

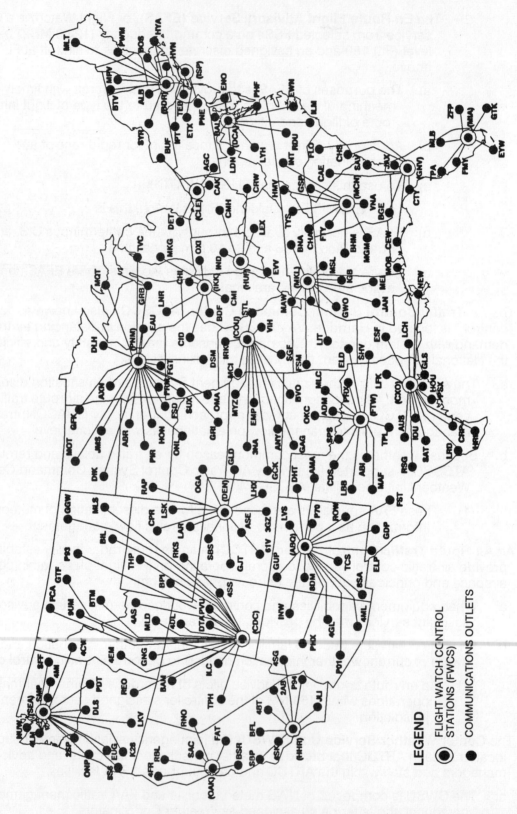

Figure 1-1.  En Route Flight Advisory (Flight Watch Facilities)

    b.    The CWSU provides FAA traffic managers with accurate and timely weather information.

        1)    This information is based on monitoring, analyzing, and interpreting real-time weather data at the ARTCC through the use of all available data sources, including radar, satellites, PIREPs, and various NWS products, such as TAFs and area forecasts, in-flight aviation weather advisories, etc.

5.    **Airport Traffic Control Tower (ATCT)**. The terminal controller informs arriving and departing aircraft of pertinent local weather conditions.

    a.    The controller must be familiar with, and remain aware of, current weather information needed to perform air traffic control duties in the vicinity of the terminal.

    b.    The responsibility for reporting visibility observations is shared with the NWS at many ATCT facilities.

        1)    If the responsibility is not shared, the controllers are properly certified and acting as official weather observers for the elements being reported.

6.    **Automatic Terminal Information Service (ATIS)** is a continuous broadcast of recorded information in selected terminal areas. Its purpose is to improve controller effectiveness and to relieve frequency congestion by automating the repetitive transmission of current weather information and other noncontrol airport information.

7.    **Direct User Access Terminal System (DUATS)** is an FAA-funded information system which enables pilots with a current medical certificate to access the system to conduct their own weather briefings.

    a.    This computer-based system receives and stores up-to-date weather and NOTAM data from the FAA's Weather Message Switching Center (WMSC).

    b.    By using a personal computer and modem, you can access the system and request specific types of weather briefings and other pertinent data for planned flights.

        1)    You can also use DUATS to file, amend, or cancel flight plans.

    c.    See Appendix D, Direct User Access Terminal System (DUATS), beginning on page 410.

8.    **Flight Information Services Data Link (FISDL)** is a system which provides properly equipped aircraft with a cockpit display of certain aeronautical weather and flight operational information.

    a.    There are two types of FISDL systems.

        1)    Broadcast FISDL system allows the pilot to passively collect weather data and to call up that data for review at an appropriate time.

            a)    The broadcast system components include a ground- or space-based transmitter, an aircraft receiver, and a cockpit display device.

        2)    Two-way FISDL system allows the pilot to make specific weather and operational requests for cockpit display.

            a)    The two-way system components include a ground- or space-based transmitter, an aircraft receiver and transmitter, and a cockpit display device.

    b.    FISDL services are provided by vendors under contract with the FAA and several commercial vendors.

        1)    The basic products under the FAA FISDL include METARs, TAFs, SIGMETs, convective SIGMETs, AIRMETs, PIREPs, and AWWs.

            a)    Value added products (e.g., radar images) may be available on a subscription service from the vendor.

c.  FISDL does not serve as the sole source of weather information, but it does augment FSS and ATC services.

    1)  FISDL may alert the pilot to specific areas of concern which will more accurately focus requests made to FSS for inflight briefings or queries made to ATC.

F.  **Weather observations** are measurements and estimates of existing weather, both at the surface and aloft. When recorded and transmitted, a weather observation becomes a report, and these reports are the basis for analyses, forecasts, advisories, and briefings.

    1.  **Aviation routine weather reports (METARs)** include weather elements pertinent to flying.

        a.  A network of airport stations provides routine up-to-date surface weather information.

        b.  Most of the stations in the network are either NWS or FAA; however, the military services and contracted civilians are also included.

        c.  Automated Surface Observing System (ASOS), Automated Weather Observing System (AWOS), and other automated weather observing systems are becoming a major part of the surface weather observing network.

    2.  **Radar Observations.** The weather radar provides detailed information about precipitation, winds, and weather systems.

        a.  The WSR-88D weather radar uses Doppler technology, which allows the radar to provide measurements of winds through a large vertical depth of the atmosphere.

            1)  This information helps support public and aviation warning and forecast programs.

            2)  Figure 7-2 on page 221 shows the WSR-88D radar network.

        b.  FAA Terminal Doppler Weather Radars (TDWR) are being installed near a number of major airports around the country.

            1)  The TDWR will be specifically used to alert and warn airport air traffic controllers of approaching wind shear, gust fronts, and heavy precipitation which could cause hazardous conditions for landing or departing aircraft.

        c.  Also installed at major airports is the FAA airport surveillance radar, ASR-9.

            1)  With the ASR-9, specific locations of six different precipitation levels will be available for the safe routing of air traffic in and about a terminal location.

            2)  However, the radar's primary function is for aircraft detection.

    3.  **Satellite Imagery.** Visible, infrared (IR), and other types of images (or pictures) of clouds are available from weather satellites in orbit.

        a.  Satellite images are then made available on a near real-time basis to NWS and FAA facilities.

        b.  Satellite images are an important source of weather information.

    4.  **Upper-Air Observations.** Other important sources of observed weather data are radiosonde balloons and pilot weather reports (PIREPs).

        a.  Upper-air observations from radiosonde, taken twice daily at specified stations, furnish temperature, humidity, pressure, and wind, often to heights above 100,000 ft.

        b.  Pilots are also a vital source of upper-air weather observations.

            1)  In fact, aircraft in flight are the only means of directly observing turbulence, icing, and height of cloud tops.

c. A new sensing system utilizing vertically oriented radars will provide increased real-time data from the upper atmosphere. These radars will provide profiles of the atmosphere. Thus, the system is known as the Profiler Network.

1) At present, upper-level winds are the only data obtained from this network, but other parameters such as temperature and moisture content at various levels eventually will be available.

2) Currently, a limited network of profilers is being tested in the central part of the country.

# COMPLETED WSR-88D INSTALLATIONS WITHIN THE CONTIGUOUS U.S.

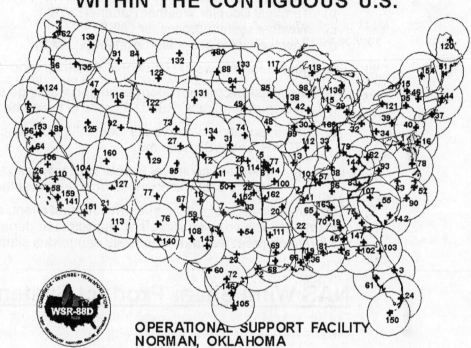

OPERATIONAL SUPPORT FACILITY
NORMAN, OKLAHOMA

# COMPLETED WSR-88D INSTALLATIONS

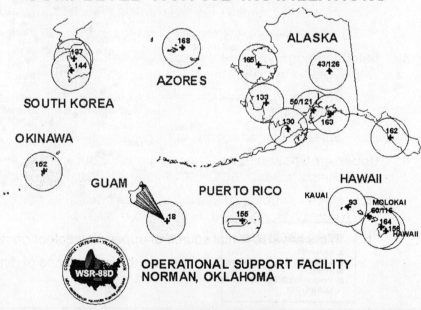

OPERATIONAL SUPPORT FACILITY
NORMAN, OKLAHOMA

Figure 7-2. WSR-88D Radar Network

5.   **Detection of Microbursts, Wind Shear, and Gust Fronts**

   a.   **The FAA's Integrated Wind Shear Detection Plan**

   1)   The FAA currently employs an integrated plan for wind shear detection that will significantly improve both the safety and the capacity of the majority of the airports served by the air carriers.

   2)   This plan integrates several programs into a single strategic concept that significantly improves the aviation weather information in the terminal area. These programs include the

   a)   Integrated terminal weather system (ITWS)
   b)   Terminal Doppler Weather Radar (TDWR)
   c)   Weather system processor (WSP)
   d)   Low-level wind shear alert system (LLWAS)

   3)   The figure below shows the locations of various wind shear systems.

   4)   The wind shear/microburst information and warnings are displayed on the ribbon display terminals (RBDT) located in the air traffic control tower.

   a)   The information displayed is identical (and standardized) in the LLWAS, TDWR, and WSP systems.

   i)   Thus, the controller is able to simply read the displayed information to the pilot without needing to interpret the data.

   5)   The early detection of a wind shear/microburst event, and the subsequent warning(s) issued to an aircraft on approach or departure, will alert the pilot/crew to be prepared for a potentially dangerous situation.

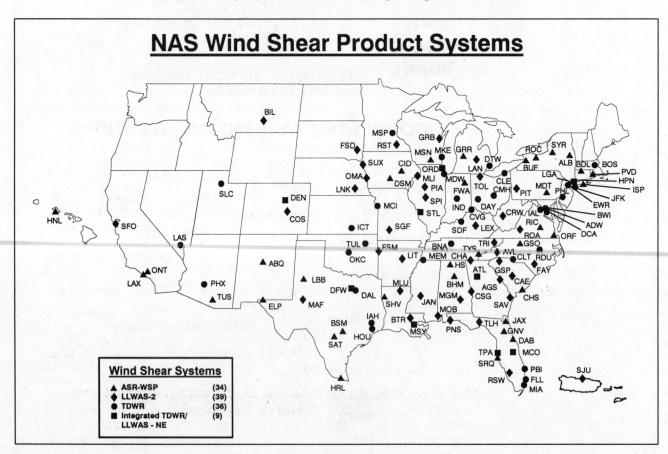

**NAS Wind Shear Product Systems**

| Wind Shear Systems | |
|---|---|
| ▲ ASR-WSP | (34) |
| ◆ LLWAS-2 | (39) |
| ● TDWR | (36) |
| ■ Integrated TDWR/ LLWAS - NE | (9) |

b. **Low-Level Wind Shear Alert System (LLWAS)**

1) The LLWAS provides wind data and software processes to detect the presence of hazardous wind shear and microbursts in the vicinity of an airport.

2) Wind sensors, mounted on poles sometimes as high as 150 ft., are ideally located 2,000 ft. to 3,500 ft., but not more than 5,000 ft., from the runway centerline.

3) Eventually, all LLWASs will be phased out; however, 39 airports will be upgraded to the LLWAS-NE (network expansion) system, which employs the very latest software and sensor technology.

4) LLWAS-NE not only will provide the controller with wind shear warnings and alerts, including wind shear/microburst detection at the centerfield wind sensor location, but also will provide the location of the hazards relative to the airport runway(s).

c. **Terminal Doppler Weather Radar (TDWR)**

1) Optimally, TDWRs are located 8 to 12 mi. off the airport proper. They are designed to look at the airspace around and over the airport to detect microbursts, gust fronts, wind shifts, and precipitation intensities.

2) TDWR products advise the controller of wind shear and microburst events impacting all runways and the areas 1/2 mi. on either side of the extended runway centerline to 3 mi. out on final approach and 2 mi. out on departure.

3) TDWR does NOT

a) Warn of wind shear outside of the alert boxes (on the arrival and departure ends of the runway)

b) Detect wind shear that is **not** a microburst or a gust front

c) Detect gusty or crosswind conditions

d) Detect turbulence

4) However, future improvements may include such areas as storm movement, improved gust front detection, storm growth and decay, microburst prediction, and turbulence detection.

d. **Weather Systems Processor (WSP)**

1) The WSP provides the controller and the pilot with the same products as the TDWR, but at a lower cost.

2) This is accomplished by utilizing new technologies to access the Weather Channel capabilities of the FAA's existing airport surveillance radar, ASR-9, located on or nearby the airport.

3) This system is currently in a developmental test status.

e. **Examples.** The following three examples include the phraseology used by the controller when issuing an alert and then its meaning to the pilot.

1) Microburst alert

a) Controller: Runway 27 arrival, microburst alert, 35-kt. loss 2 mi. final. Threshold winds 250 at 20.

b) Meaning: On approach to runway 27, there is a microburst alert; anticipate, or expect, a 35-kt. loss of airspeed at approximately 2 mi. out on final approach (where you will first encounter the microburst). Additionally, the surface winds at the threshold of runway 27 are 250° at 20 kt.

2) Wind shear alert

    a) Controller: Runway 27 arrival, wind shear alert, 20-kt. loss 3 mi. final. Threshold winds 200 at 15.

    b) Meaning: On approach to runway 27 at approximately 3 mi. out on final, expect to encounter a wind shear that will decrease airspeed by 20 kt. and possibly to encounter turbulence. Additionally, the surface winds at the threshold for runway 27 are 200° at 15 kt.

3) Multiple wind shear alerts

    a) Controller: Multiple wind shear alerts. Runway 27 arrival, wind shear alert, 20 kt. gain on runway. Runway 27 departures, wind shear alert, 20 kt. gain on runway. Runway winds 250 at 20.

    b) Meaning: Arriving and departing aircraft could encounter a wind shear condition on the runway due to a gust front with the possibility of a gain of 20 kt. in airspeed associated with the gust front. The surface winds for runway 27 are 250° at 20 kt.

f. **Terminal Weather Information for Pilots System (TWIP)**

1) With the increase in the quantity and quality of terminal weather information available through TDWR, the next step is to provide this information directly to pilots rather than to rely on voice communication with ATC.

2) The National Airspace System needs a means of delivering terminal weather information to the cockpit with a speed and an accuracy that enhance pilot awareness of weather hazards and reduce ATC workload.

3) With the TWIP capability, terminal weather information (both text and graphics) is now available directly to the cockpit on a test basis at nine locations.

4) TWIP products are generated using weather data from the TDWR or the ITWS testbed. These products include descriptions and character graphics of

    a) Microburst alerts
    b) Wind shear alerts
    c) Significant precipitation
    d) Convective activity within 30 NM of the terminal area
    e) Expected weather that will impact airport operations

5) Software has been developed to allow TDWR or ITWS to format data and send the TWIP products to a database resident at Aeronautical Radio, Inc. (ARINC).

    a) These products can then be accessed by pilots using the ARINC data link services.

## G. Communications Systems

1. High-speed communications and automated data processing have improved the flow of weather data and products through the aviation weather network.

    a. The flow of weather information within and between agencies is becoming faster as computers and satellites are being used to facilitate the exchange of data.

2. A new computer-based advanced weather interactive processing system (AWIPS) workstation is being deployed for the NWS. AWIPS is replacing the current system and will allow quicker dissemination of weather information.

    a. This system will be linked with the WSR-88D weather radar system to provide better detection, observing, and forecasting of weather systems, especially severe weather.

3. The flow of alphanumeric weather information to the FAA service outlets is accomplished through leased lines to computer-based equipment.

4.    Exchange of weather information between the NWS and FAA service outlets is generally accomplished in two ways.

    a.    Graphic products (weather maps) are received by FAA service outlets from NCEP through a private sector contractor.

    b.    Alphanumeric information is exchanged through the Weather Message Switching Center Replacement (WMSCR) in Atlanta, GA and Salt Lake City, UT.

        1)    These switching facilities serve as the gateway for the flow of alphanumeric information from one communication system to another (i.e., among the various FAA facilities, NWS, and other users).

## H.    Users

1.    The ultimate users of the aviation weather service are pilots and dispatchers.

    a.    As a user of the service, you should also contribute to it.  Send pilot weather reports (PIREPs) to help briefers, forecasters, and your fellow pilots.

    b.    The service can be no better or more complete than the information that goes into it.

2.    In the interest of safety and in compliance with Federal Aviation Regulations, you should get a complete briefing before each flight.

    a.    You can get a preliminary briefing on your telephone by listening to the PATWAS or TIBS at your home or place of business.

    b.    Many times the weather situation may be complex, and you may not completely comprehend the recorded message.

    c.    You are responsible for ensuring that you have all the information needed to make a go/no-go decision.

3.    **How to Get a Good Weather Briefing**

    a.    Call a FSS by dialing (800) WX-BRIEF [(800) 992-7433].

    b.    When requesting a briefing, identify yourself as a pilot.  Give clear and concise facts about your flight:

        1)    Type of flight, VFR or IFR
        2)    Aircraft number or your name
        3)    Aircraft type
        4)    Departure point
        5)    Proposed time of departure
        6)    Flight altitude(s)
        7)    Route of flight
        8)    Destination
        9)    Estimated time en route (ETE)

    c.    With this background, the briefer can proceed directly with the briefing and concentrate on weather relevant to your flight.

    d.    The weather briefing you receive depends on the type requested.

        1)    A **standard** briefing should include

            a)    Adverse conditions -- meteorological or aeronautical conditions reported or forecast that may influence you to alter the proposed flight

            b)    VFR flight not recommended (VNR) -- a statement issued if you indicated that your flight will be conducted under VFR and if the sky conditions or visibilities are present or forecast that, in the judgment of the FSS briefer, would make flight under VFR doubtful

        c) Synopsis -- a brief statement describing the type, location, and movement of weather systems and/or air masses which may affect your flight

        d) Current conditions -- a summary of current weather (METAR) reports-- departure, en route, and destination

        e) En route forecast -- a summary of forecast conditions for the flight

        f) Destination forecast -- forecast data about your destination, including any significant changes expected between 1 hr. before to 1 hr. after your proposed ETA

        g) Winds aloft -- forecast winds aloft for your route with temperature information available on request

        h) NOTAMs -- NOTAM information pertinent to your flight

        i) ATC delays -- information about any known ATC delays and/or flow control advisories that may affect your flight

        j) Request for PIREPs, if appropriate

        k) EFAS -- information about the availability of Flight Watch for weather updates

        l) Any other information you may request, e.g., military training routes, etc.

  2) An **abbreviated** briefing will be provided when the user requests information to

        a) Supplement mass disseminated data (e.g., TIBS, DUATS)
        b) Update a previous briefing
        c) Be limited to specific information

  3) An **outlook** briefing will be provided when the briefing is 6 or more hours in advance of proposed departure.

        a) Briefing will be limited to applicable forecast data for the proposed flight.

  e. The FSSs are here to serve you. You should not hesitate to discuss factors that you do not fully understand.

    1) You have a complete briefing only when you have a clear picture of the weather to expect.

    2) It is to your advantage to make a final weather check immediately before departure if at all possible.

### 4.  Have an Alternate Plan of Action

  a. When weather is questionable, get a picture of expected weather over a broader area.

    1) Preplan a route to take you rapidly away from the weather if it goes sour.

  b. When you fly into weather through which you cannot safely continue, you must act quickly.

    1) Without preplanning, you may not know the best direction in which to turn; a wrong turn could lead to disaster.

  c. A preplanned diversion beats panic. It is better to be safe than sorry.

# END OF CHAPTER

# CHAPTER TWO
# AVIATION ROUTINE WEATHER REPORT (METAR)

> Please take a few minutes to think about each of the concepts listed above and anticipate/imagine what they are and how they relate to the other listed concepts.
>
> Note: Appendix E, beginning on page 419, has six additional sample METARs.

An aviation routine weather report (METAR) is the weather observer's (human or automated) interpretation of the weather conditions at a given site and time. The METAR is used by the aviation community and the NWS to determine the weather conditions (VFR, MVFR, IFR) at an airport, as well as to produce a terminal aerodrome forecast (TAF), if appropriate.

Although the METAR code is being adopted worldwide, each country is allowed to make modifications or exceptions to the code for use in that particular country; e.g., the U.S. will report prevailing visibility in statute miles, runway visual range (RVR) values in feet, wind speed in knots, and altimeter setting in inches of mercury. However, temperature and dew point will be reported in degrees Celsius.

A. **Elements**. A METAR report contains the following sequence of elements in the following order:

1. Type of report
2. ICAO station identifier
3. Date and time of report
4. Modifier (as required)
5. Wind
6. Visibility
7. Runway visual range (RVR) (as required)
8. Weather phenomena
9. Sky condition
10. Temperature/dew point
11. Altimeter
12. Remarks (RMK) (as required)

NOTE: The elements in the body of a METAR report are separated by a space, except temperature and dew point, which are separated with a solidus, /. When an element does not occur or cannot be observed, that element is omitted from that particular report.

B. **Example of a METAR Report**

METAR KGNV 201953Z 24015KT 3/4SM R28/2400FT +TSRA BKN008 OVC015CB 26/25 A2985 RMK TSB32RAB32

To aid in the discussion, we have divided the report into the 12 elements:

| METAR | KGNV | 201953Z | __ | 24015KT | 3/4SM | R28/2400FT | +TSRA |
|-------|------|---------|----|---------|-------|-----------|-------|
| 1. | 2. | 3. | 4. | 5. | 6. | 7. | 8. |

| BKN008 OVC015CB | 26/25 | A2985 | RMK TSB32RAB32 |
|-----------------|-------|-------|----------------|
| 9. | 10. | 11. | 12. |

1. Aviation routine weather report
2. Gainesville, FL
3. Observation taken on the 20th day at 1953 UTC (or Zulu)
4. Modifier omitted; i.e., not required for this report
5. Wind 240° true at 15 kt.
6. Visibility 3/4 SM
7. Runway 28, runway visual range 2,400 ft.
8. Thunderstorm with heavy rain
9. Ceiling 800 ft. broken, 1,500 ft. overcast, cumulonimbus clouds
10. Temperature 26°C, dew point 25°C
11. Altimeter 29.85
12. Remarks: Thunderstorm began at 32 min. past the hour; rain began at 32 min. past the hour.

C. **Type of Report** (element 1). The type of report will always appear as the lead element of the report. There are two types of reports:

1. **METAR** -- an aviation routine weather report
2. **SPECI** -- a nonroutine aviation weather report

D. **ICAO Station Identifier** (element 2). The METAR uses the ICAO four-letter station identifier.

1. In the contiguous 48 states, the three-letter domestic location identifier is prefixed with a "K."

   a. EXAMPLE: The domestic identifier for San Francisco, CA is SFO, while the ICAO identifier is KSFO.

2. Elsewhere, the first one or two letters of the ICAO identifier indicate in which region of the world and country (or state) the station is located.

   a. Pacific locations such as Alaska and Hawaii start with "P" followed by an "A" or an "H," respectively.

      1) The last two letters reflect the specific reporting station identification.

      2) If the station's three-letter identifier begins with an "A," "H," or "G," the "P" is added to the beginning.

         a) If the station's three-letter identifier does NOT begin with an "A," "H," or "G," the last letter is dropped and the "P" is added to the beginning.

      3) EXAMPLES:

         a) ANC (Anchorage, AK) is PANC.
         b) OME (Nome, AK) is PAOM.
         c) HNL (Honolulu, HI) is PHNL.
         d) KOA (Keahole Point, HI) is PHKO.

   b. Canadian station identifiers start with "C," e.g., Toronto, Canada is CYYZ.

c.   Mexican and western Caribbean station identifiers start with "M."

d.   The identifier for the eastern Caribbean is "T" followed by the individual country's letter; e.g., San Juan, Puerto Rico is TJSJ.

E.   **Date and Time of Report** (element 3)

1.   The date and time the observation is taken are transmitted as a six-digit date/time group appended with the letter **Z** to denote Coordinated Universal Time (UTC).

a.   The first two digits are the date followed with two digits for the hour and two digits for minutes.

b.   If the report is a corrected report, the time entered shall be the same time used in the report being corrected.

F.   **Modifier** (element 4).  There are two modifiers.

1.   **AUTO** -- identifies the report as an automated weather report with no human intervention

a.   The type of sensor equipment used at the automated station will be encoded in the remarks section of the report (explained on page 237).

b.   The absence of AUTO indicates that the report was made manually or an automated station had human augmentation.

2.   **COR** -- identifies the report as a corrected report to replace an earlier report with an error

G.   **Wind** (element 5).  Wind follows the date/time or modifier element.

1.   The average 2-minute direction and speed are reported in a five- or six-digit format.

a.   The first three digits are the direction FROM which the wind is blowing.  The direction is to the nearest 10-degree increment referenced to TRUE north.

1)   On the other hand, wind direction when verbally broadcast by a control tower or advisory station is referenced to magnetic north.  This is done so the pilot can more closely relate the wind direction to the runway in use.

b.   The last two or three digits are the wind speed in knots -- two digits for speeds less than 100 kt., three digits for speeds greater than 100 kt.

c.   The abbreviation **KT** is appended to denote the use of knots for wind speed.

d.   EXAMPLE:  **24015KT** means the wind is from 240° true at 15 kt.

2.   If the wind speed is 6 kt. or less and, in the observer's opinion, the wind is varying in direction, the contraction **VRB** is used for wind direction.

a.   EXAMPLE:  **VRB04KT** means the wind is variable in direction at 4 kt.

3.   A **calm** wind is defined as a wind speed of less than 3 kt. and is coded as **00000KT**.

4.   A **gust** is a variation in wind speed of 10 kt. or more between peaks and lulls.

a.   A gust is reported by the letter **G** after the wind speed, followed by the highest reported gust.

b.   EXAMPLE:  **210103G130KT** means the wind is from 210° true at 103 kt. with gusts to 130 kt.

5.   If the wind is variable by 60° or more and the wind speed is greater than 6 kt., a variable group consisting of the extremes of the wind directions separated by **V** will follow the wind group.

a.   EXAMPLE:  **32012G22KT 280V350** means the wind direction is from 320° true at 12 kt., gusts to 22 kt., wind direction variable between 280° true and 350° true.

6.   A SPECI report will be issued if the wind direction changes by 45° or more in less than 15 min. and the sustained winds are 10 kt. or more throughout the wind shift.

H.  **Visibility** (element 6)

1.  The visibility reported is called the prevailing visibility. **Prevailing visibility** is considered representative of the visibility conditions at the observing site. This representative visibility is the greatest distance at which objects can be seen and identified through at least 180° of the horizon circle, which need not be continuous.

2.  Visibility is reported in statute miles with a space and then fractions of statute miles, as needed, with **SM** appended to it.

    a.  EXAMPLE: **1 1/2SM** means visibility is one and one-half statute miles.

3.  When visibility is less than 7 SM, the restriction to the visibility will be shown in the weather element.

    a.  The only exception is that if volcanic ash, low drifting dust, sand, or snow is observed, those restrictions to visibility are reported, even if the visibility is 7 SM or greater.

4.  If tower or surface visibility is less than 4 SM, the lesser of the two will be reported in this element; the greater will be reported in the remarks.

5.  Automated reporting stations will show visibility less than 1/4 SM as **M1/4SM** and visibility of 10 SM and greater as **10SM**.

    a.  For automated reporting stations having more than one visibility sensor, site-specific visibility that is lower than the visibility shown in the body will be shown in remarks.

6.  A SPECI report will be issued if the prevailing visibility decreases to less than or, if below, increases to equal or exceed

    a.  3 SM

    b.  2 SM

    c.  1 SM

    d.  The lowest standard instrument approach procedure (IAP) as published, or 1/2 SM if no IAP

I.  **Runway Visual Range (RVR)** (element 7)

1.  Runway visual range (RVR) is based on the measurement of a transmissometer (or the New Generation RVR system, which uses forward scatter technology) made near the touchdown point of an instrument runway which represents the horizontal distance a pilot will see down the runway from the approach end.

    a.  RVR is based on either the sighting of high-intensity runway lights or the visual contrast of other targets, whichever yields the greater visual range.

2.  RVR is reported if the airport has RVR equipment and whenever the prevailing visibility is 1 SM or less and/or the RVR for the designated instrument runway is 6,000 ft. or less.

3.  RVR is reported in the following format: **R** identifies the group, followed by the runway heading and parallel designator if needed, a solidus (/), and the visual range in feet (meters in other countries) indicated by **FT**.

    a.  EXAMPLE: **R28/1200FT** means runway 28 visual range is 1,200 ft.

4.  When the RVR varies by more than one reportable value, the lowest and highest values are shown with **V** between them.

    a.  EXAMPLE: **R19/1000V2000FT** means runway 19 visual range variable between 1,000 ft. and 2,000 ft.

5.  When the observed RVR is above the maximum value that can be determined by the system, it should be reported as **P6000**, if 6,000 ft. is the maximum value for this system.

    a.  EXAMPLE: **R04/P6000FT** means runway 4 visual range is more than 6,000 ft.

6. When the observed RVR is below the minimum value that can be determined by the system, it should be reported as **M600**, if 600 ft. is the minimum value for this system.

    a. EXAMPLE: **R32/M0600FT** means runway 32 visual range is less than 600 ft.

7. If RVR should be reported but is missing, **RVRNO** will be in the remarks.

8. A SPECI report will be issued if the highest value from the designated RVR runway decreases to less than or, if below, increases to equal or exceed 2,400 ft. during the preceding 10 min.

J. **Weather Phenomena** (element 8). The weather is reported in the following format:

1. **Intensity**. Intensity may be shown with most precipitation types including those of a showery nature.

    a. Intensity levels may be shown with obscurations such as blowing dust, sand, or snow.

    b. When more than one type of precipitation is present, the intensity refers to the first precipitation type (most predominant).

    c. The intensity symbols used are shown in Table 2-2:

| Symbol | Meaning |
|--------|---------|
| + | Heavy |
| (no symbol) | Moderate |
| − | Light |

Table 2-2. Intensity Qualifiers.

2. **Proximity**. Proximity is applied to and reported only for weather occurring in the vicinity of the airport and is denoted by **VC**.

    a. Vicinity of the airport is defined to be between 5 and 10 SM from the usual point of observation for obscurations to visibility and just beyond the point of observation to 10 SM for precipitation.

    b. VC will replace the intensity symbol; i.e., intensity and VC will never be shown in the same group.

3. **Descriptor**. The following eight descriptors shown in Table 2-3 below further identify weather phenomena and are used with certain types of precipitation and obscurations.

| Coded | Meaning | Coded | Meaning |
|-------|---------|-------|---------|
| TS | Thunderstorm | DR | Low drifting |
| SH | Showers | MI | Shallow |
| FZ | Freezing | BC | Patches |
| BL | Blowing | PR | Partial |

Table 2-3. Descriptor Qualifiers

    a. MI is used only to further describe fog that has little vertical extent (less than 6 ft.).

    b. BC is used only to further describe fog that has little vertical extent and reduces horizontal visibility.

    c. DR is used when dust, sand, or snow is raised by the wind to less than 6 ft.

    d. BL is used when dust, sand, snow, and/or spray is raised by the wind to a height of 6 ft., or more.

    e. Although **TS** and **SH** are used with precipitation and may be preceded with an intensity symbol, the intensity applies to the precipitation and not the descriptor.

1)   EXAMPLE: **+TSRA** means a thunderstorm with heavy rain, not a heavy thunderstorm with rain.

4.   **Precipitation**. Precipitation is any form of water particles, whether solid or liquid, that fall from the atmosphere and reach the ground. The nine types of precipitation are shown in Table 2-4 below.

| Coded | Meaning | Coded | Meaning |
|-------|---------|-------|---------|
| RA | Rain | GR | Hail (¼ in. in diameter or larger) |
| DZ | Drizzle | GS | Small Hail/Snow Pellets |
| SN | Snow | PL | Ice Pellets |
| SG | Snow Grains | IC | Ice Crystals |
| UP | Unknown Precipitation | | |

Table 2-4. Coded Precipitation and Meanings

a.   **UP** is used only by automated weather reporting systems to indicate that the system cannot identify the precipitation with any degree of proficiency. This situation usually occurs when rain and snow are falling at the same time.

b.   **GS** is used to indicate hail less than 1/4 in. in diameter.

c.   For **IC** to be reported, visibility must be reduced by the ice crystals to 6 SM or less.

5.   **Obscurations to Visibility**. The eight types of obscuration phenomena are listed in Table 2-4A below. Obscurations are any phenomena in the atmosphere, other than precipitation, that reduce horizontal visibility.

| Coded | Meaning | Coded | Meaning |
|-------|---------|-------|---------|
| FG | Fog (visibility less than ⅝ SM) | PY | Spray |
| BR | Mist (visibility ⅝ to 6 SM) | SA | Sand |
| FU | Smoke | DU | Dust |
| HZ | Haze | VA | Volcanic Ash |

Table 2-4A. Coded Obstructions to Vision and Meanings

a.   **FG** is used to indicate fog restricting visibility to less than 5/8 SM.

b.   **BR** is used to indicate mist restricting visibility from 5/8 to 6 SM and is never coded with a descriptor.

c.   **BCFG** and **PRFG** are used to indicate patchy fog or partial fog only if the prevailing visibility is 7 SM or greater.

d.   Precipitation types may be grouped together; obscurations are grouped separately.

1)   EXAMPLE: **RASN BR HZ**

6.   **Other**. There are six other weather phenomena that are reported when they occur, as shown in Table 2-4B below.

| Coded | Meaning | Coded | Meaning |
|-------|---------|-------|---------|
| SQ | Squall | SS | Sandstorm |
| DS | Duststorm | PO | Well-Developed Dust/Sand Whirls |
| FC | Funnel Cloud | +FC | Tornado or Waterspout |

Table 2-4B. Coded Other Weather Phenomena and Meanings

a.   A **squall** (**SQ**) is a sudden increase in wind speed of at least 16 kt., with the speed rising to 22 kt. or more and lasting at least one minute.

    b.    **+FC** is used to denote a tornado or waterspout.

        1)    The type is indicated in the remarks.

7.    **Examples of Reported Weather Phenomena**

    a.    **TSRA** means thunderstorm with moderate rain.
    b.    **+SN** means heavy snow.
    c.    **—RA FG** means light rain and fog.
    d.    **VCSH** means showers in the vicinity.

8.    If more than one significant weather phenomenon is observed, entries will be listed in the following order: tornadic activity, thunderstorms, and weather phenomena in order of decreasing predominance (i.e., the most dominant type is reported first).

    a.    If more than one significant weather phenomenon is observed, except precipitation, separate weather groups will be shown in the report.

        1)    No more than three weather groups are used to report weather phenomenon at, or in the vicinity of, the station.

    b.    If more than one type of precipitation is observed, the appropriate codes are combined into a single group with the predominant type being reported first.

        1)    Any intensity refers to the first type of precipitation in the group.

9.    The following weather phenomena require the issuance of a SPECI report:

    a.    A tornado, waterspout, or funnel cloud is first observed or disappears from sight.
    b.    A thunderstorm begins or ends.

        1)    A SPECI is not required to report the beginning of a new thunderstorm if one is currently reported.

    c.    Hail begins or ends.
    d.    Freezing precipitation begins, ends, or changes intensity.
    e.    Ice pellets begin, end, or change intensity (manual stations).
    f.    A volcanic eruption is first noted.
    g.    Squalls occur.

K.    **Sky Condition** (element 9). The sky condition is reported in the following format:

    1.    **Amount**. The amount of sky cover is reported in eighths of sky cover, using contractions in Table 2-5 below.

| Contraction | Meaning | Summation Amount |
|---|---|---|
| SKC or CLR* | Clear | 0 or 0 below 12,000 ft. |
| FEW | Few | >0 to 2/8 |
| SCT | Scattered | 3/8 to 4/8 |
| BKN | Broken | 5/8 to 7/8 |
| OVC | Overcast | 8/8 |
| VV | Vertical Visibility (indefinite ceiling) | 8/8 |
| CB | Cumulonimbus | When present |
| TCU | Towering Cumulus | When present |
| * **SKC** will be reported at manual stations. **CLR** will be used at automated stations when no clouds below 12,000 ft. are reported. | | |

Table 2-5. Reportable Contractions for Sky Cover

    a.    A **layer** is defined as clouds or obscuring phenomena aloft when the base is at approximately the same level.

    b.    A **ceiling** is defined as the height (above ground level or AGL) of the lowest broken or overcast layer aloft or vertical visibility into a surface-based obstruction.

1)    There is no provision for reporting thin layers (i.e., a layer through which blue sky or higher sky cover is visible) in the METAR code. A thin layer will be reported the same as if it were opaque.

2.    **Height**. Cloud bases are reported with three digits in hundreds of feet AGL.

   a.    Automated stations cannot report clouds above 12,000 ft.

   b.    At reporting stations located in the mountains, if the cloud layer is below the station level, the height of the layer will be shown as three solidi, ///.

   c.    When more than one layer is reported, layers are given in ascending order of height. For each layer above a lower layer(s), the sky cover contraction for that layer will be the **total sky cover**, which includes that layer and all lower layers.

   1)    EXAMPLE: **FEW010 SCT030 BKN050** reports three layers, as shown in Figure 2-2 below:

   a)    A few clouds at 1,000 ft. (2/8)
   b)    A scattered cloud layer at 3,000 ft. (3/8)
   c)    A broken cloud layer (ceiling) at 5,000 ft. (6/8)

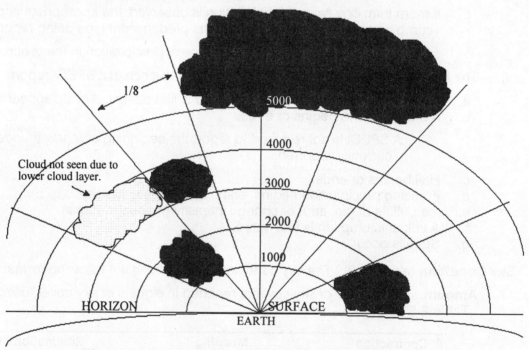

Figure 2-2.  Summation of Cloud Cover in Multiple Layers

3.    **Type or Vertical Visibility**

   a.    If **towering cumulus clouds (TCU)** or **cumulonimbus clouds (CB)** are present, they are reported after the height which represents their base.

   1)    EXAMPLES:

   a)    **SCT025TCU BKN080 BKN250** means 2,500 ft. scattered towering cumulus, ceiling 8,000 ft. broken, 25,000 ft. broken.

   b)    **SCT008 OVC012CB** means 800 ft. scattered, ceiling 1,200 ft. overcast cumulonimbus clouds.

   b.    Height into an indefinite ceiling is preceded with **VV** (vertical visibility) followed by three digits indicating the vertical visibility in hundreds of feet AGL.

   1)    The layer is spoken of as an "indefinite ceiling" and indicates total obscuration.
   2)    EXAMPLE: **1/8SM FG VV006** means visibility 1/8 SM, fog, indefinite ceiling 600 ft.

c.   Obscurations are reported when the sky is partially obscured by a ground-based phenomenon by indicating the amount of obscuration as FEW, SCT, or BKN (see Table 2-5 on page 233 for summation amount) followed by three zeros (000).

1)   In the remarks, the obscuring phenomenon precedes the amount of obscuration and three zeros.

2)   EXAMPLE:

a)   **BKN000** in the body of the report means broken cloud layer less than 50 ft.

b)   **FU BKN000** in the remarks means smoke obscuring 5/8 to 7/8 of the sky.

4.   The sky cover and ceiling, as determined from the ground, represent as nearly as possible what you should experience in flight.

a.   When at or above the reported ceiling layer aloft, you should see less than half the surface below you.

b.   When descending through a surface-based total obscuration, first view the ground directly below you from the height reported as vertical visibility into the obscuration.

c.   However, because of the differing view points of you and the observer, the observed values and what you see do not always exactly agree.

d.   Figure 2-6 below illustrates the effect of an obscured sky on the vision from a descending aircraft.

1)   Vertical visibility is the altitude above the ground from which you should first see the ground directly below the airplane (top).

2)   The real concern is slant range visibility, which most often is less than vertical visibility. Thus, you must descend to a lower altitude (bottom) before seeing a representative surface and being able to fly by visual reference to the ground.

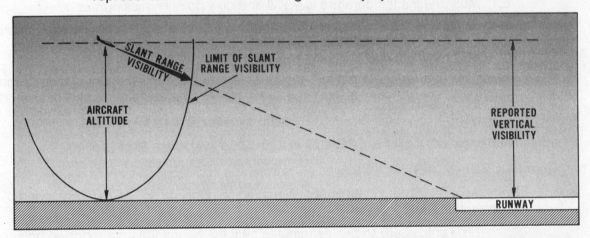

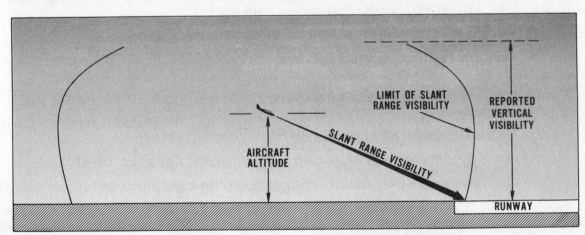

Figure 2-6.  Effect of Obscured Sky on Aircraft

5.   The following sky conditions require the issuance of a SPECI report:

   a.   The ceiling forms or dissipates below, decreases to less than, or, if below, increases to equal or exceed

   1)   3,000 ft.

   2)   1,500 ft.

   3)   500 ft.

   4)   The lowest standard IAP minimum as published (if none is published, then 200 ft.)

   b.   A layer of clouds or obscurations aloft is present below 1,000 ft., but no layer aloft was reported below 1,000 ft. in the preceding METAR or SPECI report.

L.   **Temperature/Dew Point Group** (element 10)

1.   Temperature and dew point are reported in a two-digit form in whole degrees Celsius (C) separated by a solidus, /.

   a.   Temperatures below zero are prefixed with **M**.

2.   If the temperature is available but the dew point is missing, the temperature is shown followed by a solidus.

   a.   If the temperature is missing, the group is omitted from the report.

3.   EXAMPLES:

   a.   **15/08** means temperature is 15°C and dew point is 8°C.
   b.   **00/M02** means temperature is 0°C and dew point is −2°C.
   c.   **M05/** means temperature is −5°C and dew point is missing.

4.   An air mass with a 3°C or less temperature/dew point spread is considered saturated.

M.   **Altimeter** (element 11)

1.   The altimeter is reported in a four-digit format representing tens, units, tenths, and hundredths of inches of mercury prefixed with **A**.  The decimal point is not reported.

   a.   EXAMPLE: **A2995** means the altimeter setting is 29.95 inches of mercury.

2.   Normal range of altimeter settings is from 28.00 to 31.00 inches of mercury.

N.   **Remarks** (element 12)

1.   Remarks are included in all observations, when appropriate, and are preceded by the contraction **RMK**.

   a.   Time entries are shown as minutes past the hour if the time reported occurs during the same hour the observation is taken.

   1)   If the hour is different, hours and minutes are shown.

   b.   Except for precipitation, phenomena located within 5 SM of the point of observation will be reported as at the station.

   1)   Phenomena between 5 and 10 SM are reported in the vicinity, **VC**.

   a)   Precipitation not occurring at the point of observation but within 10 SM is also reported as VC.

   2)   Phenomena beyond 10 SM are reported as distant, **DSNT**.

   c.   Direction of phenomena is indicated with the eight points of the compass (N, NE, E, SE, S, SW, W, NW).

   d.   Distance remarks are in statute miles except for automated lightning remarks, which are in nautical miles.

e.　Movement of clouds or weather is indicated by the direction toward which the phenomenon is moving.

f.　There are two categories of remarks, and our discussion will follow the order in which the remarks would be used.

2.　**Automated, Manual, and Plain Language Remarks Category**. This category of remarks may be generated from either manual or automated weather reporting stations, and they generally elaborate on parameters reported in the body of the report.

　　a.　**Volcanic eruption**. When first noted, the name of the volcano; latitude/longitude or approximate direction and distance from the reporting station; date/time; size/description/height/movement of ash cloud; and other pertinent information are entered in plain language.

　　　　1)　Pre-eruption volcanic activity is not reported.

　　　　2)　EXAMPLE:  MT. ST. HELENS VOLCANO 70 MILES NE ERUPTED 181505 LARGE ASH CLOUD EXTENDING TO APPROX 30000 FEET MOVING SE

　　b.　**Tornado, funnel cloud, waterspout**. Whenever tornadoes, funnel clouds, or waterspouts begin, are in progress, or end, the phenomena, beginning and/or ending time, location, and movement are shown.

　　　　1)　EXAMPLE: **TORNADO B13 DSNT NE** means that a tornado began 13 min. past the hour to the distant (beyond 10 SM) northeast.

　　c.　**Automated station type**. This remark is shown only if the **AUTO** modifier was used.

　　　　1)　**AO1** means the automated weather station is without a precipitation (rain/snow) discriminator.

　　　　2)　**AO2** means the automated weather station has a precipitation (rain/snow) discriminator.

　　d.　**Peak wind**. Whenever the peak wind exceeds 25 kt., **PK WND** will be included in the remarks in the next report with three digits for direction and two or three digits for speed, followed by time in hours and minutes of occurrence (only minutes if the hour can be inferred from the report time).

　　　　1)　EXAMPLES:

　　　　　　a)　**PK WND 28045/1955** means peak wind, 280° true at 45 kt., occurred at 1955 UTC.

　　　　　　b)　**PK WND 34050/38** means peak wind, 340° true at 50 kt., occurred at 38 min. past the hour.

　　e.　**Wind shift**. A wind shift is indicated by a change in wind direction of 45° or more in less than 15 min. with sustained winds of 10 kt. or more throughout the wind shift.

　　　　1)　When a wind shift occurs, **WSHFT** will be included in the remarks, followed by the time the wind shift began.

　　　　2)　The contraction, **FROPA**, indicates that the wind shift is a result of a frontal passage.

　　　　3)　EXAMPLES:

　　　　　　a)　**WSHFT 45** means the wind shift began at 45 min. past the hour.

　　　　　　b)　**WSHFT 30 FROPA** means the wind shift began at 30 min. past the hour due to a frontal passage.

　　f.　**Tower or surface visibility**. If either tower (TWR) or surface (SFC) visibility is less than 4 SM, the lesser of the two will be reported in the body of the report; the greater will be reported in the remarks.

　　　　1)　EXAMPLE: **3/4SM** (in the body) and **TWR VIS 1** (in remarks) mean that the lowest visibility is 3/4 SM, but the tower is reporting visibility of 1 SM.

g. **Variable prevailing visibility**. When the prevailing visibility rapidly increases or decreases by 1/2 SM or more during the observation, and the average prevailing visibility is less than 3 SM, the visibility is variable.

    1) Variable visibility is shown in the remarks with the minimum and maximum visibility values.

    2) EXAMPLE: **VIS 1V2** means the visibility is variable between 1 and 2 SM.

h. **Sector visibility**. Sector visibility is shown in remarks when it differs from the prevailing visibility and either the prevailing or the sector visibility is less than 3 SM.

    1) EXAMPLE: **VIS N 2** means the visibility to the north is 2 SM.

i. **Visibility at second site**. For automated reporting stations having more than one visibility sensor, site-specific visibility that is lower than the visibility shown in the body will be shown in the remarks.

    1) EXAMPLE: **VIS 2 RY11** means the visibility is 2 SM at runway 11.

j. **Lightning**

    1) When lightning is seen by the weather observer, it will be included in remarks. The frequency of occurrence, type of lightning when observed, and the location will be indicated. The location is determined in reference to point of observation.

       a) The following contractions are used for frequency of lightning:

| Contraction | Meaning |
|---|---|
| OCNL | Occasional (less than 1 flash/min.) |
| FRQ | Frequent (about 1 to 6 flashes/min.) |
| CONS | Continuous (more than 6 flashes/min.) |

       b) The following contractions are used for type of lightning:

| Contraction | Meaning |
|---|---|
| CG | Cloud-to-ground |
| IC | In-cloud |
| CC | Cloud-to-cloud |
| CA | Cloud-to-air |

       c) EXAMPLE. **OCNL LTGICCG 2 NW** means occasional lightning in-cloud and cloud-to-ground 2 SM to the northwest.

    2) When lightning is detected by an automated weather system

       a) Within 5 NM of the airport location point (ALP), it will be reported as TS in the body of the report with no remarks.

       b) Between 5 and 10 NM of the ALP, it will be reported as VCTS in the body of the report with no remarks.

       c) Beyond 10 but within 30 NM of the ALP, it will be reported in remarks only as LTG DSNT followed by the direction from the ALP.

k. **Beginning and/or ending times for precipitation and thunderstorms**

    1) When precipitation begins or ends, remarks will show the type of precipitation as well as the beginning and/or ending time(s) of occurrence.

        a) Types of precipitation may be combined if beginning or ending at the same time.

        b) These remarks are not required in SPECI reports but will be shown in the next METAR report.

        c) EXAMPLE: **RAB05E30SNB20E55** means that rain began at 5 min. past the hour and ended at 30 min. past the hour; snow began at 20 min. past the hour and ended at 55 min. past the hour.

    2) When thunderstorms begin or end, remarks will show the thunderstorm as well as the beginning and/or ending time(s) of occurrence.

        a) These remarks are required in SPECI reports and will also be shown in the next METAR report.

        b) EXAMPLE: **TSB05E40** means the thunderstorm began at 5 min. past the hour and ended at 40 min. past the hour.

l. **Thunderstorm location**. Thunderstorm location and movement will be in remarks as TS followed by location and movement.

    1) EXAMPLES:

        a) **TS OHD MOV N** means thunderstorm overhead, moving north.

        b) **TS VC NE MOV NE** means thunderstorm in the vicinity northeast, moving northeast.

m. **Hailstone size**. The size of the largest hailstone, in 1/4-in. increments, is shown in remarks preceded with the contraction for hail, **GR**.

    1) EXAMPLE: **GR 3/4** means hailstones 3/4 in. in diameter.

    2) If small hail or snow pellets, GS, are encoded in the body of the report, no size remark will be shown.

n. **Virga**. When precipitation is observed but does not reach the ground, **VIRGA** will be shown in remarks. The direction from the station may also be reported.

    1) EXAMPLES:

        a) **VIRGA** means virga at the station.
        b) **VIRGA SW** means virga southwest of the station.

o. **Variable ceiling height**. Whenever the ceiling is below 3,000 ft. and is variable, the remark **CIG** will be shown, followed with the lowest and highest ceiling heights.

    1) EXAMPLE: **CIG 005V010** means the ceiling is variable between 500 ft. and 1,000 ft.

p. **Obscurations**. When an obscuration (surface or aloft) is observed, the obscuring phenomenon, followed by the amount of obscuration (FEW, SCT, or BKN) and three zeros, is shown in remarks.

    1) EXAMPLES:

        a) **FG FEW000** means fog is obscuring 1/8 to 2/8 of the sky.
        b) **FU BKN020** means a broken layer of smoke aloft, with bases at 2,000 ft.

q.  **Variable sky condition**.  When a layer that is 3,000 ft. or less is varying in sky cover, remarks will show the variability range.

1)  If there is more than one cloud layer, the variable layer is identified by including the layer height.

2)  EXAMPLES:

a)  **SCT V BKN** means scattered variable broken.
b)  **BKN025 V OVC** means 2,500 ft. broken variable overcast.

r.  **Significant cloud types**.  When significant cloud types are observed, they are shown in remarks.

1)  **Cumulonimbus (CB)** or **cumulonimbus mammatus (CBMAM)**, direction from the station, and direction of movement (if known).  If the cloud is beyond 10 SM from the airport, DSNT will indicate distance.

a)  EXAMPLES:

i)  **CB W MOV E** means cumulonimbus clouds west of the station moving east.

ii)  **CBMAM DSNT S** means a cumulonimbus mammatus cloud distant south.

2)  **Towering cumulus (TCU)**, location (if known), and direction from station

a)  **TCU OHD** means towering cumulus overhead.

3)  Altocumulus castellanus (ACC), stratocumulus standing lenticular (SCSL), altocumulus standing lenticular (ACSL), cirrocumulus standing lenticular (CCSL), or rotor clouds (ROTOR CLD).  Remarks describe the cloud and the direction from the station.

a)  EXAMPLES:

i)  **ACSL SW-S** means altocumulus standing lenticular clouds southwest through south.

ii)  **APRNT ROTOR CLD S** means an apparent rotor cloud to the south.

s.  **Ceiling height at second site**.  When an automated station uses meteorological discontinuity sensors, remarks will be shown to identify site-specific ceiling heights.

1)  EXAMPLE:  **CIG 020 RY11** means ceiling 2,000 ft. at runway 11.

t.  **Pressure rising/falling rapidly**.  At designated stations, when the pressure is rising or falling at a rate of at least 0.06 in. per hour and the pressure change totals 0.02 or more at time of observation, remarks will show **PRESRR** or **PRESFR**, respectively.

u.  **Sea-level pressure**.  At designated stations that report sea-level pressure, this remark begins with **SLP** and is coded using tens, units, and tenths of sea-level pressure in hectoPascals (hPa), which are equivalent to millibars.

1)  If no sea-level pressure is available, it will be coded **SLPNO**.

a)  SLPNO is not entered in SPECI reports taken at manual stations.

2)  EXAMPLE:  **SLP132** means the sea-level pressure is 1013.2 hPa.

v.  **Aircraft mishap**.  If a report is taken to document weather conditions when notified of an aircraft mishap, the remark, **(ACFT MSHP)**, including the parentheses, is included in the report.

1)  The remark is not transmitted.

w.  **No SPECI available**. At manual weather observing stations that do not take SPECI reports, **NOSPECI** will be shown in remarks of all METAR reports.

x.  **Snow increasing rapidly**. The rapid accumulation of snow is reported at designated stations whenever the snow depth increases by one inch or more in the past hour.

  1)  The remark, **SNINCR**, is followed with the depth increase in the past hour, a solidus (/), and the total snow depth on the ground at the time of the report.

   a)  EXAMPLE: **SNINCR 2/10** means the snow depth increased 2 in. during the past hour and the current total snow depth on the ground is 10 in.

y.  **Other significant information**. Operationally significant information such as fog dispersal operations, runway conditions, or first and last reports may be added to remarks here.

3.  **Additive and Maintenance Data Remarks Category**. Additive data groups are reported only at designated stations, and the maintenance data groups are reported only from automated weather reporting stations.

  a.  **Hourly precipitation**. An hourly precipitation group will be reported only at designated weather reporting stations, when precipitation has occurred since the last hour.

   1)  This element begins with **P** followed by four digits indicating precipitation amount in hundredths of an inch.

    a)  Four zeros indicate that less than one-hundredth of an inch of precipitation fell in the past hour.

  b.  **3- and 6-hour precipitation amount**. A precipitation amount group will be reported at designated stations in the 3- and 6-hr. reports when precipitation has occurred since the previous 3- or 6-hr. report.

   1)  This element begins with **6** followed by four digits indicating precipitation amount using tens, units, tenths, and hundredths of an inch.

  c.  **24-hour precipitation**. A 24-hr. precipitation group will be reported at designated stations in the 1200Z observation (or other designated time) when more than a trace of precipitation has fallen in the past 24-hr. period.

   1)  This element begins with a **7** followed by four digits indicating precipitation amount using tens, units, tenths, and hundredths of an inch.

  d.  **Snow depth on ground**. A snow depth on the ground report will be reported at designated stations in the 0000Z and 1200Z observation, whenever more than a trace of snow is on the ground, and included in the 0600Z and 1800Z observation, whenever more than a trace of snow is on the ground and more than a trace of precipitation (water equivalent) has occurred in the past 6 hr.

   1)  This element begins with **4/** followed by three digits indicating snow depth in inches.

  e.  **Water equivalence of snow on the ground**. Water equivalence is reported at designated stations in the 1800Z report if the average snow, snow pellets, ice pellets, ice crystals, or hail depth is 2 in. or more.

   1)  This remark is not used if the water equivalent consists entirely of hail.

   2)  The remark is preceded with **933** followed by the water equivalent in tens, units, and tenths of an inch.

  f.  **Cloud types**. Cloud types are reported at designated stations in 3- and 6-hr. reports. The predominant low, middle, and high cloud types are coded.

   1)  This element begins with **8/** followed with coding used by the World Meteorological Organization for cloud identification.

   2)  FAA reporting stations will not encode this group.

g.  **Duration of sunshine**. This remark is included in the 0800Z report at designated sunshine duration measuring stations.

  1)  If the station is closed at 0800Z, the remark is included in the first 6-hr. METAR report after the station opens.

  2)  The remark begins with **98** followed by three digits of the number of minutes of sunshine in hundreds, tens, and units taken from a sunshine sensor.

h.  **Hourly temperature/dew point**. An hourly temperature/dew point group will be reported at designated stations to show hourly temperature and dew point to the tenth of a degree Celsius.

  1)  The element begins with **T** followed with **0**, if the temperature is 0°C or higher, or **1** to denote the temperature is less than 0°C, followed by the temperature in tens, units, and tenths; then 0 or 1 again; and then the dew point in tens, units, and tenths.

i.  **6-hr. maximum temperature**. A 6-hr. maximum temperature group will be reported at designated stations to show the maximum temperature for the last 6 hr.

  1)  This element begins with **1** to identify maximum temperature, followed with **0** or **1** to denote whether the temperature is higher or lower than 0°C, respectively, and then the temperature in tens, units, and tenths.

j.  **6-hr. minimum temperature**. A 6-hr. minimum temperature group will be reported at designated stations to show the minimum temperature for the last 6 hr.

  1)  This element begins with 2 to identify minimum temperature. Then it is coded in the same manner as maximum temperature.

k.  **24-hr. maximum/minimum temperature**. A maximum/minimum temperature group will be reported at designated stations to show the maximum and minimum temperatures for the last 24 hr.

  1)  This element begins with **4** and is coded in the same manner as the maximum temperature element.

l.  **3-hr. pressure tendency**. At designated stations, the 3- and 6-hr. reports include a remark regarding pressure tendency.

  1)  This element begins with **5** followed with 0 through 9 from the pressure characteristic and then the change in pressure over the past 3 hr.

  2)  The change in pressure is based on either the station pressure or the altimeter and is reported in tens, units, and tenths.

m.  **Sensor status indicators**

  1)  If the RVR element in the body should be reported but is missing, **RVRNO** will be shown here.

    a)  **RVRNO** means that runway visual range information is not available.

  2)  If an automated weather reporting station is equipped with the following sensors and they are not working, the following remarks will appear:

    a)  **PWINO** -- present weather identifier not available
    b)  **PNO** -- precipitation amount not available
    c)  **FZRANO** -- freezing rain information indicator not available
    d)  **TSNO** -- lightning information not available
    e)  **VISNO** -- visibility sensor information not available
    f)  **CHINO** -- cloud height indicator information not available

n.  **Maintenance indicator**. A maintenance indicator (dollar) sign, **$**, is included when an automated weather reporting system detects that maintenance is needed on the system.

O.  **Automated Surface Observing System (ASOS)**. The ASOS is the primary surface weather observing system of the U.S.

   1.  ASOS is designed to support aviation operations and weather forecast activities.

      a.  The program to install and operate up to 1,700 systems throughout the U.S. is a joint effort of the NWS, the FAA, and the Department of Defense.

      b.  The ASOS will provide continuous minute-by-minute observations and perform the basic observing functions necessary to generate a METAR report and other aviation weather information.

   2.  ASOS weather information may be transmitted over a discrete VHF radio frequency or the voice portion of a local navigation aid (NAVAID).

      a.  ASOS transmissions on a discrete VHF frequency are engineered to be receivable to a maximum of 25 NM from the ASOS site and a maximum altitude of 10,000 ft. AGL.

         1)  Local conditions may limit the maximum reception distance and/or altitude.

   3.  While the automated system and the human observer may differ in their methods of data collection and interpretation, both produce an observation that is similar in form and content.

      a.  For the "objective" elements (i.e., pressure, ambient temperature, dew point temperature, wind, and precipitation accumulation), both the ASOS and the observer use a fixed location and time-averaging technique.

      b.  For the "subjective" elements, observers use a fixed-time, spatial-averaging technique to describe the visual elements (i.e., sky condition, visibility, and present weather), while the ASOS uses a fixed-location, time-averaging technique.

         1)  Although this is a fundamental change, the manual and automated techniques yield remarkably similar results within the limits of their respective capabilities.

      c.  Among the basic strengths of the ASOS observation is the fact that critical weather parameters (e.g., sky condition and visibility) are measured at specific locations where they are needed most.

   4.  **System Description**

      a.  The ASOS at each airport location consists of four main components:

         1)  Individual weather sensors
         2)  Data collection package(s) (DCP)
         3)  Acquisition control unit
         4)  Peripherals and displays

      b.  The ASOS sensors perform the basic function of data acquisition.

         1)  They continuously sample and measure the ambient environment, derive raw sensor data, and make them available to the collocated DCP.

   5.  Every ASOS will contain the following basic set of sensors:

      a.  Cloud height indicator (one or possibly three)
      b.  Visibility sensor (one or possibly three)
      c.  Precipitation identification sensor
      d.  Freezing rain sensor (at select sites)
      e.  Pressure sensors (two sensors at small airports; three sensors at large airports)
      f.  Ambient temperature/dew point temperature sensor
      g.  Anemometer (wind direction and speed sensor)
      h.  Rainfall accumulation sensor

6. The ASOS data outlets include

    a. Those necessary for on-site airport users

    b. National communications networks

    c. Computer-generated voice (available through FAA radio broadcast to pilots and/or dial-in telephone line)

7. An ASOS report without human intervention will contain only that weather data capable of being reported automatically.

    a. The modifier for this METAR report is **AUTO**.

        1) When an observer backs up an ASOS site, the **AUTO** modifier is not used.

8. There are two types of automated stations:

    a. **AO1** for automated stations without a precipitation discriminator

    b. **AO2** for automated stations with a precipitation discriminator

        1) A precipitation discriminator can determine the difference between liquid and frozen/freezing precipitation.

P. **Automated Weather Observing System (AWOS)**. AWOS is the predecessor to ASOS; i.e., ASOS is replacing AWOS.

1. ASOS will always prepare a METAR report, while only some AWOS sites will prepare METAR reports.

    a. Generally, AWOS is operated by the FAA or state government agencies.

2. Some AWOS locations will be augmented by certified observers who will provide weather and obstruction-to-vision information in the remarks of the report when the reported visibility is less than 7 SM.

    a. These sites, along with the hours of augmentation, are to be published in the *Airport/Facility Directory* (A/FD).

    b. Augmentation is identified in the observation as "Observer Weather."

3. The AWOS wind speed, direction and gusts, temperature, dew point, and altimeter setting are exactly the same as for manual observations.

    a. AWOS will also report density altitude when it exceeds the field elevation by more than 1,000 ft.

    b. The reported visibility is derived from a sensor near the touchdown point of the primary runway.

        1) The visibility is a runway visibility value (RVV), using a 10-min. harmonic average.

        2) The AWOS sensors have been calibrated against the FAA transmissometer standards used for runway visual range values.

        3) Since the AWOS visibility is an extrapolation of a measurement at the touchdown point of the runway, it may differ from the standard prevailing visibility.

4. The reported sky condition/ceiling is derived from a ceilometer located next to the visibility sensor.

    a. The AWOS algorithm integrates the last 30 min. of ceilometer data to derive cloud layers and heights.

    b. This output may also differ from the observer sky condition in that the AWOS is totally dependent upon the cloud advection over the sensor site.

5.  AWOS sites are operationally classified into the following four basic levels:

    a.  **AWOS-A** -- only reports altimeter setting

    b.  **AWOS-1** -- usually reports altimeter setting, wind data, temperature, dew point, and density altitude

    c.  **AWOS-2** -- provides the information provided by AWOS-1 plus visibility

    d.  **AWOS-3** -- provides the information provided by AWOS-2 plus cloud/ceiling data

6.  AWOS information (system level, frequency, telephone number, etc.) concerning specific locations is published, as the systems become operational, in the *A/FD* and, if applicable, in published instrument approach procedures (IAP).

    a.  Selected AWOS sites may be incorporated into nationwide data collection and dissemination networks in the future.

7.  **AWOS Broadcasts**

    a.  A computer-generated voice is used in AWOS to automate the broadcast of the minute-by-minute weather observations via VHF radio and/or telephone.

        1)  Some AWOS sites are configured to permit the addition of an operator-generated voice message, e.g., weather remarks following the automated parameters.

    b.  AWOS information is transmitted over a discrete VHF radio frequency or the voice portion of a local NAVAID.

        1)  AWOS VHF radio transmissions are engineered to be receivable to a maximum of 25 NM from the AWOS site to a maximum altitude of 10,000 ft. AGL.

            a)  Local conditions may limit AWOS reception.

        2)  The system transmits a 20- to 30-sec. weather message updated each minute.

    c.  Most AWOS sites have a dial-up capability so that the weather message can be accessed via telephone.

8.  The phraseology used generally follows that used for other weather broadcasts. Following are explanations of the exceptions.

    a.  The location/name of the airport and the phrase, "Automated Weather Observation," followed by the time, are announced.

        1)  The word "Test" is added following "Observation" when the system is not in commissioned (operational) status.

        2)  The phrase, "Temporarily Inoperative," is added when the system is inoperative.

    b.  The lowest reported visibility value in AWOS is "less than 1/4 (SM)."

        1)  The phrase, "Visibility Missing," is announced only if the AWOS site has a visibility sensor and the information is not available.

    c.  In the future, some AWOS sites are to be configured to determine the occurrence of precipitation.

    d.  Ceiling and sky cover

        1)  With the exception of an indefinite ceiling, all ceiling heights are measured (by a ceilometer).

        2)  "No clouds detected" is announced as "No clouds below XXX" or, in newer systems, as "Sky clear below XXX," where XXX is the range limit of the sensor.

        3)  "Sky condition missing" is announced only if the system is configured with a ceilometer and the ceiling and sky cover information is not available.

e.   Remarks

    1)   Automated remarks include

        a)   Density altitude

        b)   Variable visibility

        c)   Variable wind direction

    2)   Manual input remarks are prefaced with the phrase, "Observer Weather," and are generally limited to

        a)   Type and intensity of precipitation

        b)   Thunderstorm and direction

        c)   Obstruction to vision when the visibility is less than 3 SM

9.   If an AWOS is capable of producing a METAR report, the modifier **AUTO** will be used, if it is done without human intervention.

    a.   The remarks section will contain either **AO1** or **AO2**, depending on if the site has a precipitation discriminator.

10.  **Weather Observing Programs Summary Table**

| WEATHER OBSERVING PROGRAMS | | | | | | |
|---|---|---|---|---|---|---|
| **Element Reported** | **AWOS-A** | **AWOS-1** | **AWOS-2** | **AWOS-3** | **ASOS** | **Manual** |
| Altimeter | X | X | X | X | X | X |
| Wind | | X | X | X | X | X |
| Temperature/ Dew Point | | X | X | X | X | X |
| Density Altitude | | X | X | X | X | |
| Visibility | | | X | X | X | X |
| Clouds/Ceiling | | | | X | X | X |
| Precipitation | | | | | X | X |
| Remarks | | | | | X | X |

# END OF CHAPTER

# CHAPTER THREE
# PILOT WEATHER REPORTS (PIREPs)

A. **Purpose.** No observation is more timely than the one made from the cockpit. In fact, aircraft in flight are the **only** means of directly observing cloud tops, icing, and turbulence.

1. PIREPs are also valuable in that they help fill the gaps between reporting stations.

2. You should report any observation that may be of concern to other pilots.

3. Also, if conditions were forecasted but were not encountered, you should provide a PIREP.

   a. This will help the NWS to verify forecast products and create accurate products for the aviation community.

4. Pilots should help themselves, the aviation public, and the aviation weather forecasters by sending pilot reports.

B. **Format.** A PIREP is usually transmitted in a prescribed format (shown in Table 3-2 on page 248).

1. Required elements for all PIREPs are type of report, location, time, flight level, type of aircraft, and at least one weather element encountered.

2. When not required, elements without reported data are omitted.

3. All altitude references are MSL unless otherwise noted.

4. Distances are in NM, and time is in UTC.

5.  To lessen the chance of misinterpretation, icing and turbulence reports state intensities using standard terminology when possible.

    a.  If a PIREP stated, "Pretty rough at 6,500...," the description of the strength of the turbulence at 6,500 ft. could have many interpretations.

        1)  A report of "light," "moderate," or "severe" turbulence would have been more concise and understandable.

    b.  If a pilot's description of an icing or turbulence encounter cannot readily be translated into standard terminology, the pilot's description should be transmitted verbatim.

| | |
|---|---|
| UUA/UA | Type of report:<br>   URGENT (UUA) - Any PIREP that contains any of the following weather phenomena: tornadoes, funnel clouds, or waterspouts; severe or extreme turbulence, including clear air turbulence (CAT); severe icing; hail; volcanic ash; low-level wind shear (LLWS) (pilot reports air speed fluctuations of 10 knots or more within 2,000 feet of the surface); any other weather phenomena reported which are considered by the controller to be hazardous, or potentially hazardous, to flight operations.<br>   ROUTINE (UA) - Any PIREP that contains weather phenomena not listed above, including low-level wind shear reports with air speed fluctuations of less than 10 knots. |
| /OV | Location: Use VHF NAVAID(s) or an airport using the three- or four-letter location identifier. Position can be over a site, at some location relative to a site, or along a route. Ex: /OV KABC; /OV KABC090025; /OV KABC045020-DEF; /OV KABC-KDEF |
| /TM | Time: Four digits in UTC. Ex: /TM 0915 |
| /FL | Altitude/Flight level: Three digits for hundreds of feet with no space between FL and altitude. If not known, use UNKN. Ex: /FL095; /FL310; /FLUNKN |
| /TP | Aircraft type: Four digits maximum; if not known, use UNKN. Ex: /TP L329; /TP B737; /TP UNKN |
| /SK | Sky cover: Describes cloud amount, height of cloud bases, and height of cloud tops. If unknown, use UNKN. Ex: /SK SCT040-TOP080; /SK BKNUNKN-TOP075; /SK BKN-OVC050-TOPUNKN; /SK OVCUNKN-TOP085 |
| /WX | Flight visibility and weather: Flight visibility (FV) reported first and use standard METAR weather symbols. Intensity (– for light, no qualifier for moderate, and + for heavy) shall be coded for all precipitation types except ice crystals and hail. Ex: /WX FV05SM – RA; /WX FV01SM SN BR; /WX RA |
| /TA | Temperature (Celsius): If below zero, prefix with an "M." Temperature should also be reported if icing is reported. Ex: /TA 15; /TA M06 |
| /WV | Wind: Direction from which the wind is blowing coded in tens of degrees using three digits. Directions of less than 100 degrees shall be preceded by a zero. The wind speed shall be entered as a two- or three-digit group immediately following the direction, coded in whole knots using the hundreds, tens, and units digits.<br>Ex: /WV 27045KT; /WV 280110KT |
| /TB | Turbulence: Use standard contractions for intensity and type (CAT or CHOP when appropriate). Include altitude only if different from FL. (See Table 3-3 on page 249.)<br>Ex: /TB EXTRM; /TB OCNL LGT-MDT BLO 090; /TB MOD-SEV CHOP 080-110 |
| /IC | Icing: Describe using standard intensity and type contractions. Include altitude only if different from FL. (See Table 3-4 on page 251.) Ex: /IC LGT-MDT RIME; /IC SEV CLR 028-045 |
| /RM | Remarks: Use free form to clarify the report putting hazardous elements first<br>Ex: /RM LLWS –15 KT SFC-030 DURGC RY 22 JFK |

Table 3-2. Encoding PIREPs

C. **Turbulence.** Table 3-3 below classifies each turbulence intensity according to its effects on aircraft control, structural integrity, and articles and occupants within the aircraft.

| Intensity | Aircraft Reaction | Reaction Inside Aircraft |
|---|---|---|
| Light | Turbulence that momentarily causes slight, erratic changes in altitude and/or attitude (pitch, roll, yaw). Report as **light turbulence** or **clear air turbulence (CAT)**. or Turbulence that causes slight, rapid, and somewhat rhythmic bumpiness without appreciable changes in altitude or attitude. Report as **light chop**. | Occupants may feel a slight strain against belts or shoulder straps. Unsecured objects may be displaced slightly. Food service may be conducted and little or no difficulty is encountered in walking. |
| Moderate | Turbulence that causes changes in altitude and/or attitude occurs but the aircraft remains in positive control at all times. It usually causes variations in indicated airspeed. Report as **moderate turbulence** or **moderate CAT**. or Turbulence that is similar to light chop but of greater intensity. It causes rapid bumps or jolts without appreciable changes in aircraft or attitude. Report as **moderate chop**. | Occupants feel definite strains against seat belts or shoulder straps. Unsecured objects are dislodged. Food service and walking are difficult. |
| Severe | Turbulence that causes large, abrupt changes in altitude and/or attitude. It usually causes large variations in indicated airspeed. Aircraft may be momentarily out of control. Report as **severe turbulence** or **severe CAT**. | Occupants are forced violently against seat belts or shoulder straps. Unsecured objects are tossed about. Food service and walking are impossible. |
| Extreme | Turbulence in which the aircraft is violently tossed about and is practically impossible to control. It may cause structural damage. Report as **extreme turbulence** or **extreme CAT**. | |

Table 3-3. Turbulence Reporting Criteria.

1.  Pilots should report location(s), time (UTC), altitude, aircraft type, proximity to clouds, intensity, and, when applicable, type (chop or CAT) and duration (occasional, intermittent, or continuous) of turbulence.

    a.  Definition of duration terms

        1)  Occasional means less than 1/3 of the time.
        2)  Intermittent means 1/3 to 2/3 of the time.
        3)  Continuous means more than 2/3 of the time.

2.  High-level turbulence (normally above 15,000 ft. AGL) that is not associated with cumuliform clouds (including thunderstorms) shall be reported as clear air turbulence (CAT).

D.  **Locations of Probable Turbulence by Intensities** (as it relates to weather and terrain features)

    1.  **Light Turbulence**

        a.  In hilly and mountainous areas, even with light winds
        b.  In and near small cumulus clouds
        c.  In clear-air convective currents over heated surfaces
        d.  With weak wind shears in the vicinity of

            1)  Troughs aloft
            2)  Lows aloft
            3)  Jet streams
            4)  The tropopause

        e.  In the lower 5,000 ft. of the atmosphere

            1)  When winds are near 15 kt.
            2)  Where the air is colder than the underlying surfaces

2. **Moderate Turbulence**

    a.   In mountainous areas with a wind component of 25 to 50 kt. perpendicular to and near the level of the ridge

        1)   At all levels from the surface to 5,000 ft. above the tropopause with preference for altitudes

            a)   Within 5,000 ft. of the ridge level
            b)   At the base of relatively stable layers below the base of the tropopause
            c)   Within the tropopause layer

        2)   Extending downstream from the lee of the ridge for 150 to 300 NM

    b.   In and near thunderstorms in the dissipating stage

    c.   In and near other towering cumuliform clouds

    d.   In the lower 5,000 ft. of the troposphere

        1)   When surface winds are 30 kt. or more
        2)   Where heating of the underlying surface is unusually strong
        3)   Where there is an invasion of very cold air

    e.   In fronts aloft

    f.   In areas where the vertical wind shear exceeds 6 kt. per 1,000 ft., and/or horizontal wind shear exceeds 18 kt. per 150 NM

3. **Severe Turbulence**

    a.   In mountainous areas with a wind component exceeding 50 kt. perpendicular to and near the level of the ridge

        1)   In 5,000-ft. layers

            a)   At and below the ridge level in rotor clouds or rotor action
            b)   At the tropopause
            c)   Sometimes at the base of other stable layers below the tropopause

        2)   Extending downstream from the lee of the ridge for 50 to 150 NM

    b.   In and near growing and mature thunderstorms

    c.   Occasionally in other towering cumuliform clouds

    d.   50 to 100 NM on the cold side of the center of the jet stream, in troughs aloft, and in lows aloft where

        1)   Vertical wind shear exceeds 10 kt. per 1,000 ft., and
        2)   Horizontal wind shear exceeds 40 kt. per 150 NM.

4. **Extreme Turbulence**

    a.   In mountain wave situations, in and below the level of well-developed rotor clouds. Sometimes it extends to the ground.

    b.   In severe thunderstorms. A severe thunderstorm is indicated by

        1)   Large hailstones (diameter ¾ in. or greater),
        2)   Strong radar echoes, or
        3)   Almost continuous lightning.

E.  **Icing.** Table 3-4 below classifies each intensity according to its operational effects on the aircraft.

| Intensity | Airframe Ice Accumulation | Pilot Report |
|---|---|---|
| Trace | Ice becomes perceptible. Rate of accumulation slightly greater than sublimation. Deicing/anti-icing equipment is not used unless encountered for an extended period of time (over one hour). | Location, time, altitude/FL, aircraft type, temperature, and icing intensity and type |
| Light | The rate of accumulation may create a problem if flight is prolonged in this environment (over one hour). Occasional use of deicing/anti-icing equipment removes/prevents accumulation. It does not present a problem if the deicing/anti-icing equipment is used. | Location, time, altitude/FL, aircraft type, temperature, and icing intensity and type |
| Moderate | The rate of accumulation is such that even short encounters become potentially hazardous and use of deicing/anti-icing equipment or diversion is necessary. | Location, time, altitude/FL, aircraft type, temperature, and icing intensity and type |
| Severe | The rate of accumulation is such that deicing/anti-icing equipment fails to reduce or control the hazard. Immediate diversion is necessary. | Location, time, altitude/FL, aircraft type, temperature, and icing intensity and type |

Table 3-4. Icing Intensities, Airframe Ice Accumulation, and Pilot Report

1.  **Types of Icing Reported**

    a.  **Rime ice** is rough, milky, opaque ice formed by the instantaneous freezing of small supercooled water droplets.

    b.  **Clear ice** is a glossy, clear or translucent ice formed by the relatively slow freezing of large supercooled water drops.

    c.  **Mixed ice** is a combination of rime and clear ice.

F.  **Wind Shear.** Because unexpected changes in wind speed and direction can be hazardous to aircraft operations at low altitudes on approach to, and departing from, airports, pilots are urged to promptly volunteer reports of wind shear conditions they encounter to controllers.

1.  An advanced warning of this information will assist other pilots in avoiding or coping with a wind shear on approach or departure.

    a.  See Part III, Chapter 1, The Aviation Weather Service Program, for a discussion on microburst, wind shear, and gust front detection programs beginning on page 222.

2.  When describing conditions, pilots should avoid the use of the terms "negative" or "positive" wind shear.

    a.  PIREPs of "negative wind shear on final," intended to describe loss of airspeed and lift, have been interpreted to mean that no wind shear was encountered.

3.  The pilot should report the loss or gain of airspeed and the altitudes at which it was encountered.

    a.  Pilots who are not able to report wind shear in these specific terms are encouraged to make reports in terms of the effects on their aircraft.

G.  **Examples** (Refer to Table 3-2 on page 248.)

1.  KORD UUA/OV ORD/TM 1235/FLUNKN/TP B727/TB MOD/RM LLWS +/– 20KT BLW 003 DURGD 27L.

    a.  Urgent PIREP (due to a report of wind shear) over Chicago O'Hare at 1235 UTC, flight level unknown.  The pilot of a Boeing 727 reported moderate turbulence and low-level wind shear (LLWS) with airspeed fluctuations of plus to minus 20 kt. below 300 ft. during descent to runway 27 left.

2.  KOKC UA/OV KOKC090064/TM 1522/FL080/TP C172/SK SCT090-TOPUNKN/WX FV05SM HZ/TA M04/WV 24540KT/TB LGT/RM IN CLR.

    a.  Routine PIREP, 64 NM east (090 radial) of the Oklahoma City VOR at 1522 UTC, flight level 8,000 ft. MSL.  The pilot of a Cessna 172 reported a scattered cloud layer with bases at 9,000 ft. MSL and top height unknown.  Flight visibility is 5 SM in haze, outside air temperature is –04°C, wind is 245° at 40 kt., light turbulence, and the airplane is in clear skies.

3.  Nonmeteorological PIREPs sometimes help air traffic controllers.  A plain language report stated the following:

    ...3N PNS LARGE FLOCK OF BIRDS HDG GEN N MAY BE SEAGULLS FRMN...

    a.  This PIREP alerted pilots and controllers to a bird hazard.

# END OF CHAPTER

# CHAPTER FOUR
# RADAR WEATHER REPORT (SD)

A. **Introduction**. General areas of precipitation, including rain, snow, and thunderstorms, can be observed by radar. Radar stations report each hour at H+35 (35 min. past the hour).

　1. The radar weather report (SD) includes the type, intensity, and location of the echo top of the precipitation.

　　a. The intensity trend of the precipitation is not coded on the SD.

　2. All heights are reported above mean sea level (MSL).

　3. Table 3-5 below explains symbols denoting intensity.

| Symbol | Intensity |
|--------|-----------|
| – | Light |
| (none) | Moderate |
| + | Heavy |
| ++ | Very Heavy |
| X | Intense |
| XX | Extreme |

Table 3-5. Precipitation Intensity

　4. Table 3-6 below lists the symbols used to denote types of precipitation.

　　a. Note that these symbols do not reflect the METAR code.

　　b. The SD is automatically generated from WSR-88D weather radar data, and the type of precipitation is determined by computer models and is limited to the ones listed below.

| Symbol | Meaning |
|--------|---------|
| R | Rain |
| RW | Rain Shower |
| S | Snow |
| SW | Snow Shower |
| T | Thunderstorm |

Table 3-6. Symbols Used in SD

B.    **Example of an SD**

TLX 1935 LN 8TRW++ 86/40 164/60 20W C2425 MTS 570 AT 159/65
AUTO ⌃ MO1 NO2 ON3 PM34 QM3 RL2=

To aid in the discussion, we have divided the SD into the following elements numbered 1
through 10:

| TLX 1935 | LN | 8 | TRW++ | 86/40  164/60 |
|----------|----|---|-------|---------------|
| 1.       | 2. | 3.| 4.    | 5.            |

| 20W | C2425 | MTS 570 AT 159/65 | AUTO |
|-----|-------|-------------------|------|
| 6.  | 7.    | 8.                | 9.   |

⌃ MO1 NO2 ON3 PM34 QM3 RL2=
        10.

1.    Location identifier and time of radar observation (Oklahoma City at 1935 UTC)

2.    Echo pattern (LN, or line, in the example).  The following echo pattern, or configuration,
      codes may be used.

      a.    LN (line) -- a line of convective echoes with precipitation intensities that are heavy or
            greater, at least 30 NM long, at least four times as long as it is wide, and at least 25%
            coverage within the line

      b.    AREA (area) -- a group of echoes of similar type and not classified as a line

      c.    CELL (cell) -- a single isolated convective echo, such as a rain shower

3.    Coverage, in tenths, of precipitation in the defined area.  In this example, 8 means
      8/10 coverage in the line.

4.    Type and intensity of weather.  In this example, the line of echoes contains thunderstorms
      (T) with very heavy rain showers (RW++).

5.    Azimuth, referenced to true north, and range in NM from the radar site of points defining the
      echo pattern.  In this example, the center points of the line are located at 86° at 40 NM and
      164° at 60 NM.

      a.    For lines and areas, two azimuth and range sets define the pattern.
      b.    For cells, there will be only one azimuth and range set.

6.    Dimension of echo pattern.  The dimensions of an echo pattern are given when azimuth and
      range define only the center line of the pattern.

      a.    In this example, 20W means the line has a total width of 20 NM, 10 NM either side of a
            center line drawn from the points provided in item 5. above.

7.    Cell movement.  Movement is coded only for cells; it will not be coded for lines or areas.

      a.    In this example, C2425 means the cells within the line are moving from 240° at 25 kt.

8.    Maximum top and location.  Maximum top may be coded with the symbols "MT" or "MTS."

      a.    MTS means that satellite data as well as radar information were used to measure the
            top of the precipitation.

      b.    In this example, MTS 570 AT 159/65 means the maximum precipitation top (measured
            by satellite and radar data) is 57,000 ft. located 159° at 65 NM from the radar site.

9.    AUTO means the SD is automated from WSR-88D weather radar data.

10.   Digital section. This information is used for preparing the radar summary chart.

C.  **Blank Reports**.  When a radar weather report is transmitted but does not contain any encoded weather observation, a contraction is sent that indicates the operational status of the radar.

| Contraction | Operational Status |
|---|---|
| PPINE | Radar is operating normally, but no echoes are being detected. |
| PPINA | Radar observation is not available. |
| PPIOM | Radar is inoperative or out of service. |
| AUTO | Automated radar report from WSR-88D. |

Table 3-6.  Contractions of Radar Operational Status.

1.  EXAMPLE:  TLX 1135 PPINE AUTO means that the Oklahoma City, OK radar at 1135 UTC detects no echoes.

D.  **Using SDs**.  When using hourly and special SDs in preflight planning, note the location and coverage of echoes, the type of weather reported, the intensity, and especially the direction of movement.

1.  A WORD OF CAUTION:  Remember that the WSR-88D (or any weather radar) detects particles in the atmosphere that are of precipitation size or larger.

     a.  It does **not** display locations of cloud-size particles and, thus, neither ceilings nor restrictions to visibility.

2.  An area may be blanketed with fog or low stratus, but the SD would not include information about it.

3.  You should use SDs along with METARs, satellite photos, and forecasts when planning a flight.

4.  SDs will help you to plan ahead to avoid thunderstorm areas.  Once airborne, however, you must depend on contact with Flight Watch (which has the capability to view current radar images), airborne radar, or visual sighting to evade individual storms.

# END OF CHAPTER

# CHAPTER FIVE
# SATELLITE WEATHER PICTURES

Please take a few minutes to study each of the concepts listed above and anticipate/imagine what they are and how they relate to the other listed concepts.

A. **Types of Satellites**

1. Two types of weather satellites are in use today by the U.S.

    a. Geostationary operational environmental satellites (GOES) are used for short-range warning and "now-casting."

    b. Polar-orbiting operational environmental satellites (POES) are used for longer-term forecasting.

2. Two GOES satellites are used for imaging.

    a. One GOES is stationed over the equator at 75°W longitude and is referred to as GOES EAST since it covers the eastern U.S.

    b. The other GOES is stationed over the equator at 135°W longitude and is referred to as GOES WEST since it covers the western U.S.

    c. Together, these satellites cover North and South America and the surrounding waters.

    d. Each satellite transmits an image of the earth, pole to pole, every 15 min.

    e. When disastrous weather threatens the U.S., the satellites can scan small areas rapidly so that an image can be received as often as every 1 min.

       1) Data from these rapid scans are used at NWS offices.

3. Complimenting the GOES satellites are two POES satellites.

    a. Each satellite is a polar orbiter and orbits the Earth on a track that nearly crosses the north and south poles.

    b. These satellites are able to monitor the entire Earth, tracking atmospheric variables and providing atmospheric data and cloud images. They track weather conditions that eventually affect the weather and climate of the U.S.

    c. Operating as a pair, these satellites ensure that data from any region of the Earth are no more than 6 hr. old.

B. **Types of Imagery**. Two types of imagery are available, and when combined, they give a great deal of information about clouds.

1. Through interpretation, the analyst can determine the type of cloud, the temperature of cloud tops (from this the approximate height of the cloud can be determined), and the thickness of cloud layers.

    a. From this information, the analyst gets a good idea of the associated weather.

2. One type of imagery is visible imagery. With a visible image, we are looking at clouds and the Earth reflecting sunlight to the satellite sensor.

    a. The greater the reflected sunlight reaching the sensor, the whiter the object is on the image.

        1) The amount of reflectivity reaching the sensor depends upon the height, thickness, and ability of the object to reflect sunlight.

        2) Since clouds are much more reflective than the Earth, clouds will usually show up white on the image, especially thick clouds.

        3) Thus, the visible image is primarily used to determine the presence of clouds and the type of cloud from shape and texture.

3. The second type of imagery is infrared (IR) imagery. With an IR image, we are looking at heat radiation being emitted by the Earth and clouds.

    a. The images show temperature differences between cloud tops and the ground, as well as temperature gradations of cloud tops and along the Earth's surface.

        1) Ordinarily, cold temperatures are displayed light gray or white.

        2) High clouds appear whitest.

        3) However, various computer-generated enhancements are sometimes used to sharply illustrate important temperature contrasts.

    b. IR images are used to determine cloud top temperatures, which can approximate the height of the cloud.

4. From this, you can see the importance of using visible and IR imagery together when interpreting clouds.

    a. Visible images are available only during the day.

    b. IR images are available both day and night.

C. **Using Satellite Images.** Satellite images are processed by the NWS as well as private companies. Thus, satellite images can be viewed through the use of many different sources.

1. Depending on the source of the satellite images, they may be updated anywhere from every 15 min. to every hour.

    a. Thus, it is important to note the time on the images when interpreting them.

2. By viewing satellite images, the development and dissipation of weather can be seen and followed over the entire country and coastal regions.

# END OF CHAPTER

# CHAPTER SIX
# RADIOSONDE ADDITIONAL DATA (RADAT)

A.  **Introduction.** Radiosonde additional data (RADAT) is obtained from the radiosonde observations which are conducted at 0000Z and 1200Z. A radiosonde is attached to a balloon and has instruments to measure specific information and a radio transmitter to transmit the information back to a ground station. The information contained in a RADAT includes the observed freezing level and the relative humidity associated with the freezing level.

B.  **Format.** The format associated with a RADAT is "Stn ID Time RADAT UU (D) (hphphp)(/n)."

1.  Explanation:

    a.  Stn ID Time -- three-letter station identifier and the observation time (UTC)

    b.  RADAT -- a contraction identifying the data as "freezing-level data"

    c.  UU -- relative humidity (a percentage) at the freezing level

        1)  When more than one level is sent, this value is the highest relative humidity observed at any of the levels transmitted.

    d.  (D) -- a coded letter, "L," "M," or "H," indicating that the relative humidity is for the lowest, middle, or highest level coded. This letter is omitted when only one level is sent.

    e.  (hphphp) -- height in hundreds of feet MSL at which the sounding (or radiosonde) crossed the 0°C isotherm. No more than three levels are coded.

        1)  If the sounding crosses the 0°C isotherm more than three times, the levels are coded for the lowest, the highest, and the intermediate crossing with the highest relative humidity.

    f.  (/n) -- indicator to show the number of crossings of the 0°C isotherm, other than those coded. The indicator is omitted when all levels are coded.

C.  **Examples**

1.  SJU 1200 RADAT 87160

    a.  San Juan, PR, 1200 UTC freezing-level data. Relative humidity at the freezing level was 87%, and the freezing level was at 16,000 ft. MSL.

2.  ALB 1200 RADAT 84M019045051/1

    a.  Albany, NY, 1200 UTC freezing level data. Relative humidity was 84% at the middle crossing. Three crossings of the 0°C isotherm occurred at 1,900 ft. MSL, 4,500 ft. MSL, and 5,100 ft. MSL. The "/1" indicates one additional crossing that was not coded.

3.  DNR 1200 RADAT ZERO

    a.  Denver, CO, 1200 UTC freezing-level data. The entire sounding was below 0°C.

4.  ABR 0000 RADAT MISG

    a.  Aberdeen, SD, 0000 UTC freezing level. The sounding was terminated before the first crossing of the 0°C isotherm. All temperatures were above freezing.

# END OF CHAPTER

# CHAPTER SEVEN
# TERMINAL AERODROME FORECAST (TAF)

Please take a few minutes to study each of the concepts listed above and anticipate/imagine what they are and how they relate to the other listed concepts.

## A. Introduction

1. The terminal aerodrome forecast (TAF) is a concise statement of the expected meteorological conditions at an airport during a specified period (usually 24 hr.).

   a. The TAF covers a geographic area within a 5-SM radius of the airport's center.

2. TAFs are scheduled four times daily for 24-hr. periods beginning at 0000Z, 0600Z, 1200Z, and 1800Z.

   a. Figures 4-1 and 4-2 on pages 268 and 269 depict the locations where TAFs are issued.

3. TAFs use the same codes as the METAR reports. See Part III, Chapter 2, Aviation Routine Weather Report (METAR), beginning on page 227, for a detailed explanation of those codes.

4. All times in a TAF refer to Coordinated Universal Time (UTC).

   a. The term Zulu (Z) may be used to denote UTC.

5. In order for a TAF to be issued for an airport, a minimum of two consecutive, complete METAR (routine) observations (manual or automated), not less than 30 min. apart nor more than 1 hr. apart, is required for that airport.

   a. After a TAF has been issued, the forecaster will use all available weather data sources to maintain the TAF.

   b. However, if during this time a METAR is missing or part of the METAR is missing, and the forecaster feels that other weather sources cannot provide the necessary information, the forecaster will discontinue the TAF.

B.  **Elements**.  A TAF contains the following sequence of elements in the following order (items 1-9). Forecast change indicators (items 10-12) and probability forecast (item 13) are used as appropriate.

| Communications Header | Forecast of Meteorological Conditions | Time Elements |
|---|---|---|
| 1.  Type of report | 5.  Wind | 10.  Temporary (TEMPO) |
| 2.  ICAO station identifier | 6.  Visibility | 11.  From (FM) |
| 3.  Date and time of origin | 7.  Significant weather | 12.  Becoming (BECMG) |
| 4.  Valid period date and time | 8.  Sky condition | 13.  Probability (PROB) |
|  | 9.  Wind shear (optional) |  |

C.  **Example of a TAF**

    TAF
    KOKC 051130Z 051212 14008KT 5SM BR BKN030 WS018/32030KT TEMPO 1316 1 1/2SM BR
        FM1600 16010KT P6SM SKC BECMG 2224 20013G20KT 4SM SHRA OVC020
        PROB40 0006 2SM TSRA OVC008CB=

To aid in the discussion, we have divided the TAF above into the following elements numbered 1. through 13.

| TAF | KOKC | 051130Z | 051212 | 14008KT | 5SM | BR | BKN030 | WS018/32030KT |
|---|---|---|---|---|---|---|---|---|
| 1. | 2. | 3. | 4. | 5. | 6. | 7. | 8. | 9. |

TEMPO 1316 1 1/2SM BR          FM1600 16010KT P6SM SKC
         10.                              11.

BECMG 2224 20013G20KT 4SM SHRA OVC020
         12.

PROB40 0006 2SM TSRA OVC008CB=
         13.

1.  Routine terminal aerodrome forecast

2.  Oklahoma City, OK

3.  Forecast prepared on the 5th day at 1130 UTC (or Z)

4.  Forecast valid from the 5th day at 1200 UTC until 1200 UTC on the 6th

5.  Wind 140° true at 8 kt.

6.  Visibility 5 SM

7.  Visibility obscured by mist

8.  Ceiling 3,000 ft. broken

9.  Low-level wind shear at 1,800 ft., wind 320° true at 30 kt.

10.  Temporary (spoken as occasional) visibility 1½ SM in mist between 1300 UTC and 1600 UTC

11.  From (or after) 1600 UTC, wind 160° true at 10 kt., visibility more than 6 SM, sky clear

12.  Becoming (gradual change) wind 200° true at 13 kt., gusts to 20 kt., visibility 4 SM in moderate rain showers, ceiling 2,000 ft. overcast between 2200 UTC and 2400 UTC

13.  Probability (40% chance) between 0000 UTC and 0600 UTC of visibility 2 SM, thunderstorm, moderate rain, ceiling 800 ft. overcast, cumulonimbus clouds (The = sign indicates end of forecast.)

D.  **Type of Report** (element 1).  The report type header will always appear as the first element in the TAF.

   1.  There are two types of TAF issuances:

      a.  **TAF** means a routine forecast.
      b.  **TAF AMD** means an amended forecast.

         1)  An amended forecast is issued when the current TAF no longer adequately describes the ongoing weather or the forecaster feels the TAF is not representative of the current or expected weather.

   2.  Corrected (COR) or routine delayed (RTD) TAFs are identified only in the communications header which precedes the actual forecast.

      a.  A **TAF COR** is issued when a TAF contains typographical or other errors.

      b.  A **TAF RTD** is issued for those airports not having a 24-hr. observing schedule.  When the first two complete METARs of the day are received, a routine delayed TAF is issued.

         1)  After this, the TAF is issued at the normal times.

E.  **ICAO Station Identifier** (element 2).  The TAF uses the ICAO four-letter station identifier as described in Part III, Chapter 2, Aviation Routine Weather Report (METAR), on page 228.

   1.  EXAMPLE:  **KOKC** is the identifier for Will Rogers World Airport in Oklahoma City, OK.

F.  **Date and Time of Origin** (element 3).  This element provides the date and time (UTC) that the forecast is actually prepared.

   1.  The format is a two-digit date and a four-digit time followed by the letter **Z**.

      a.  EXAMPLE:  **051130Z** means the TAF was prepared on the 5th day at 1130 UTC (or Z )

   2.  Routine TAFs are prepared and filed roughly one-half hour prior to the scheduled issuance time.

G.  **Valid Period Date and Time** (element 4)

   1.  The valid period of the forecast is a two-digit date followed by the two-digit beginning hour and two-digit ending hour in UTC.

      a.  EXAMPLE:  **051212** means the TAF is valid from the 5th day at 1200 UTC to 1200 UTC on the 6th.

   2.  Valid periods beginning at 0000 UTC will be indicated as **00**, and valid periods ending at 0000 UTC will be indicated as **24**.

      a.  EXAMPLE:  **110024** means the TAF is valid from the 11th at 0000 UTC to 0000 UTC on the 12th.

   3.  In the case of an amended, corrected, or delayed forecast, the valid period may be less than 24 hr.

   4.  For airports with less than 24-hr. observational coverage for which part-time TAFs are provided, the TAF will be valid until the end of the scheduled forecast even if the observations have ceased before that time.

      a.  **AMD NOT SKED** (amendment not scheduled) or **NIL AMD** (no amendment) will be issued after the forecast information.

      b.  **AMD NOT SKED AFT (closing time) Z** will be used if the times of the observations are known and judged reliable.

         1)  EXAMPLE:  **AMD NOT SKED AFT 0300Z** means amendment not scheduled after 0300 UTC.

c. For the TAFs issued while the location is closed, the word **NIL** (no TAF) will appear in place of the forecast text.

   1) A delayed (RTD) forecast will then be issued for the location after two complete observations are received.

d. **AMD LTD TO CLD VIS AND WIND** (amendment limited to clouds, visibility, and wind) is used at observation sites that have part-time augmentation. This remark means that there will be amendments only for clouds, visibility, and wind.

   1) There will be no amendments for thunderstorms or freezing/frozen precipitation.

H. **Forecast Meteorological Conditions.** This is the body of the TAF.

   1. The wind, visibility, and sky condition elements are always included in the initial time group of the forecast. Weather is included only if it is significant to aviation.

   2. If a significant, lasting change in any of the elements is expected during the valid period, a new time period with the changes is included.

      a. It should be noted that, with the exception of a FM group (see From (FM) Group, item P. on page 264), the new time period will include only those elements expected to change; i.e., if a lowering of the visibility is expected but the wind is expected to remain the same, the new time period reflecting the lower visibility would not include a forecast wind.

         1) The forecast wind would remain the same as in the previous time period.

   3. Temporary conditions expected during a specific time period are included with that period.

I. **Wind** (element 5). This element is a five- or six-digit group with the forecast surface wind direction (first three digits) referenced to true north and the average speed (last two digits or three digits if 100 kt. or greater).

   1. The abbreviation **KT** is appended to denote the use of knots for wind speed.

   2. Wind gusts are denoted by the letter **G** appended to the wind speed, followed by the highest expected gust.

   3. Variable winds are encoded when it is impossible to forecast a wind direction due to winds associated with convective activity or low wind speeds (3 kt. or less).

      a. Variable wind direction is noted by **VRB** where the three-digit direction usually appears.

   4. A calm wind (0 kt.) is forecast as **00000KT**.

   5. EXAMPLES:

      a. **14008KT** means wind from 140° (true) at 8 kt.
      b. **35012G20KT** means wind from 350° (true) at 12 kt. with gusts to 20 kt.
      c. **VRB20KT** means wind direction variable at 20 kt.

J. **Visibility** (element 6)

   1. The expected prevailing visibility is forecast in statute miles with a space and then fractions of statute miles, as needed, with **SM** appended to it.

      a. EXAMPLE: **2 1/2SM** means visibility is forecast to be two and one-half statute miles.

   2. Sector or variable visibility is not forecast, and the visibility group is omitted if missing.

   3. Forecast visibility greater than 6 SM is coded **P6SM**.

   4. If the prevailing visibility is forecast to be 6 SM or less, one or more weather phenomena (e.g., RA, SN, TS, etc.) must be included.

   5. If volcanic ash is forecasted, the visibility must also be forecasted even if it is more than 6 SM.

K. **Significant Weather** (element 7). The expected weather phenomenon or phenomena are coded in TAF reports using the same format, qualifiers, and phenomena contractions as METAR reports, except **UP** (unknown precipitation). See Part III, Chapter 2, Aviation Routine Weather Report (METAR), beginning on page 231.

    1. Obscurations to vision will be forecast whenever the prevailing visibility is forecast to be 6 SM or less.

        a. Precipitation and volcanic ash will be forecast regardless of the visibility forecast.

    2. If no significant weather is expected to occur during a specific time period in the forecast, the weather group is omitted for that time period.

        a. If, after a time period in which significant weather has been forecast, a change to a forecast of no significant weather occurs, the contraction **NSW** (no significant weather) will appear as the weather included in becoming (BECMG) or temporary (TEMPO) groups.

            1) NSW will not be used in the initial time period of a TAF or in from (FM) groups.

    3. If the forecaster determines that there could be weather that impacts aviation in the vicinity of the airport, the forecaster will include those conditions after the weather group.

        a. The letters **VC** describe conditions that will occur within the vicinity of an airport (5 to 10 SM) and will be used only with FG (fog), SH (showers), or TS (thunderstorm).

L. **Sky Conditions** (element 8). TAF sky conditions use the METAR format as described in Part III, Chapter 2, Aviation Routine Weather Report (METAR), beginning on page 233, except that cumulonimbus clouds (CB) are the only cloud type forecast in TAFs.

    1. When the sky is obscured due to a surface-based phenomenon, vertical visibility (VV) into the obscuration is forecast.

        a. Partial obscurations are not forecast.

    2. For aviation purposes, the ceiling is the lowest broken or overcast layer or vertical visibility.

    3. EXAMPLES:

        a. **BKN030** means ceiling 3,000 ft. broken.
        b. **VV008** means indefinite ceiling (vertical visibility) 800 ft.

M. **Wind Shear** (optional data) (element 9). Wind shear is the forecast of nonconvective low-level winds (up to and including 2,000 ft. AGL) and is entered after the sky conditions when wind shear is expected. The element is omitted if wind shear is not expected to occur.

    1. The forecast includes the height of the wind shear followed by the wind direction and wind speed at the indicated height.

    2. Wind shear is encoded with the contraction **WS**, followed by a three-digit height solidus (/), and winds at the height indicated in the same format as surface winds. Height is given in hundreds of feet AGL up to and including 2,000 ft.

    3. EXAMPLE: **WS018/32030KT** means low-level wind shear at 1,800 ft., wind (at 1,800 ft. AGL) 320° true at 30 kt.

N. **Time Elements**

    1. Three forecast change indicators are used when a rapid, a gradual, or a temporary change is expected in some or all of the forecast meteorological conditions.

        a. The forecast change indicators are

            1) From (FM)
            2) Becoming (BECMG)
            3) Temporary (TEMPO)

        b. Each change indicator marks a time group within the TAF.

    2. The probability (PROB) forecast also indicates a time of forecast weather events.

O.  **Temporary (TEMPO) Group** (element 10)

   1.  The **TEMPO** group is used for temporary fluctuations in wind, visibility, weather, or sky condition that are expected to last for generally less than 1 hour at a time (occasional) and to occur during less than half the time period.

   2.  The **TEMPO** indicator is followed by a four-digit group giving the beginning and ending hour of the time period during which the temporary conditions are expected.

   3.  Only the changing forecast meteorological conditions are included in **TEMPO** groups.

      a.  The omitted conditions are carried over from the previous time group.

   4.  EXAMPLE:  **5SM BR BKN030 TEMPO 1316 1 1/2SM BR** means visibility 5 SM in mist, ceiling 3,000 ft. broken with occasional (temporary conditions) visibility 1 1/2 SM in mist between 1300 UTC and 1600 UTC.  Note that the sky condition (BKN030) does not change; i.e., only the visibility is forecast to change.

P.  **From (FM) Group** (element 11)

   1.  The **FM** group is used when a rapid and significant change, usually occurring in less than 1 hour in prevailing conditions, is expected.

      a.  Typically, a rapid change of prevailing conditions to more or less a completely new set of conditions is associated with a synoptic feature (i.e., cold or warm front) passing through the terminal area.

   2.  Appended to the **FM** indicator is the four-digit hour and minute when the change is expected to begin and continue until the next change group or until the end of the current forecast.

   3.  A **FM** group will always mark the beginning of a new line in a TAF report.

   4.  Each **FM** group contains all the required elements, i.e., wind, visibility, weather, sky condition, and wind shear.

      a.  Weather and wind shear will be omitted in **FM** groups when they are not significant to aviation.

      b.  **FM** groups will not include the contraction **NSW**.

   5.  EXAMPLE:  **FM1600 16010KT P6SM SKC** means the forecast for the hours from 1600 UTC to 2200 UTC (beginning of the next forecast group in the example TAF on page 260), wind 160° true at 10 kt., visibility greater than 6 SM, sky clear.

Q.  **Becoming (BECMG) Group** (element 12)

   1.  The **BECMG** group is used when a gradual change in conditions is expected over a longer time period, but no longer than 2 hours.

   2.  Appended to the **BECMG** indicator is a four-digit group with the beginning hour and ending hour of the change period.  The gradual change will occur within this time period.

   3.  Only the changing forecast meteorological conditions are included in **BECMG** groups.

      a.  The omitted conditions are carried over from the previous time group.

   4.  EXAMPLE:  **BECMG 2224 20013G20KT 4SM SHRA OVC020** means that, between 2200Z and 2400Z, conditions are forecast to become wind 200° true at 13 kt. with gusts to 20 kt., visibility 4 SM, moderate rain showers, ceiling 2,000 ft. overcast.

R.  **Probability (PROB) Forecast** (element 13)

   1.  The **PROB** indicates the chance of thunderstorms or other precipitation events occurring, along with associated weather conditions (wind, visibility, and sky conditions).

      a.  A probability forecast will not be used during the first 6 hr. of a TAF.

   2.  Appended to the **PROB** contraction is the probability value.

a. EXAMPLES:

1) **PROB40** means there is a 40-49% probability.
2) **PROB30** means there is a 30-39% probability.

b. If the chance is 50% or greater, it is considered a prevailing weather condition and is included in the weather section or the TEMPO change indicator group.

3. Appended to the **PROB** indicator is a four-digit group giving the beginning and ending hour of the time period during which the precipitation or thunderstorms are expected.

4. EXAMPLE: **PROB40 0006 2SM TSRA OVC008CB** means that, between 0000 UTC and 0600 UTC, there is a chance (40-49% probability) of visibility 2 SM, thunderstorm with moderate rain, ceiling 800 ft. overcast cumulonimbus clouds.

## S. Sample TAFs

```
TAF
KOMA 010530Z 010606 24009KT P6SM SCT050 BKN250 BECMG 1214 BKN040
          OVC250 PROB40 1822 25012G18KT 4SM –TSRA BKN030CB BECMG 2224
          SCT030=
```

| | |
|---|---|
| TAF | Routine terminal aerodrome forecast |
| KOMA | Omaha, NE |
| 010530Z | Prepared on the 1st day at 0530 UTC |
| 010606 | Forecast valid from 0600 UTC on the 1st until 0600 UTC on the 2nd |
| 24009KT P6SM SCT050 BKN250 | Wind 240° at 9 kt., visibility greater than 6 SM, 5,000 ft. scattered, ceiling 25,000 ft. broken |
| BECMG 1214 BKN040 OVC250 | A gradual change to ceiling 4,000 ft. broken, 25,000 ft. overcast between 1200 UTC and 1400 UTC (forecast wind and visibility carried over from previous time group |
| PROB40 1822 25012G18KT 4SM –TSRA BKN030CB | Chance (40-49% probability) between 1800 UTC and 2200 UTC of wind 250° at 12 kt., gusts to 18 kt., visibility 4 SM, thunderstorm, light rain, ceiling 3,000 ft. broken cumulonimbus |
| BECMG 2224 SCT030= | A gradual change to 3,000 ft. scattered between 2200 UTC and 2400 UTC, end of forecast (forecast wind and visibility carried over from previous time group) |

```
TAF
KSPI   251730Z 251818 01015KT P6SM BKN050 PROB40 1819 –RA BECMG
          1921 33009KT SCT050
          FM0200 VRB04KT P6SM SKC PROB30 1115 4SM MIFG=
```

| | |
|---|---|
| TAF | Routine terminal aerodrome forecast |
| KSPI | Springfield, IL |
| 251730Z | Prepared on the 25th day at 1730 UTC |
| 251818 | Valid from 1800 UTC on the 25th to 1800 UTC on the 26th |
| 01015KT P6SM BKN050 | Wind 010° at 15 kt., visibility greater than 6 SM, ceiling 5,000 ft. broken |
| PROB40 1819 –RA | Chance (40-49% probability) of light rain between 1800 UTC and 1900 UTC |
| BECMG 1921 33009KT SCT050 | A gradual change to wind 330° at 9 kt., 5,000 ft. scattered between 1900 UTC and 2100 UTC (visibility carried over from previous time group) |
| FM0200 VRB04KT P6SM SKC | After 0200 UTC, wind variable in direction at 4 kt., visibility greater than 6 SM, sky clear |
| PROB30 1115 4SM MIFG= | Chance (30-39% probability) visibility 4 SM in shallow fog between 1100 UTC and 1500 UTC, end of forecast |

TAF AMD
KOTH 270740Z 270806 VRB04KT 3SM BR SCT003 OVC015 TEMPO 1114 1SM
        —RA OVC005
        FM1400 34014KT 2SM -RA BR SCT005 OVC035 PROB40 1416 –FZRA
        BECMG 1618 34010KT P6SM BKN040 BECMG  2301  00000KT NSW SKC=

| | |
|---|---|
| TAF AMD | Amended terminal aerodrome forecast |
| 270740Z | Prepared on the 27th day at 0740 UTC |
| 270806 | Valid time from 0800 UTC on the 27th to 0600 UTC on the 28th |
| VRB04KT 3SM BR SCT003 OVC015 | Wind variable in direction at 4 kt., visibility 3 SM  in mist, 300 ft. scattered, ceiling 1,500 ft. overcast |
| TEMPO 1114 1SM –RA OVC005 | Occasional (temporary) visibility 1 SM in light  rain, ceiling 500 ft. overcast between 1100 UTC  and 1400 UTC |
| FM1400 34014KT 2SM –RA BR SCT005 OVC035 | From (after) 1400 UTC, wind 340° at 14 kt., visibility 2 SM, light rain, mist, 500 ft. scattered, ceiling 3,500 ft. overcast |
| PROB40 1416 –FZRA | Chance (40-49% probability) between 1400 UTC and 1600 UTC of light freezing rain |
| BECMG 1618 34010KT P6SM BKN040 | A gradual change to wind 340° at 10 kt., visibility greater than 6 SM, ceiling 4,000 ft. broken between 1600 UTC and 1800 UTC |
| BECMG 2301 00000KT NSW SKC= | A gradual change to a calm wind, no significant weather, sky clear (forecast visibility carried over from previous time group); these conditions continuing until 0600 UTC, the end of the forecast period |

TAF AMD
KSEA 161440Z 161512 23012KT P6SM SCT014 OVC020 TEMPO 1518 5SM –DZ BR
        BKN012 BECMG 1820 27008KT BKN022 BECMG 2123 30010KT SCT025
        BECMG 0406 21012KT OVC015=

| | |
|---|---|
| TAF AMD | Amended terminal aerodrome forecast |
| KSEA | Seattle-Tacoma, WA |
| 161440Z | Prepared on the 16th day at 1440 UTC |
| 161512 | Valid from 1500 UTC on the 16th to 1200 UTC on the 17th |
| 23012KT P6SM  SCT014 OVC020 | Wind 230° at 12 kt., visibility greater than 6 SM, 1,400 ft. scattered, ceiling 2,000 ft. overcast |
| TEMPO 1518 5SM –DZ BR BKN012 | Occasional (temporary) visibility 5 SM in light drizzle and mist, ceiling 1,200 ft. broken between 1500 UTC and 1800 UTC |
| BECMG 1820 27008KT BKN022 | A gradual change to wind 270° at 8 kt., ceiling 2,200 ft. broken between 1800 UTC and 2000 UTC (visibility carried over from previous time group) |
| BECMG 2123 30010KT SCT025 | A gradual change to wind 300° at 10 kt., 2,500 ft. scattered between 2100 UTC and 2300 UTC (visibility carried over from previous time group) |
| BECMG 0406 21012KT OVC015= | A gradual change to wind 210° at 12 kt., ceiling 1,600 ft. overcast between 0400 UTC and 0600 UTC (visibility carried over from previous time group), end of forecast |

T.  **International Differences in METAR/TAF**

    1.  **Altimeter Setting**

        a.  In the METAR report, the U.S. reports the altimeter setting in inches of mercury (e.g., A2992), but internationally it will be reported in hectoPascal/millibar (e.g., Q1013).

    2.  **Wind**

        a.  Internationally, wind speed may be reported in knots (KT), kilometers per hour (KMH), or meters per second (MPS).

        b.  Appropriate units are indicated in both METAR and TAF.

        c.  Wind direction variability is reported only for wind speeds of 3 kt. or less internationally, but wind direction variability is reported for wind speeds up to 6 kt. in the U.S.

    3.  **Visibility**

        a.  Internationally, visibility is reported in four digits using meters with the direction of the lowest visibility sector.

            1)  EXAMPLE: **6000SW** means visibility is lowest at 6,000 meters to the southwest.
            2)  International visibility reports also contain a trend code.

                a)  **D** means down.
                b)  **U** means up.
                c)  **N** means no change.
                d)  **V** means variable.

        b.  In the U.S., we use prevailing visibility.

    4.  **Wind Shear**

        a.  Low-level wind shear (below 2,000 ft. AGL) not associated with convective activity will appear in TAFs produced only in the U.S. and Canada.

    5.  **Other**

        a.  The contraction **CAVOK** (ceiling and visibility OK) replaces visibility, weather, and sky conditions on international METARs and TAFs if

            1)  Visibility is 10 kilometers or more.

            2)  No clouds are below 1,500 meters (5,000 ft.) or below the highest minimum sector altitude, whichever is greater.

            3)  There are no cumulonimbus clouds.

            4)  There are no significant weather phenomena.

        b.  International and U.S. military TAFs also contain forecasts of maximum and minimum temperature, icing, and turbulence.

            1)  The U.S. has no requirement to forecast temperatures in the TAF, and the National Weather Service will continue to forecast icing and turbulence in AIRMETs and SIGMETs.

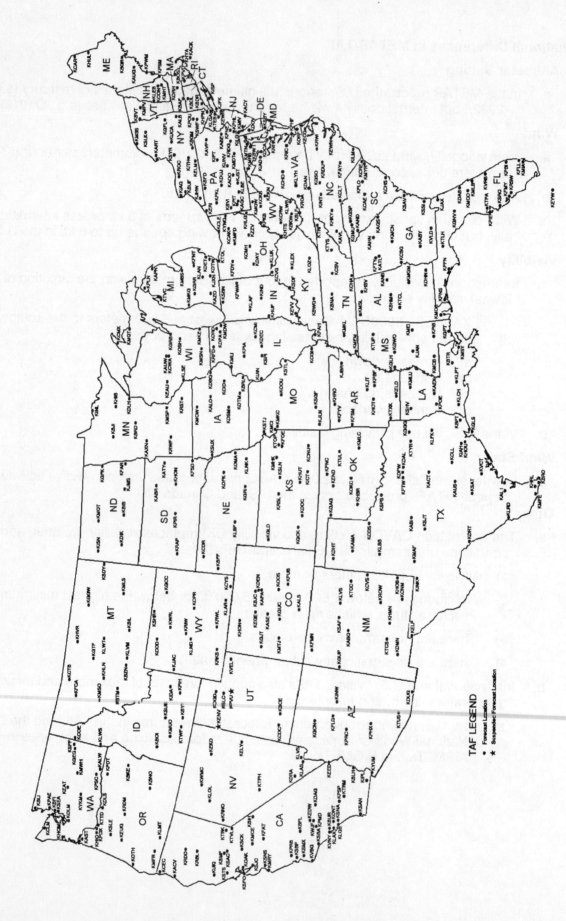

Figure 4-1.  Aviation Terminal Forecast Locations - Contiguous United States

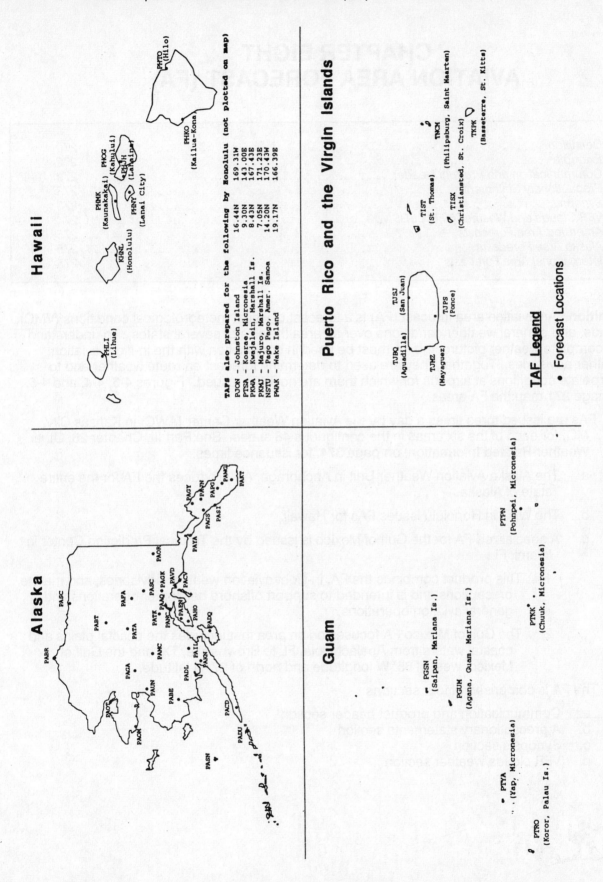

Figure 4-2.  Aviation Terminal Forecast Locations - Alaska, Pacific (Hawaii and Guam), Puerto Rico

**END OF CHAPTER**

# CHAPTER EIGHT
# AVIATION AREA FORECAST (FA)

A. **Definition.** An aviation area forecast (FA) is a forecast of visual meteorological conditions (VMC), clouds, and general weather conditions over an area the size of several states. To understand the complete weather picture, the FA must be used in conjunction with the in-flight aviation weather advisories. Together, they are used to determine forecast en route weather and to interpolate conditions at airports for which there are no TAFs issued. Figures 4-3, 4-4, and 4-5 on page 271 map the FA areas.

1. FAs are issued three times a day by the Aviation Weather Center (AWC) in Kansas City, MO, for each of the six areas in the contiguous 48 states. See Part III, Chapter 26, Other Weather-Related Information, on page 374, for issuance times.

   a. The Alaska Aviation Weather Unit in Anchorage, AK produces the FA for the entire state of Alaska.

   b. The WFO in Honolulu issues FAs for Hawaii.

   c. A specialized FA for the Gulf of Mexico is issued by the Tropical Prediction Center in Miami, FL.

      1) This product combines the FA, in-flight aviation weather advisories, and marine precautions and is intended to support offshore helicopter operations and general aviation operations.

      2) The Gulf of Mexico FA focuses on an area that includes the coastal plains and coastal waters from Apalachicola, FL to Brownsville, TX, and the Gulf of Mexico, west of 85°W longitude and north of 27°N latitude.

2. The FA is comprised of four sections:

   a. Communication and product header section
   b. A precautionary statements section
   c. Synopsis section
   d. VFR clouds/weather section

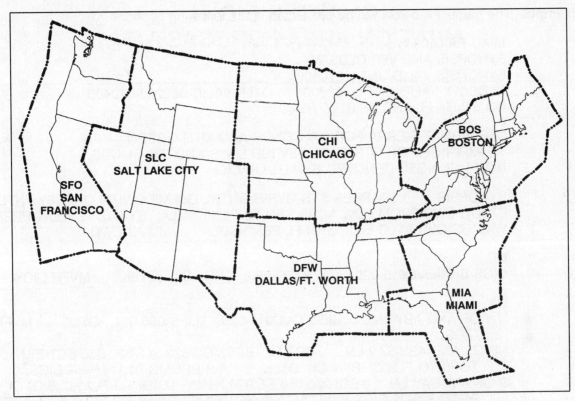

Figure 4-3. FA Locations -- Contiguous United States

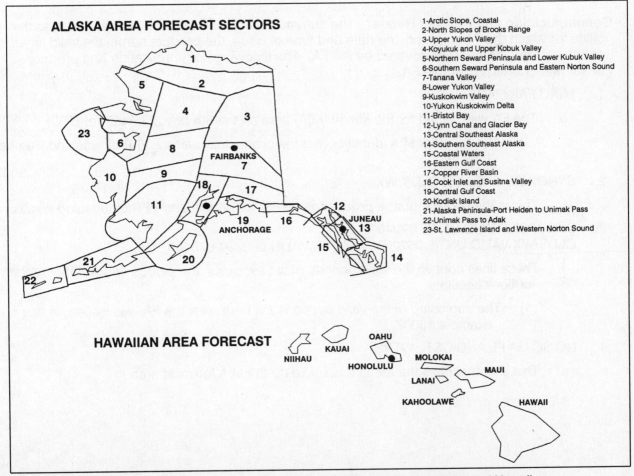

Figures 4-4 and 4-5. Locations of the Area Forecasts in Alaska and Hawaii

B. **Example**. The partial FA below will be discussed in the following sections.

```
MIAC FA 060945
SYNOPSIS AND VFR CLDS/WX
SYNOPSIS VALID UNTIL 070400
CLDS/WX VALID UNTIL 062200 . . . OTLK VALID 062200-070400
NC SC GA FL AND CSTL WTRS
.
SEE AIRMET SIERRA FOR IFR CONDS AND MTN OBSCN.
TS IMPLY SEV OR GTR TURB SEV ICE LLWS AND IFR CONDS.
NON MSL HGTS DENOTED BY AGL OR CIG.
.
SYNOPSIS . . . LOW PRES SYS OVR S CNTRL GLFMEX WITH QSTNRY FNTL SYS
EXTDG EWD FROM LOW ACRS SRN FL PENINSULA.  BY 04Z . . . LOW PRES SYS
WILL BE OVR ATLC E OF SRN FL PENINSULA . . . MOVG EWD.
.
NC
CIGS BKN-OVC015-025.  TOPS 060-080.  CHC –DZ.  OTLK . . . MVFR CIGS.
.
GA
NW OF AND-ABY LN . . . CIGS OVC015-025.  TOPS 060-080.  OTLK . . . MVFR
     CIGS.
NE OF AND-ABY-SAV LN . . . SCT CI.  BECMG 1622 SCT030-050 SCT-BKN100.
     TOPS TO FL200.  BKN CI.  OTLK . . . VFR BECMG 0104 MVFR CIGS.
S OF ABY-SAV LN . . . SCT030-050 SCT-BKN100.  TOPS TO FL200.  BKN CI.
     BECMG 1622 CIGS BKN-OVC030-050.  CHC –RA LWRG VIS 3-5SM.  OTLK . . .
     MVFR CIGS.
```

C. **Communication and Product Header**. The communication and product header identifies the office for which the FA is issued, the date and time of issue, the product name, the valid times, and the states and/or areas covered by the FA. The following communication and product headers explain the example FA.

    1.    MIAC FA 060945

        a.    The FA was issued for the Miami (MIA) area on the 6th day at 0945Z (or UTC).

            1)    The "C" after MIA identifies that the product contains a VFR clouds and weather forecast.

    2.    SYNOPSIS AND VFR CLDS/WX

        a.    This is the plain language product name, i.e., synopsis and VFR clouds and weather.

    3.    SYNOPSIS VALID UNTIL 070400
          CLDS/WX VALID UNTIL 062200 . . . OTLK VALID 062200-070400

        a.    These lines contain the valid periods of the synopsis, the clouds and weather, and the outlook sections.

            1)    The beginning of the valid period is the hour after the FA was issued, in this example 1000Z.

    4.    NC SC GA FL AND CSTL WTRS

        a.    This line describes the areas that make up the MIA forecast area.

D. **Precautionary Statements**. Three statements are in all FAs.

1. SEE AIRMET SIERRA FOR IFR CONDS AND MTN OBSCN.

   a. This statement is to alert you that IFR weather conditions and/or mountain obscurations may be occurring or may be forecast to occur in a portion of the FA area.

      1) You should always check the latest AIRMET Sierra for the FA area.

      2) See Part III, Chapter 10, In-Flight Aviation Weather Advisories, beginning on page 279, for information on AIRMETs.

2. TS IMPLY SEV OR GTR TURB SEV ICE LLWS AND IFR CONDS.

   a. The above statement is included as a reminder of the hazards existing in all thunderstorms.

      1) Thunderstorms imply severe or greater turbulence, severe icing, low-level wind shear, and IFR weather conditions.

   b. Thus, these thunderstorm-associated hazards are not spelled out within the body of the FA.

3. NON MSL HGTS ARE DENOTED BY AGL OR CIG.

   a. The purpose of this statement is to alert you that the heights, for the most part, are above mean sea level (MSL). All heights are in hundreds of feet.

      1) EXAMPLE: BKN030. TOPS 100. HYR TRRN OBSCD means bases of the broken clouds are 3,000 ft. MSL with tops 10,000 ft. MSL. Terrain above 3,000 ft. MSL will be obscured.

   b. The tops of the clouds are **always** MSL.

   c. Heights above ground level (AGL) are noted in either of two ways:

      1) Ceilings by definition are above ground. Therefore, the contraction **CIG** indicates above ground.

         a) EXAMPLE: CIG BKN-OVC015 means that ceilings are expected to be broken to overcast sky cover with bases at 1,500 ft. AGL.

      2) AGL SCT020 means scattered clouds with bases 2,000 ft. AGL.

   d. Thus, if the contraction "AGL" or "CIG" is **not** denoted, height is automatically above MSL.

E. **Synopsis**

1. The synopsis is a brief summary of the location and movements of fronts, pressure systems, and circulation patterns for an 18-hr. period.

   a. References to low ceilings and/or visibilities, strong winds or any other phenomena the forecaster considers useful may also be included.

2. EXAMPLE (from the FA on page 272):

   SYNOPSIS . . . LOW PRES SYS OVR S CNTRL GLFMEX WITH QSTNRY FNTL SYS EXTDG EWD FROM LOW ACRS SRN FL PENINSULA. BY 04Z . . . LOW PRES SYS WILL BE OVR ATLC E OF SRN FL PENINSULA . . . MOVG EWD.

   a. This paragraph states that a low pressure system over the south-central Gulf of Mexico with a quasistationary frontal system extending eastward from the low across the southern Florida (FL) peninsula. By 0400Z, the low pressure system will be over the Atlantic, east of the southern FL peninsula, moving eastward.

F.  **VFR Clouds and Weather (VFR CLDS/WX)**

1.  The VFR CLDS/WX section contains a 12-hr. specific forecast, followed by a 6-hr. (18-hr. in Alaska) categorical outlook giving a total forecast period of 18 hr. (30 hr. in Alaska).

2.  The VFR CLDS/WX section is usually several paragraphs long. The breakdown may be by states or by well-known geographical areas. (See Figure 4-11, Geographical Areas and Terrain Features, on page 281.)

3.  The specific forecast section gives a general description of clouds and weather covering an area greater than 3,000 square miles and significant to VFR flight operations.

    a.  Hazardous weather conditions, i.e., instrument meteorological conditions (IMC), icing conditions, and turbulent conditions, are not included in the FA but are included in the in-flight aviation weather advisories (convective SIGMET, SIGMET, or AIRMET).

4.  Surface visibility and obstructions to vision are included when the forecast visibility is 3 to 5 SM. Precipitation, thunderstorms, and sustained winds of 20 kt. or greater are always included when forecast.

5.  The conditional term OCNL (occasional) is used to describe clouds and visibilities that may affect VFR flights.

    a.  The term OCNL is used when there is a greater-than-50% probability of a phenomenon occurring but the probability is for less than half of the forecast period.

6.  The area coverage terms, ISOL (isolated), WDLY SCT (widely scattered), SCT or AREAS (scattered or areas), and NMRS or WDSPRD (numerous or widespread), are used to indicate the area coverage of thunderstorms or showers.

    a.  ISOL may also be used to describe areas of ceilings or visibilities that are expected to affect areas of less than 3,000 square miles.

    b.  The area coverage terms are defined in Table 4-1 below.

| Terms | Coverage |
|---|---|
| Isolated (ISOL) | Single cells (no percentage) |
| Widely scattered (WDLY SCT) | Less than 25% of area affected |
| Scattered or Areas (SCT or AREAS) | 25 to 54% of area affected |
| Numerous or Widespread (NMRS or WDSPRD) | 55% or more of area affected |

Table 4-1. Area Coverage of Showers and Thunderstorms

7.  EXAMPLE (from the FA on page 272):

    NC
    CIGS BKN-OVC015-025. TOPS 060-080. CHC –DZ. OTLK . . . MVFR CIGS.

    a.  This part of the VFR CLDS/WX section is the forecast for the entire state of North Carolina (NC).

    b.  For NC, the ceilings are expected to be broken to overcast sky cover with bases between 1,500 ft. and 2,500 ft. AGL with tops between 6,000 ft. and 8,000 ft. MSL. Also during this time, there is a chance of light drizzle. The categorical outlook is marginal VFR (MVFR) due to ceiling heights.

8. A categorical outlook, identified by OTLK, is included for each area breakdown. A categorical outlook of IFR and MVFR can be due to ceilings only (CIG), restriction to visibility only (TS, FG, etc.), or a combination of both.

   a. EXAMPLE:  OTLK . . . VFR BCMG MVFR CIG BR AFT 09Z means the weather is expected to be VFR, becoming MVFR due to low ceiling and visibility restricted by mist after 0900Z.

   b. WND is included in the outlook if winds, sustained or gusty, are expected to be 20 kt. or greater.

   c. The criteria for each category is listed below.

      1) IFR -- ceiling less than 1,000 ft. and/or visibility less than 3 SM

      2) MVFR (marginal VFR) -- ceiling 1,000 to 3,000 ft. inclusive and/or visibility 3 to 5 SM inclusive

      3) VFR -- no ceiling or ceiling greater than 3,000 ft. and visibility greater than 5 SM

## G.  Amended Area Forecasts

1. Amendments to the FA are issued as necessary.  An amended FA is identified by **AMD**, which is located on the first line after the date and time.

   a. The entire FA is transmitted again with the letters **UPDT** after the state to indicate what sections have been amended/updated.

2. Area forecasts are also amended and updated by in-flight advisories (AIRMETs, SIGMETs, and convective SIGMETs).

3. A corrected FA is identified by **COR**, and a delayed FA is identified by **RTD**, which are located in the first line after the time and date.

## H.  Alaska Area Forecasts

1. The Alaska FA is in a specialized format.

2. The Alaska FA has a SYNOPSIS section similar to the FAs for the 48 contiguous states and Hawaii.

3. The primary difference is that the forecasts and advisories for each individual geographic area are combined into a single product.

   a. Forecast weather elements are sky condition, cloud height, mountain obscuration, visibility, weather and/or obstructions to visibility, strong surface winds (direction and speed), icing, freezing level, and mountain pass conditions. Hazards and flight precautions, including AIRMETs and SIGMETs, may be found in their respective geographic area.

## I.  International Area Forecasts

1. Area forecasts from the surface to 25,000 ft. are also prepared in the international format for areas in the Atlantic Ocean, the Caribbean Sea, and the Gulf of Mexico.

2. Significant weather forecasts for 25,000 to 60,000 ft. are prepared in chart form and international text format for the northern and western hemisphere.

3. Latitude and longitude coordinates are used to identify the boundary points of the area forecasts and significant weather.

# END OF CHAPTER

# CHAPTER NINE
# TRANSCRIBED WEATHER BROADCASTS (TWEB) TEXT PRODUCTS

A. **Introduction**. The three TWEB text products -- a route forecast, a vicinity forecast, and a synopsis -- are prepared by NWS forecast offices for more than 200 selected routes over the contiguous U.S. (see Figure 4-6 on page 278). Not all NWS forecast offices issue all three of these products.

1. The TWEB products may be used in the TIBS transcriptions described in Part III, Chapter 1, The Aviation Weather Service Program, beginning on page 215.

2. The TWEB products are valid for 12 hr. and are issued four times a day (0200Z, 0800Z, 1400Z, and 2000Z).

3. A TWEB route forecast or vicinity forecast will not be issued if the TAF for that airport has not been issued.

   a. A **NIL TWEB** will be issued instead.

4. Because of their varied accessibility and route formats, TWEB products are important and useful weather information available to the pilot for flight operations.

B. **Route and Vicinity Forecasts.** The route (a 50-NM-wide corridor along the route) and vicinity (the area within a 50-NM radius) forecasts provide specific information about the following:

1. Sustained surface winds (25 kt. or greater)
2. Visibility
3. Weather and obscuration to vision

   a. If visibility of 6 SM or less is forecast, obscurations to vision/weather will be included.
   b. Thunderstorms and volcanic ash will always be included regardless of the visibility.

4. Sky conditions (coverage and ceiling/cloud heights)

   a. Cloud bases can be described in either MSL or AGL (CIGS or BASES), depending on which of the following statements is used:

      1) ALL HGTS MSL XCP CIGS means all heights are MSL except ceilings.
      2) ALL HGTS AGL XCP TOPS means all heights are AGL except tops.

   b. Use of AGL, CIGS, and BASES should be limited to cloud bases within 4,000 ft. AGL.

   c. Cloud tops, referenced to MSL, should also be forecast following the sky cover, when expected to be below 15,000 ft. MSL, using the sky cover contractions FEW, SCT, or BKN.

5. Mountain obscurement
6. Nonconvective low-level wind shear

   a. Nonconvective low-level wind shear is included when the TAF for the airport involved has issued a nonconvective low-level wind shear forecast.

NOTE: Icing and turbulence will NOT be issued.

C. **Synopsis**

1. A TWEB synopsis is a brief description of the weather systems affecting the route during the forecast valid period.

2. The synopsis describes movement of pressure systems, movement of fronts, upper-air disturbances, or air flow.

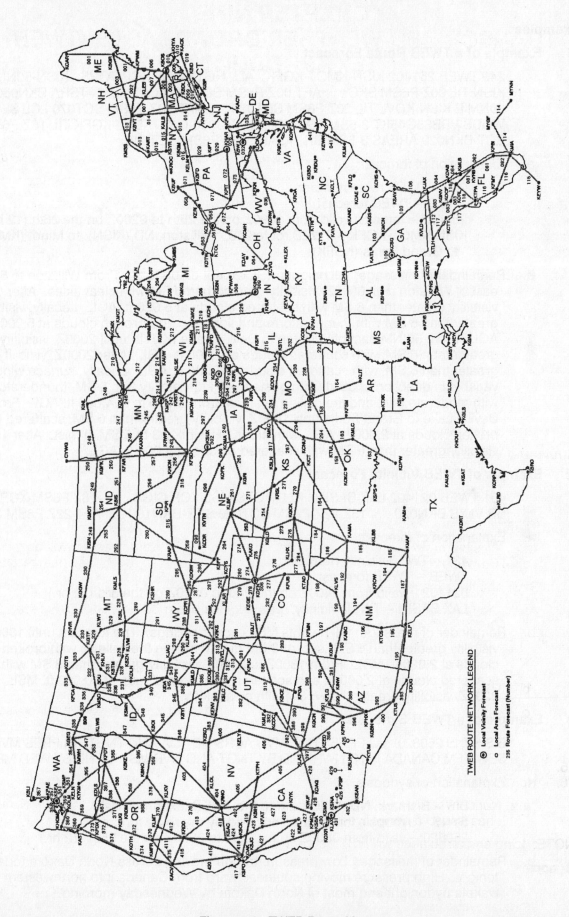

Figure 4-6.  TWEB Route Map

## D.   Examples

### 1.   Example of a TWEB Route Forecast

> 249 TWEB 251402 KISN-KMOT-KGFK.  ALL HGTS AGL XCP TOPS.  KISN-50NM E KISN TIL 00Z P6SM SKC . . . AFT 00Z P6SM SCT050 LCL P6SM –TSRA BKN050. 50NM E KISN-KDVL TIL 20Z P6SM SCT070 . . . AFT 20Z P6SM SCT070 LCL SFC WNDS VRB35G45KT 3-5SM TSRA CIGS OVC030-040.  KDVL-KGFK TIL 16Z P6SM SCT-BKN020 AREAS 3-5SM BR . . . AFT 16Z P6SM SCT040.

a.   Explanation of forecast:

> 249 -- TWEB route number
> TWEB -- TWEB forecast
> 251402 -- forecast valid from 1400Z on the 25th to 0200Z on the 26th (12 hr.)
> KISN-KMOT-KGFK -- TWEB route 249:  Williston, ND (KISN), to Minot (KMOT), to Grand Forks (KGFK)

b.   Remainder of message:  All heights AGL except cloud tops.  From Williston to 50 NM east of Williston until 0000Z, visibility greater than 6 SM with clear skies.  After 0000Z, visibility greater than 6 SM with scattered clouds at 5,000 ft. AGL.  Locally, visibility greater than 6 SM with thunderstorm and light rain and broken clouds at 5,000 ft. AGL.  From 50 NM east of Williston to Devil's Lake (KDVL) until 2000Z, visibility greater than 6 SM, with scattered clouds at 7,000 ft. AGL.  After 2000Z, visibility greater than 6 SM, with scattered clouds at 7,000 ft. AGL.  Locally, surface winds variable in direction at 35 kt. with gusts to 45 kt., visibility 3 to 5 SM, thunderstorm with moderate rain, and the ceiling overcast from 3,000 ft. to 4,000 ft. AGL.  From Devil's Lake to Grand Forks until 1600Z, visibility greater than 6 SM, scattered to broken clouds at 2,000 ft. AGL, with areas of visibility 3 to 5 SM in mist.  After 1600Z, visibility greater than 6 SM with scattered clouds at 4,000 ft. AGL.

### 2.   Example of TWEB Vicinity Forecast

> 431 TWEB 021402 LAX BASIN.  ALL HGTS MSL XCP CIGS.  TIL 18Z P6SM XCP 3SM BR VLYS BKN020 . . . 18Z-22Z P6SM SCT020 SCT-BKN100 . . . AFT 22Z P6SM SKC.

a.   Explanation of forecast:

> 431 -- TWEB vicinity number
> TWEB -- TWEB forecast
> 021402 -- valid from 1400Z on the 2nd to 0200Z on the 3rd (12 hr.)
> LAX BASIN -- TWEB vicinity:  Los Angeles basin

b.   Remainder of message:  All heights MSL except ceilings.  The forecast until 1800Z, visibility greater than 6 SM, except 3 SM due to mist in the valleys, with broken clouds at 2,000 ft. MSL.  From 1800Z to 2200Z, visibility greater than 6 SM with scattered clouds at 2,000 ft. and scattered to broken clouds at 10,000 ft. MSL.  After 2200Z, visibility greater than 6 SM with clear skies.

### 3.   Example of a TWEB Synopsis

> BIS SYNS 250820.  LO PRES TROF MVG ACRS ND TDA AND TNGT.  HI PRES MVG SEWD FM CANADA INTO NWRN ND BY TNGT AND OVR MST OF ND BY WED MRNG.

a.   Explanation of synopsis:

> BIS -- Bismark, ND (WFO issuing the TWEB text products)
> SYNS -- synopsis for the area covered by the route forecast
> 250820 -- valid from 0800Z on the 25th to 2000Z on the 25th (12 hr.)

b.   Remainder of message:  Low pressure trough moving across North Dakota today and tonight.  High pressure moving southeastward from Canada into northwestern North Dakota by tonight and most of North Dakota by Wednesday morning.

# END OF CHAPTER

# CHAPTER TEN
# IN-FLIGHT AVIATION WEATHER ADVISORIES
# (WST, WS, WA)

> Please take a few minutes to study each of the concepts listed above and anticipate/imagine what they are and how they relate to the other listed concepts.

A. **Definition**. In-flight aviation weather advisories are forecasts that advise en route aircraft of the development of potentially hazardous weather.

1. All in-flight advisories in the contiguous 48 states are issued by the Aviation Weather Center (AWC) in Kansas City, MO.

   a. In Alaska, the Alaska Aviation Weather Unit (AAWU) issues in-flight aviation weather advisories.

   b. The weather forecast office (WFO) in Honolulu issues advisories for the Hawaiian Islands.

2. All heights are referenced to MSL, except in the case of ceilings (CIG), which indicate above ground level (AGL).

3. The advisories are of three types -- convective SIGMET (WST), SIGMET (WS), and AIRMET (WA).

   a. All in-flight advisories use the same location identifiers (either VORs, airports, or well-known geographic areas) to describe the hazardous weather areas as shown in Figures 4-12 and 4-11 on pages 280 and 281.

B. **Convective SIGMET (WST)**

1. Convective SIGMETs are issued in the contiguous 48 states (i.e., none for Alaska and Hawaii) for any of the following:

   a. Severe thunderstorm due to

      1) Surface winds greater than or equal to 50 kt.
      2) Hail at the surface greater than or equal to ¾ in. in diameter
      3) Tornadoes

   b. Embedded thunderstorms

   c. A line of thunderstorms

   d. Thunderstorms producing precipitation greater than or equal to heavy precipitation affecting 40% or more of an area of at least 3,000 square mi.

2. Any convective SIGMET implies severe or greater turbulence, severe icing, and low-level wind shear.

   a. A convective SIGMET may be issued for any convective situation that the forecaster feels is hazardous to all categories of aircraft.

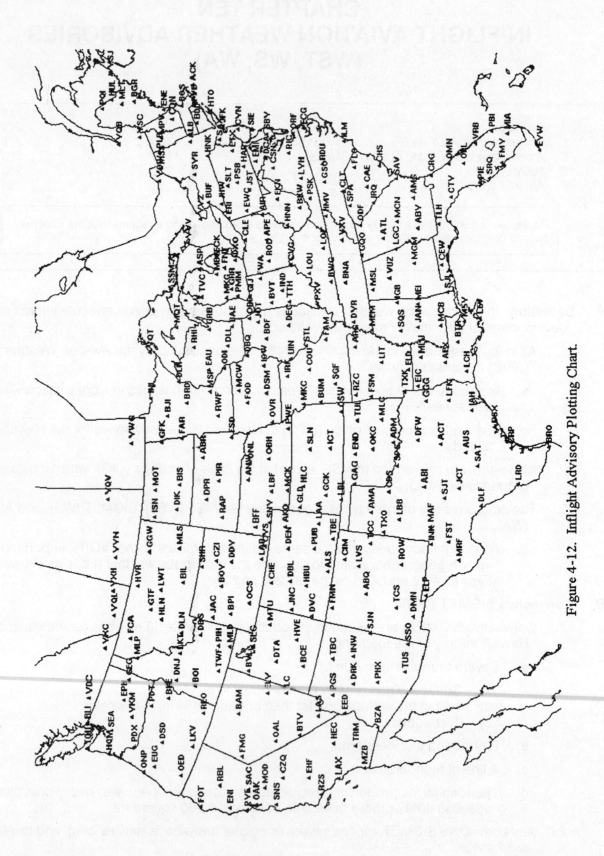

Figure 4-12.  Inflight Advisory Plotting Chart.

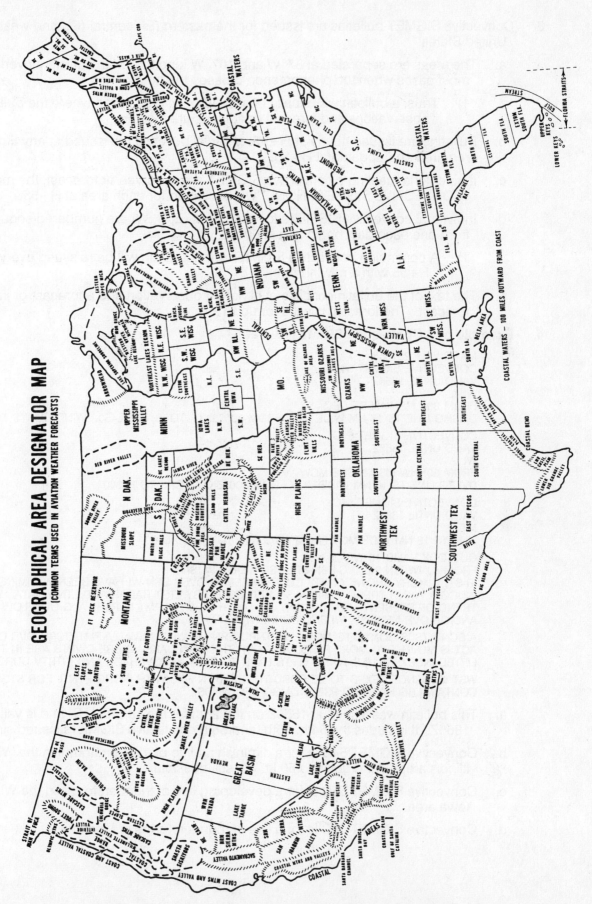

Figure 4-11.  Geographical Areas and Terrain Features

3. Convective SIGMET bulletins are issued for the eastern (E), central (C), and western (W) United States.

   a. The areas are separated at 87°W and 107°W longitude with sufficient overlap to cover most cases when the phenomenon crosses the boundaries.

      1) Thus, a bulletin will usually be issued only for the area where the bulk of observations and forecast conditions are located.

   b. Bulletins are issued hourly at H+55. Special bulletins are issued at any time as required and updated at H+55.

   c. If no criteria meeting a convective SIGMET are observed or forecast, the message "CONVECTIVE SIGMET...NONE" will be issued for each area at H+55.

   d. Individual convective SIGMETs for each area (E, C, W) are numbered sequentially from one each day, beginning at 0000Z.

      1) A continuing convective SIGMET phenomenon will be reissued every hour at H+55 with a new number.

   e. The text of the bulletin consists of either an observation and a forecast or just a forecast. The forecast is valid for up to 2 hr.

4. **Example of a Convective SIGMET Bulletin**

   ```
   MKCC WST 251655
   CONVECTIVE SIGMET 54C
   VALID UNTIL 1855Z
   WI IL
   FROM 30E MSN-40ESE DBQ
   DMSHG LINE TS 15 NM WIDE MOV FROM 30025KT. TOPS TO FL450. WIND GUSTS TO 50 KT POSS.

   CONVECTIVE SIGMET 55C
   VALID UNTIL 1855Z
   WI IA
   FROM 30NNW MSN-30SSE MCW
   DVLPG LINE TS 10 NM WIDE MOV FROM 30015KT. TOPS TO FL300.

   CONVECTIVE SIGMET 56C
   VALID UNTIL 1855Z
   MT
   LINE TS 15 NM WIDE MOV FROM 27020KT. TOPS TO FL380.

   OUTLOOK VALID 151855-252255
   FROM 60NW ISN-INL-TVC-SBN-BRL-FSD-BIL-60NW ISN
   IR STLT IMGRY SHOWS CNVTV CLD TOP TEMPS OVER SRN WI HAVE BEEN WARMING STEADILY
   INDCG A WKNG TREND. THIS ALSO REFLECTED BY LTST RADAR AND LTNG DATA. WKNG
   TREND OF PRESENT LN MAY CONT . . . HWVR NEW DVLPMT IS PSBL ALG OUTFLOW BDRY
   AND/OR OVR NE IA/SW WI BHD CURRENT ACT.

   A SCND TS IS CONTG TO MOV EWD THRU ERN MT WITH NEW DVLPMT OCRG OVR CNTRL ND. MT
   ACT IS MOVG TWD MORE FVRBL AMS OVR THE WRN DAKS WHERE DWPTS ARE IN THE UPR 60S WITH
   LIFTED INDEX VALUES TO MS 6. TS EXPD TO INCR IN COVERAGE AND INTSTY DURG AFTN HRS.

   WST ISSUANCES EXPD TO BE RQRD THRUT AFTN HRS WITH INCRG PTNTL FOR STGR CELLS TO
   CONTAIN LRG HAIL AND PSBLY DMGG SFC WNDS.
   ```

   a. This bulletin was issued at 1655Z on the 25th for the central U.S. and is valid until 1855Z. It includes the 54th, 55th, and 56th convective SIGMETs issued since 0000Z.

   b. Convective SIGMET 54C is for a diminishing line of thunderstorms in the Wisconsin-Illinois area. Wind gusts to 50 kt. are still possible.

   c. Convective SIGMET 55C is for a developing line of thunderstorms in the Wisconsin-Iowa area.

   d. Convective SIGMET 56C is for a line of thunderstorms in Montana.

e.  The outlook is valid from 1855Z to 2255Z on the 25th.

1)  It is a forecast of convective activity in the central U.S.

2)  The outlook forecasts a weakening trend of thunderstorm activity over southern Wisconsin, but new development is possible over northeast Iowa and southwest Wisconsin.

3)  Additionally, thunderstorms are moving into North Dakota, where dewpoints are high and lifted index values (to −6) are favorable for thunderstorm activity.

C.  **Area Covered by SIGMETs and AIRMETs**

1.  SIGMETs and AIRMETs are issued corresponding to the FA areas (Figures 4-3 and 4-4 on page 271).  The maximum forecast period is 4 hr. for SIGMETs and 6 hr. for AIRMETs.

a.  Both advisories are considered "widespread" because they must be either affecting or forecast to affect an area of at least 3,000 square mi. at any one time.

b.  However, if the area to be affected during the forecast period is very large, it could be that only a small portion of the total area would be affected at any one time.

1)  An example would be a 3,000-square-mi. phenomenon forecast to move across an area totaling 25,000 square mi. during a forecast period.

D.  **SIGMET (WS)**

1.  A SIGMET advises of nonconvective weather that is potentially hazardous to all aircraft.

a.  In the conterminous U.S., SIGMETs are issued when the following phenomena occur or are expected to occur:

1)  Severe icing not associated with thunderstorms

2)  Severe or extreme turbulence or clear air turbulence (CAT) not associated with thunderstorms

3)  Duststorms, sandstorms, or volcanic ash lowering surface or in-flight visibilities to below 3 SM

4)  Volcanic eruption

b.  In Alaska and Hawaii, SIGMETs are also issued for

1)  Tornadoes
2)  Lines of thunderstorms
3)  Embedded thunderstorms
4)  Hail greater than or equal to ¾ in. in diameter

2.  SIGMETs are unscheduled products that are valid for 4 hr. unless conditions are associated with a hurricane.  Then the SIGMETs are valid for 6 hr.

a.  Unscheduled updates and corrections are issued as necessary.

3.  A SIGMET is identified by an alphabetic designator from NOVEMBER through YANKEE, excluding SIERRA and TANGO.

a.  The first issuance of a SIGMET is labeled UWS (urgent weather SIGMET), and subsequent issuances are at the forecaster's discretion.

b.  Issuances for the same phenomenon are sequentially numbered, using the original designator until the phenomenon ends.

1)  EXAMPLE:  The first issuance in the CHI FA area for a phenomenon moving from the SLC FA area will be SIGMET PAPA 3, if the previous two issuances, PAPA 1 and PAPA 2, were in the SLC FA area.

2)  Note that no two different phenomena across the country can have the same alphabetic designator at the same time.

4.  **Example of a Domestic SIGMET**

    BOSR WS 050600
    SIGMET ROMEO 2 VALID UNTIL 051000
    ME NH VT
    FROM CAR TO YSJ TO CON TO MPV TO CAR
    MOD TO OCNL SEV TURB BLW 080 EXP DUE TO STG NWLY FLOW.  CONDS CONTG BYD 1000Z.

    a.  This SIGMET bulletin was issued for the BOS FA area at 0600Z on the 5th day of the month and is valid until 1000Z (maximum valid period for a SIGMET is 4 hr.).

    b.  The designator ROMEO 2 means this is the second issuance for this phenomenon (severe turbulence).

    c.  The third line identifies the affected states in the BOS FA area, and the fourth line defines the area.

    d.  The text reads:  Moderate to occasional severe turbulence below 8,000 ft. MSL expected due to strong northwesterly flow.  Conditions continuing beyond 1000Z.

5.  Some NWS offices have been designated by the International Civil Aviation Organization (ICAO) as Meteorological Watch Offices (MWOs).  These offices are responsible for issuing international SIGMETs for designated areas, which include Alaska, Hawaii, portions of the Atlantic and Pacific Oceans, and the Gulf of Mexico.

    a.  The NWS offices that issue international SIGMETs are the AAWU in Anchorage, AK; the Tropical Prediction Center in Miami, FL; the WFO in Honolulu, HI; the Aviation Weather Center in Kansas City, MO; and the WFO on Guam Island.

    b.  The international SIGMET is issued for 12 hr. for volcanic ash events, 6 hr. for hurricanes and tropical storms, and 4 hr. for all other events.

    c.  International SIGMETs are identified by an alphabetic designator from Alpha through Mike and are numbered sequentially until that weather phenomenon ends.

    d.  The criteria for an international SIGMET are

        1)  Thunderstorms occurring in lines, embedded in clouds, or occurring in large areas that produce tornadoes or large hail

        2)  Tropical cyclones

        3)  Severe icing

        4)  Severe or extreme turbulence

        5)  Dust storms and sandstorms lowering visibilities to less than 3 SM

        6)  Volcanic ash

6.  **Example of an International SIGMET**

    ZCZC MIASIGA1L
    TTAA00 KNHC 121600
    KZNY SIGMET LIMA 5 VALID 121600/122000 UTC KNHC-
    ACT TS OBS BY SATELLITE WI AREA BOUNDED BY 30N69W 31N64.6W 26.4N66.4W 27.5N69.4W
    30N69W.  CB TOPS TO FL480.  MOV ENE 15 KT.  INTSF.

    a.  This is SIGMET LIMA 5 issued by the Tropical Prediction Center at 1600 UTC (Z) on the 12th.  It is valid from 1600 to 2000 UTC (Z) on the 12th.

    b.  The text reads:  Active thunderstorms observed by satellite within an area bounded by ... (latitude/longitude coordinates)...Cumulonimbus tops to FL480.  Moving east-northeast at 15 kt.  The storms are intensifying.

E.  **AIRMET (WA)**

1.  AIRMETs are advisories of significant weather phenomena but describe conditions at intensities lower than those requiring SIGMETs to be issued.  AIRMETs are intended for dissemination to all pilots in the preflight and en route phase of flight to enhance safety.

    a.  AIRMET bulletins are issued on a scheduled basis every 6 hr.

        1)  Unscheduled updates and corrections are issued as necessary.

    b.  Each AIRMET bulletin contains

        1)  Any current AIRMETs in effect
        2)  An outlook for conditions expected after the AIRMET valid period

2.  There are three AIRMETs:

    a.  AIRMET Sierra describes

        1)  IFR weather conditions -- ceilings less  than 1,000 ft. and/or visibility less than 3 SM affecting over 50% of the area at one time

        2)  Extensive mountain obscuration

        NOTE:  AIRMET Sierra is referenced in the area forecast.

    b.  AIRMET Tango describes

        1)  Moderate turbulence
        2)  Sustained surface winds of 30 kt. or greater
        3)  Nonconvective low-level wind shear

    c.  AIRMET Zulu describes

        1)  Moderate icing
        2)  Freezing-level heights

3.  After the first issuance each day, scheduled or unscheduled bulletins are numbered sequentially for easier identification.

4.  **Example of an AIRMET Tango**

    ```
    SLCT WA 121345
    AIRMET TANGO UPDT 2 FOR TURB VALID UNTIL 122000.
    AIRMET TURB. . .NV UT CO AZ NM
    FROM LKV TO CHE TO ELP TO 60S TUS TO YUM TO EED TO RNO TO LKV OCNL MOD TURB BLW
    FL180 DUE TO MOD SWLY/WLY WNDS.  CONDS CONTG BYD 20Z THRU 02Z.
    .
    AIRMET TURB. . .NV WA OR CA CSTL WTRS
    FROM BLI TO REO TO BTY TO DAG TO SBA TO 120W FOT TO 120W TOU TO BLI OCNL MOD TURB
    BTWN FL180 AND FL400 DUE TO WNDSHR ASSOCD WITH JTST.  CONDS CONTG BYD 20Z THRU 02Z.
    ....
    ```

    a.  This is the second AIRMET Tango bulletin issued for the SLC FA area on the 12th day at 1345Z.  It is for turbulence and is valid from 1400Z to 2000Z (6 hr.).

        1)  This AIRMET identifies two areas of turbulence.  We will discuss only the first.

    b.  The first area of turbulence affects the states of Nevada, Utah, Colorado, Arizona, and New Mexico within the SLC FA area.

    c.  The text reads:  From Lakeview, OR to Hayden, CO to El Paso, TX to 60 NM south of Tucson, AZ to Yuma, AZ to Needles, CA, to Reno, NV to Lakeview, OR, occasional moderate turbulence below FL180 due to moderate southwesterly/westerly winds. Conditions continuing beyond 2000Z thru 0200Z.

        1)  Notice that Lakeview, OR and Needles, CA are points used to identify the area. However, the states of Oregon and California are not included because they are not in the SLC FA area.

# END OF CHAPTER

# CHAPTER ELEVEN
# WINDS AND TEMPERATURES ALOFT FORECAST (FD)

## A. Type and Time of Forecast

1. Winds and temperatures aloft are forecast for specific locations in the contiguous U.S., as shown in Figure 4-9 on page 287. FD forecasts are also prepared for a network of locations in Alaska and Hawaii, as shown in Figure 4-10 on page 288.

   a. Forecasts are made twice daily based on the 0000Z and 1200Z radiosonde data for use during specific time intervals.

2. Note that charts are available for winds and temperatures aloft for both (1) forecasts and (2) observations. See Section III, Chapter 21, Winds and Temperatures Aloft Charts, beginning on page 337.

## B. Example of an FD Forecast

1. Below is a sample FD message containing a heading and six FD locations. The heading always includes the time during which the FD may be used (1700-2100Z in the example) and a notation "TEMPS NEG ABV 24000," which means that, since temperatures above 24,000 ft. are always negative, the minus sign is omitted.

DATA BASED ON 151200Z

VALID 151800Z FOR USE 1700-2100Z TEMPS NEG ABV 24000

| FT | 3000 | 6000 | 9000 | 12000 | 18000 | 24000 | 30000 | 34000 | 39000 |
|-----|------|--------|--------|--------|--------|--------|--------|--------|--------|
| ALA |      |        | 2420   | 2635–08 | 2535–18 | 2444–30 | 245945 | 246755 | 246862 |
| AMA |      | 2714   | 2725+00 | 2625–04 | 2531–15 | 2542–27 | 265842 | 256352 | 256762 |
| DEN |      |        | 2321–04 | 2532–08 | 2434–19 | 2441–31 | 235347 | 236056 | 236262 |
| HLC |      | 1707–01 | 2113–03 | 2219–07 | 2330–17 | 2435–30 | 244145 | 244854 | 245561 |
| MKC | 0507 | 2006+03 | 2215–01 | 2322–06 | 2338–17 | 2348–29 | 236143 | 237252 | 238160 |
| STL | 2113 | 2325+07 | 2332+02 | 2339–04 | 2356–16 | 2373–27 | 239440 | 730649 | 731960 |

   a. The line labeled "FT" (forecast levels) shows 9 standard FD levels.

   1) The 45,000-ft. and 53,000-ft. levels are electronically transmitted and are available in the communications system.

      a) The pilot may request these levels from the FSS briefer.

   2) Through 12,000 ft., the levels are true altitude, and from 18,000 ft. and above, the levels are pressure altitude.

   b. Note that some lower-level wind groups are omitted.

   1) No winds are forecast within 1,500 ft. of station elevation.

   2) Also, no temperatures are forecast for any level within 2,500 ft. of station elevation.

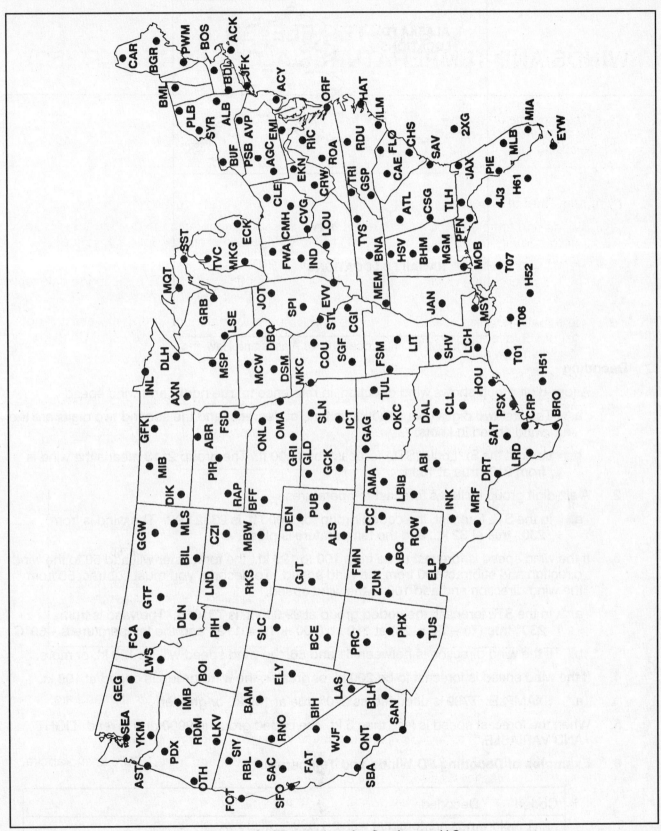

Figure 4-9.  FD Locations for Contiguous U.S.

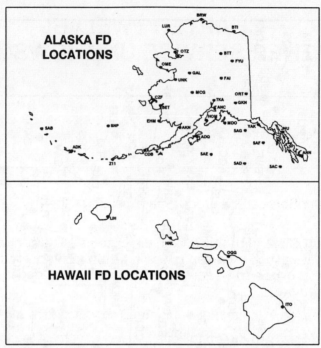

Figure 4-10.  FD Locations for Alaska and Hawaii

## C.  Decoding

1.  A four-digit group shows wind direction, in reference to true north, and wind speed.

    a.  The first two digits give direction in tens of degrees, and the second two digits are the wind speed in knots.

    b.  Look at the St. Louis (STL) forecast for 3,000 ft.  The group 2113 means the wind is from 210° true at 13 kt.

2.  A six-digit group includes forecast temperatures.

    a.  In the STL forecast, the coded group for 9,000 ft. is 2332+02.  The wind is from 230° true at 32 kt., and the temperature is plus 2°C.

3.  If the wind speed is forecast to be from 100 to 199 kt., the forecaster will add 50 to the wind direction and subtract 100 from the wind speed.  To decode, you must subtract 50 from the wind direction and add 100 to the wind speed.

    a.  In the STL forecast, the coded group at 39,000 ft. is 731960.  The wind is from 230° true (73 − 50 = 23) at 119 kt. (100 + 19 = 119), and the temperature is −60°C.

    b.  If the wind direction is between 51 and 86, the wind speed will be 100 kt. or more.

4.  If the wind speed is forecast to be 200 kt. or greater, the wind group is coded at 199 kt.

    a.  EXAMPLE:  7799 is decoded as 270° true at 199 kt. or greater.

5.  When the forecast speed is less than 5 kt., the coded group is 9900 and is read "LIGHT AND VARIABLE."

6.  **Examples of Decoding FD Winds and Temperatures**

| Coded | Decoded |
|---|---|
| 9900+00 | Wind light and variable, temperature 0°C. |
| 2707 | 270° at 7 kt. |
| 850552 | 350° (85 − 50 = 35) at 105 kt. (100 + 05 = 105), temperature −52°C |

# END OF CHAPTER

# CHAPTER TWELVE
# CENTER WEATHER SERVICE UNIT (CWSU) PRODUCTS

A. **Origin**. Center Weather Service Unit products are issued by the CWSU meteorologist located in the ARTCCs.

B. **Meteorological Impact Statement (MIS)** is an unscheduled flow control and flight operations planning forecast of conditions expected to begin beyond 2 hr. but within 12 hr. after issuance. It enables the impact of expected weather conditions to be included in air traffic control decisions in the near future.

   1. An MIS will be issued when the following three conditions are met:

      a. Any one of the following conditions occurs, is forecast to occur, and, if previously forecast, is no longer expected to occur:

         1) Convective SIGMET criteria
         2) Moderate or greater icing and/or turbulence
         3) Heavy or freezing precipitation
         4) Low IFR conditions (ceiling less than 500 ft. and/or visibility less than 1 SM)
         5) Surface winds/gusts 30 kt. or greater
         6) Low-level wind shear within 2,000 ft. of the surface
         7) Volcanic ash, dust, or sandstorm

      b. The impact occurs on air traffic flow within the ARTCC area of responsibility.

      c. The forecast lead time (the time between issuance and onset of a phenomenon), in the forecaster's judgment, is sufficient to make issuance of a Center Weather Advisory (CWA) unnecessary.

   2. **An Example of an MIS**

      ZOA MIS 01 VALID 041415-041900
      ...FOR ATC PLANNING PURPOSES ONLY...
      FOR SFO BAY AREA
      DNS BR/FG WITH CIG BLO 005 AND VIS OCNL BLO 1SM TIL 19Z.

      a. This MIS from the Fremont, CA, ARTCC is the first issuance of the day. It was issued at 1415Z on the 4th and is valid until 1900Z on the 4th. This forecast is for the San Francisco Bay area. The forecast is for dense mist/fog with ceilings below 500 ft. and visibilities occasionally below 1 SM until 1900Z.

C.  **Center Weather Advisory (CWA)** is an advisory for use by air crews to anticipate and avoid adverse weather conditions in the en route and terminal environments.

   1.  The CWA is **not** a flight planning forecast but a **nowcast** for conditions beginning within the next 2 hr.

      a.  Maximum valid time of a CWA is 2 hr. from the time of issuance.

      b.  If conditions are expected to continue beyond the valid period (2 hr.), a statement will be included in the advisory.

   2.  A CWA may be issued for the following three situations:

      a.  As a supplement to an existing in-flight aviation weather advisory for the purpose of improving or updating the definition of the phenomenon in terms of location, movement, extent, or intensity relevant to the ARTCC area of responsibility.  This is important for the following reason:

         1)  A SIGMET for severe turbulence is issued by the AWC, and the outline covers the entire ARTCC area for the total 4-hr. valid time period.  However, the forecaster may issue a CWA covering only a relatively small portion of the ARTCC area at any one time during the 4-hr. period.

      b.  When an in-flight aviation weather advisory has not yet been issued but conditions meet the criteria based on current pilot reports, and the information must be disseminated sooner than the AWC can issue the in-flight aviation weather advisory

         1)  In the case of an impending SIGMET, the CWA will be issued as urgent (UCWA) to allow the fastest possible dissemination.

      c.  When in-flight aviation weather advisory criteria are not met but conditions are or will shortly be adversely affecting the safe flow of air traffic within the ARTCC area of responsibility

   3.  **An Example of a CWA**

```
ZME1 CWA 081300
ZME CWA 101 VALID UNTIL 081500
FROM MEM TO JAN TO LIT TO MEM
AREA SCT VIP 5-6 (INTENSE/EXTREME) TS MOV FROM 26025KT.  TOPS TO
FL450.
```

      a.  This CWA was issued by the Memphis, TN, ARTCC on the 8th at 1300Z and is valid until the 8th at 1500Z.

         1)  The "1" after the ZME in the first line denotes that this CWA has been issued for the first weather phenomenon to occur for the day.

         2)  The "101" in the second line denotes the phenomenon number again (1) and the issuance number (01) for this phenomenon.

      b.  The CWA covers an area from Memphis to Jackson, MS to Little Rock, AR to Memphis.  An area of scattered intense to extreme thunderstorms are moving from 260° at 25 kt.  Tops to flight level (FL) 450 (45,000 ft.).

# END OF CHAPTER

# CHAPTER THIRTEEN
# HURRICANE ADVISORY (WH)

A. **Purpose.** When a hurricane threatens a coastline but is located at least 300 NM offshore, an abbreviated hurricane advisory (WH) is issued to alert aviation interests.

   1. The advisory gives the location of the storm center, its expected movement, and the maximum winds in and near the storm center.

   2. It does not contain details of associated weather, such as specific ceilings, visibilities, weather, and hazards that are found in the FAs, TAFs, and in-flight aviation weather advisories.

B. **Example of an Abbreviated Hurricane Advisory**

   ZCZC MIATCPAT4
   TTAA00 KNHC 190841
   BULLETIN
   HURRICANE DANNY ADVISORY NUMBER 13
   NATIONAL WEATHER SERVICE MIAMI FL
   4 AM CDT SAT JUL 19 1997

   ...DANNY STILL MOVING LITTLE...ANY NORTHWARD DRIFT WOULD BRING THE CENTER ONSHORE...

   HURRICANE WARNINGS ARE IN EFFECT FROM GULFPORT MISSISSIPPI TO APALACHICOLA FLORIDA. SMALL CRAFT SOUTHWEST OF GULFPORT SHOULD REMAIN IN PORT UNTIL WINDS AND SEAS SUBSIDE.

   AT 4 AM CDT...0900Z...THE CENTER OF HURRICANE DANNY WAS LOCATED BY NATIONAL WEATHER SERVICE RADAR AND RECONNAISSANCE AIRCRAFT NEAR LATITUDE 30.2 NORTH...LONGITUDE 88.0 WEST...VERY NEAR THE COAST ABOUT 25 MILES SOUTH-SOUTHEAST OF MOBILE ALABAMA.

   DANNY HAS MOVED LITTLE DURING THE PAST FEW HOURS. WHILE SOME ERRATIC MOTION CAN BE EXPECTED DURING THE NEXT FEW HOURS...A GRADUAL TURN TOWARD THE NORTHEAST IS EXPECTED TODAY. ON THIS COURSE...THE CENTER IS EXPECTED TO MAKE LANDFALL IN THE WARNING AREA TODAY. HOWEVER ANY DEVIATION TO THE NORTH OF THE TRACK WOULD BRING THE CENTER ONSHORE WITHIN THE WARNING AREA AT ANYTIME.

   MAXIMUM SUSTAINED WINDS ARE NEAR 75 MPH WITH HIGHER GUSTS. SOME STRENGTHENING IS STILL POSSIBLE PRIOR TO LANDFALL. DAUPHIN ISLAND RECENTLY REPORTED GUSTS TO 66 MPH AND THE PRESSURE DROPPED TO 989MB...29.20 INCHES.

   DANNY HAS A RELATIVELY SMALL WIND FIELD. HURRICANE FORCE WINDS EXTEND OUTWARD UP TO 25 MPH FROM THE CENTER AND TROPICAL STORM FORCE WINDS EXTEND OUTWARD UP TO 70 MILES.

   LATEST MINIMUM CENTRAL PRESSURE REPORTED BY A RECONNAISSANCE AIRCRAFT WAS 986 MB...29.11 INCHES.

   RADAR SHOWS RAIN BANDS AFFECTING THE AREA FROM SOUTHERN MISSISSIPPI TO THE FLORIDA PANHANDLE. TOTALS OF 10 TO 20 INCHES...LOCALLY HIGHER...COULD OCCUR NEAR THE TRACK OF DANNY DURING THE NEXT FEW DAYS.

   STORM SURGE FLOODING OF 4 TO 5 FEET ABOVE NORMAL TIDES IS POSSIBLE ALONG THE GULF COAST EAST OF THE CENTER.

# END OF CHAPTER

# CHAPTER FOURTEEN
# CONVECTIVE OUTLOOK (AC)

A.  **Purpose**. A convective outlook (AC) is a forecast of severe and nonsevere thunderstorms across the nation.

   1.  The AC is a tool for planning flights later in the day to avoid thunderstorms.

B.  **Forecast**. An AC consists of two forecasts, Day 1 (24 hr.) and Day 2 (next 24 hr.).

   1.  These forecasts describe areas where a high, moderate, or slight risk of severe thunderstorms exists.

      a.  For the definitions of high, moderate, and slight risks, see Part III, Chapter 24, Convective Outlook Chart, on page 359.

   2.  The **severe thunderstorm** criteria are

      a.  Winds equal to or greater than 50 kt. at the surface,
      b.  Hail equal to or greater than ¾ in. in diameter at the surface, or
      c.  Tornadoes.

   3.  Forecast reasoning is also included in all ACs.

   4.  This outlook is also produced as a chart (see Part III, Chapter 24, Convective Outlook Chart, on page 359).

C.  **Schedule**. ACs are produced by the Storm Prediction Center (SPC) located in Norman, OK.

   1.  The issuance times for Day 1 are 0600Z, 1300Z, 1630Z, 2000Z, and 0100Z.
   2.  Day 2 is issued at 0830Z (standard time) or 0730Z (daylight savings time), and 1730Z.

D.  **Example of a Convective Outlook**

   2ND DAY CONVECTIVE OUTLOOK . . . REF AFOS NMCGPH980.

   VALID 121200Z - 131200Z

   NO SVR TSTMS FCST.

   GEN TSTMS ARE FCST TO THE RIGHT OF A LINE FROM OTH 45 ENE MER 20 S MRY.

   ...GENERAL THUNDERSTORM FORECAST DISCUSSION . . .
   NARROW HIGH AMPLITUDE UPPER TROUGH IS FORECAST TO MOVE INTO MID PACIFIC COASTAL REGION LATE IN PERIOD. PROGGED SOUNDINGS SUGGEST THAT SUFFICIENT INSTABILITY FOR LOW TOPPED CONVECTION WILL BE PRESENT AHEAD OF TROUGH. ISOLATED THUNDERSTORMS ARE EXPECTED . . . PRIMARILY AFTER 13/00Z . . . IN AREAS OF INCREASING UVV FROM CENTRAL CA NWD TO THE SWRN OREGON COASTAL REGION.

   . . JOHNS . . 03/11/98

# END OF CHAPTER

# CHAPTER FIFTEEN
# SEVERE WEATHER WATCH BULLETIN (WW)

> Please take a few minutes to study each of the concepts listed above and anticipate/imagine what they are and how they relate to the other listed concepts.

A. **Purpose**. A Severe Weather Watch Bulletin (WW) defines areas of possible severe thunderstorms or tornado activity.

    1. The bulletins are issued by the Storms Prediction Center at Norman, OK.

    2. WWs are unscheduled and are issued as required.

B. **Scope**. A severe thunderstorm watch describes areas of expected severe thunderstorms, and a tornado watch describes areas where the threat of tornadoes exists.

    1. Severe thunderstorm criteria are 3/4-in. hail or larger and/or wind gusts of 50 kt. (58 mph) or greater.

C. **Alert Severe Weather Watch (AWW)**. In order to alert the WFOs, CWSUs, FSSs, and other users, a preliminary message called the Alert Severe Weather Watch (AWW) message is sent before the main bulletin.

    1. **Example of an AWW**

        MKC AWW 011734
        WW 75 TORNADO TX OK AR 011800Z-020000Z
        AXIS..80 STATUTE MILES EAST AND WEST OF A LINE..60ESE DAL/DALLAS TX/ - 30 NW ARG/ WALNUT RIDGE AR/
        ..AVIATION COORDS.. 70NM E/W /58W GGG - 25NW ARG/
        HAIL SURFACE AND ALOFT..1¾ INCHES. WIND GUSTS..70 KNOTS. MAX TOPS TO 450. MEAN WIND VECTOR 24045.

    2. Soon after the preliminary message goes out, the actual watch bulletin itself is issued.

D. **Severe Weather Watch Bulletin Format**

    1. Type of severe weather watch, watch area, valid time period, type of severe weather possible, watch axis, meaning of a watch, and a statement that persons should be on the lookout for severe weather

    2. Other watch information, i.e., references to previous watches

    3. Phenomena, intensities, hail size, wind speeds (knots), maximum CB tops, and estimated cell movement (mean wind vector)

    4. Cause of severe weather

    5. Information on updating ACs (convective outlooks)

E.   **Example of a Severe Weather Watch Bulletin**

BULLETIN - IMMEDIATE BROADCAST REQUESTED
TORNADO WATCH NUMBER 381
STORM PREDICTION CENTER NORMAN OK
556 PM CDT MON JUN 2 1997

THE STORM PREDICTION CENTER HAS ISSUED A TORNADO WATCH FOR PORTIONS OF

NORTHEAST NEW MEXICO
TEXAS PANHANDLE

EFFECTIVE THIS MONDAY NIGHT AND TUESDAY MORNING FROM 630 PM UNTIL
MIDNIGHT CDT.

TORNADOES...HAIL TO 2¾ INCHES IN DIAMETER...THUNDERSTORM WIND GUSTS TO 80
MPH...AND DANGEROUS LIGHTNING ARE POSSIBLE IN THESE AREAS.

THE TORNADO WATCH AREA IS ALONG AND 60 STATUTE MILES NORTH AND SOUTH OF
A LINE FROM 50 MILES SOUTHWEST OF RATON NEW MEXICO TO 50 MILES EAST OF
AMARILLO TEXAS.

REMEMBER...A TORNADO WATCH MEANS CONDITIONS ARE FAVORABLE FOR
TORNADOES AND SEVERE THUNDERSTORMS IN AND CLOSE TO THE WATCH AREA.
PERSONS IN THESE AREAS SHOULD BE ON THE LOOKOUT FOR THREATENING WEATHER
CONDITIONS AND LISTEN FOR LATER STATEMENTS AND POSSIBLE WARNINGS.

OTHER WATCH INFORMATION... CONTINUE... WW 378... WW 379... WW 380

DISCUSSION...THUNDERSTORMS ARE INCREASING OVER NE NM IN MOIST
SOUTHEASTERLY UPSLOPE FLOW.  OUTFLOW BOUNDARY EXTENDS EASTWARD INTO THE
TEXAS PANHANDLE AND EXPECT STORMS TO MOVE ESE ALONG AND NORTH OF THE
BOUNDARY ON THE N EDGE OF THE CAP.  VEERING WINDS WITH HEIGHT ALONG WITH
INCREASING MID LVL FLOW INDICATE A THREAT FOR SUPERCELLS.

AVIATION...TORNADOES AND A FEW SEVERE THUNDERSTORMS WITH HAIL SURFACE AND
ALOFT TO 2¾ INCHES.  EXTREME TURBULENCE AND SURFACE WIND GUSTS TO 70
KNOTS.  A FEW CUMULONIMBI WITH MAXIMUM TOPS TO 550.  MEANS STORM MOTION
VECTOR 28025.

F.   **Status Reports**.  Additional reports are issued as needed to show progress of storms and to
delineate areas no longer under the threat of severe storm activity.

1.   Cancellation bulletins are issued when it becomes evident that no severe weather will
develop or that storms have subsided and are no longer severe.

G.   **Local Warnings**.  When tornadoes or severe thunderstorms develop, the local WSO office will
issue the warnings covering those areas.

# END OF CHAPTER

# CHAPTER SIXTEEN
# SURFACE ANALYSIS CHART

Please take a few minutes to study each of the concepts listed above and anticipate/imagine what they are and how they relate to the other listed concepts.

## A. Introduction

1. The surface analysis chart is a computer-generated chart, with frontal analysis by forecasters at the Hydrometeorological Prediction Center (HPC). The surface analysis chart covers the contiguous 48 states and adjacent areas.

   a. The chart is transmitted every 3 hr.

   b. Figure 5-1 on page 296 is a surface analysis chart, and Figure 5-2 on page 297 illustrates the symbols depicting fronts and pressure centers.

## B. Valid Time.
The chart's valid time corresponds to the time of the plotted observations. A date-time group in Universal Coordinated Time (UTC) tells the user when conditions portrayed on the chart were occurring.

## C. Isobars
are solid lines depicting the sea-level pressure pattern and are usually spaced at 4-mb (millibar), or 4-hPa (hectoPascal) intervals. HectoPascal (hPa) is a metric unit of pressure (1 mb = 1 hPa).

1. When the pressure gradient is weak, dashed isobars are sometimes inserted at 2-mb/hPa intervals to define the pressure pattern more clearly.

2. Each isobar is labeled.

   a. EXAMPLES: 1032 signifies 1032.0 mb/hPa; 1000 signifies 1000.0 mb/hPa; and 992 signifies 992.0 mb/hPa.

## D. Pressure Systems.
The letter "L" denotes a low pressure center, and the letter "H" denotes a high pressure center.

1. The pressure at each center is indicated by a three- or four-digit number, which is the central pressure in millibars or hectoPascals.

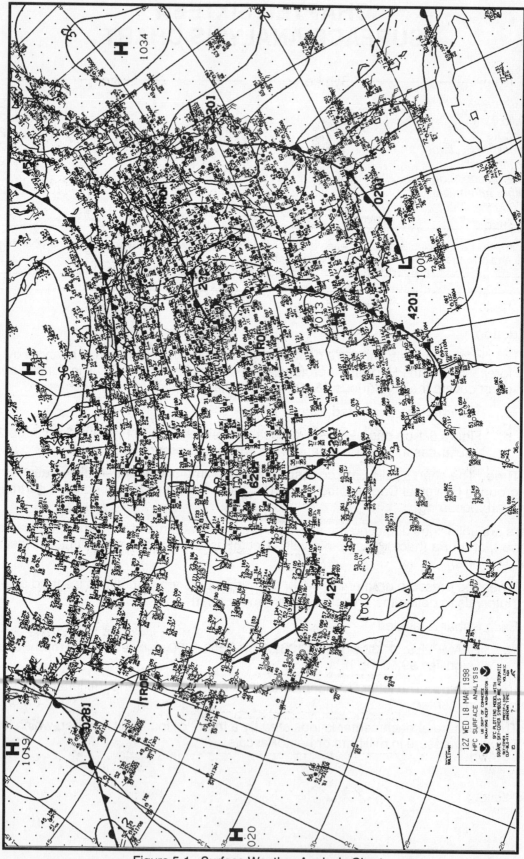

Figure 5-1.  Surface Weather Analysis Chart

| Color | Symbol | Description |
|-------|--------|-------------|
| Blue | **H** | High Pressure Center |
| Red | **L** | Low Pressure Center |
| Blue | | Cold Front |
| Red | | Warm Front |
| Red/Blue | | Stationary Front |
| Purple | | Occluded Front |
| Blue | | Cold Frontogenesis |
| Red | | Warm Frontogenesis |
| Red/Blue | | Stationary Frontogenesis |
| Blue | | Cold Frontolysis |
| Red | | Warm Frontolysis |
| Red/Blue | | Stationary Frontolysis |
| Purple | | Occluded Frontolysis |
| Purple | | Squall Line |
| Brown | | Dryline |
| Brown | | Trough |
| Yellow | | Ridge |

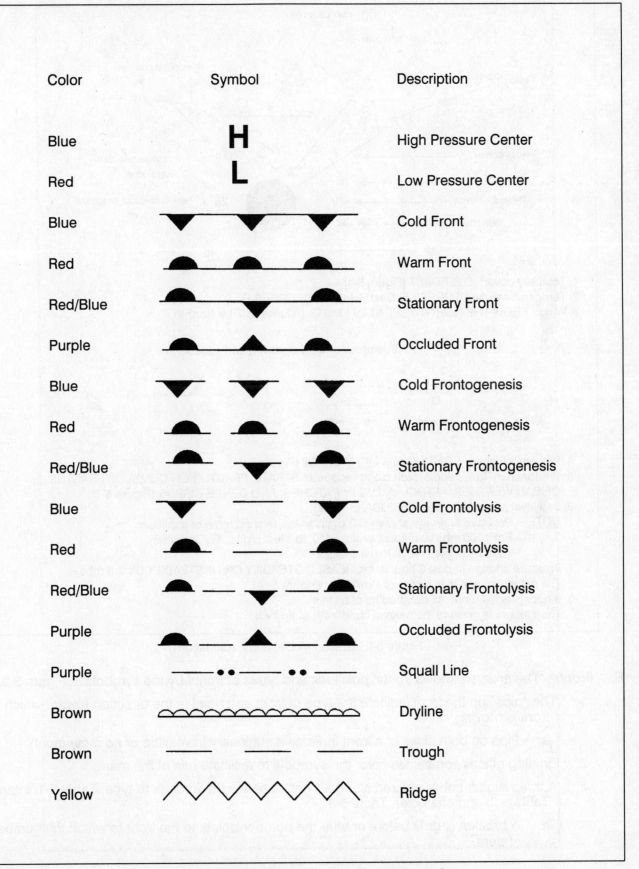

Figure 5-2. Symbols Used on the Surface Analysis

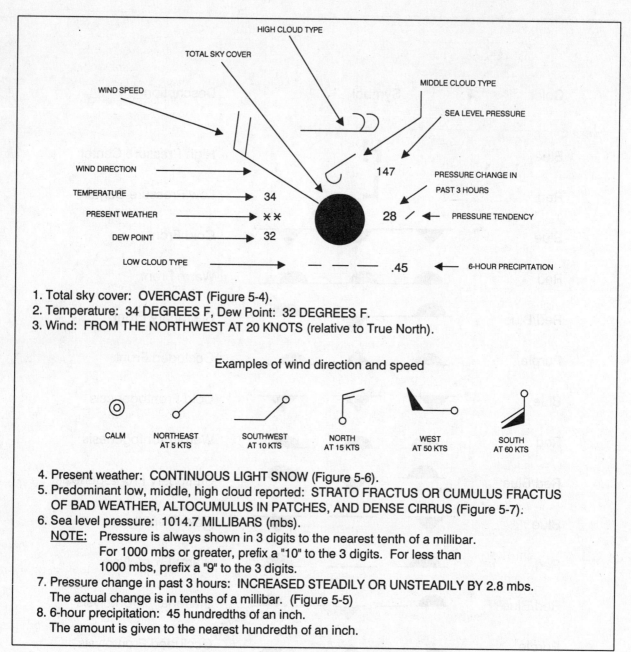

1. Total sky cover:  OVERCAST (Figure 5-4).
2. Temperature:  34 DEGREES F, Dew Point:  32 DEGREES F.
3. Wind:  FROM THE NORTHWEST AT 20 KNOTS (relative to True North).

Examples of wind direction and speed

| CALM | NORTHEAST AT 5 KTS | SOUTHWEST AT 10 KTS | NORTH AT 15 KTS | WEST AT 50 KTS | SOUTH AT 60 KTS |

4. Present weather:  CONTINUOUS LIGHT SNOW (Figure 5-6).
5. Predominant low, middle, high cloud reported:  STRATO FRACTUS OR CUMULUS FRACTUS OF BAD WEATHER, ALTOCUMULUS IN PATCHES, AND DENSE CIRRUS (Figure 5-7).
6. Sea level pressure:  1014.7 MILLIBARS (mbs).
   NOTE:   Pressure is always shown in 3 digits to the nearest tenth of a millibar. For 1000 mbs or greater, prefix a "10" to the 3 digits.  For less than 1000 mbs, prefix a "9" to the 3 digits.
7. Pressure change in past 3 hours:  INCREASED STEADILY OR UNSTEADILY BY 2.8 mbs. The actual change is in tenths of a millibar.  (Figure 5-5)
8. 6-hour precipitation:  45 hundredths of an inch. The amount is given to the nearest hundredth of an inch.

Figure 5-3.  Station Model and Explanation

E.   **Fronts**.  The analysis shows frontal positions and types of fronts by the symbols in Figure 5-2.

   1.   The "pips" on the front indicate the type of front and point in the direction toward which the front is moving.

      a.   Pips on both sides of a front indicate a stationary front (little or no movement).

   2.   Briefing offices sometimes color the symbols to facilitate use of the map.

   3.   A three-digit number entered along a front classifies the front as to type, Table 5-1; intensity, Table 5-2; and character, Table 5-3.

      a.   A bracket ([ or ]) before or after the number points to the front to which the number refers.

| Code Figure | Description |
|:---:|:---|
| 0 | Quasi-stationary at surface |
| 2 | Warm front at surface |
| 4 | Cold front at surface |
| 6 | Occlusion |
| 7 | Instability line |

Table 5-1. Type of Front

| Code Figure | Description |
|:---:|:---|
| 0 | No specification |
| 1 | Weak, decreasing |
| 2 | Weak, little or no change |
| 3 | Weak, increasing |
| 4 | Moderate, decreasing |
| 5 | Moderate, little or no change |
| 6 | Moderate, increasing |
| 7 | Strong, decreasing |
| 8 | Strong, little or no change |
| 9 | Strong, increasing |

Table 5-2. Intensity of Front

| Code Figure | Description |
|:---:|:---|
| 0 | No specification |
| 5 | Forming or existence expected |
| 6 | Quasi-stationary |
| 7 | With waves |
| 8 | Diffuse |

Table 5-3. Character of Front

    b.    EXAMPLE: In Figure 5-1, on page 296, the front extending from the coast of South Carolina across Florida to the Gulf of Mexico is labeled "020." This number means a quasi-stationary front at the surface ("0" in Table 5-1), weak with little or no change ("2" in Table 5-2) and with no specific character ("0" in Table 5-3).

        1)    Pips on both sides also identify this front as a stationary front.

    4.    Two short lines across a front indicate a change in classification.

## F. Trough and Ridge

    1.    A trough of low pressure with significant weather will be depicted as a thick, dashed line running through the center of the trough and identified with the word "TROF."

    2.    The symbol for a ridge of high pressure is very rarely, if at all, depicted (Figure 5-2).

## G. Other Information

    1.    The observations from a number of stations are plotted on the chart to aid in analyzing and interpreting the surface weather features.

        a.    These plotted observations are referred to as **station models**.

    2.    Various types of station models are plotted on the chart.

        a.    A round station symbol represents a location where the observations are taken by observers.

        b.    A square station symbol represents a location where the sky cover was determined by an automated system.

        c.    Other plotted station models that appear over water on the chart are data from ships, buoys, and offshore oil platforms.

    3.    Figure 5-3 on page 298 is an example of a station model that shows where the weather information is plotted.

        a.    Figures 5-4 through 5-7 on pages 300 through 302 help explain the decoding of the station model.

4.  An outflow boundary will be depicted as a thick, dashed line with the word "OUTBNDY."

5.  A dry line will be depicted as a line with unshaded scallops. It will also be identified with the words "DRY LINE."

6.  Each chart has a legend (see Figure 5-1 on page 296) identifying the chart as the surface analysis and giving the date and time of the chart, with additional information regarding the automated observation sites and a volcanic ash symbol.

## H.  Using the Chart

1.  The surface analysis chart provides a ready means of locating pressure systems and fronts.

    a.  It also gives an overview of winds, temperatures, and dew point temperatures at chart time.

2.  When using the chart, keep in mind that weather moves and conditions change.

    a.  The surface analysis chart must be used in conjunction with other information to give a more complete weather picture.

| Symbol | Total sky cover |
|--------|-----------------|
| ○ | Clear (0/8) |
| ◔ | FEW (less than 1/8 to 2/8) |
| ◑ | SCT (3/8 to 4/8) |
| ◕ | BKN (5/8 to 7/8) |
| ◍ | Breaks in overcast |
| ● | OVC (8/8) |
| ⊗ | Total sky obscuration |
| Ⓜ | Missing cloud (or sky cover) observation or partial obscuration |

Figure 5-4.  Sky Cover Symbols

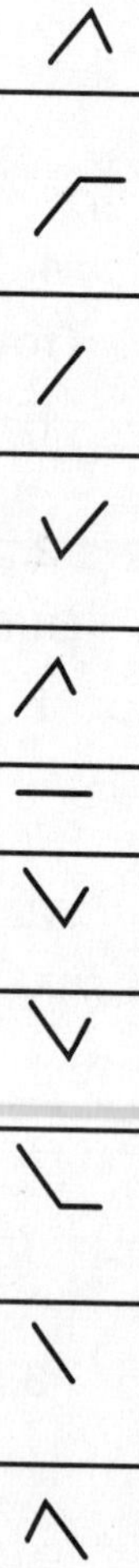

| Description of Characteristic | | Graphic |
|---|---|---|
| Primary Unqualified Requirement | Additional Requirements | |
| **HIGHER**<br>Atmospheric pressure now higher than 3 hours ago | Increasing then decreasing | ∧ |
| | Increasing then steady; or | ⌐ |
| | increasing then increasing more slowly | |
| | Increasing — Steadily | / |
| | Increasing — Unsteadily | |
| | Decreasing or steady then increasing; or | ✓ |
| | increasing then increasing more rapidly | |
| **THE SAME**<br>Atmospheric pressure now same as 3 hours ago | Increasing then decreasing | ∧ |
| | Steady | — |
| | Decreasing then increasing | ∨ |
| **LOWER**<br>Atmospheric pressure now lower than 3 hours ago | Decreasing then increasing | ∨ |
| | Decreasing then steady; or | \ |
| | decreasing then decreasing more slowly | |
| | Decreasing — Steadily | \ |
| | Decreasing — Unsteadily | |
| | Steady or increasing then decreasing; or | ∧ |
| | decreasing then decreasing more rapidly | |

Figure 5-5.  Barometer Tendencies

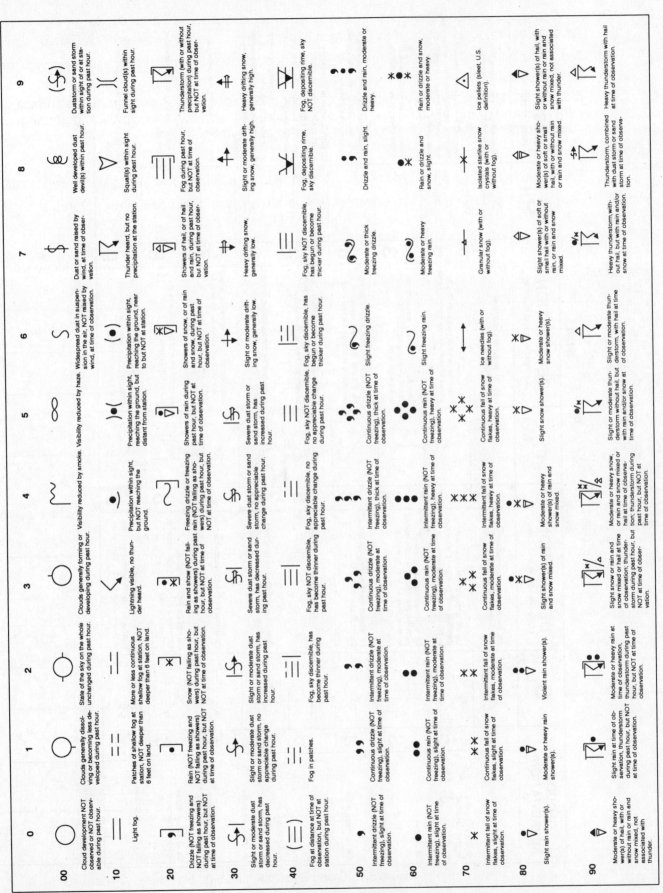

Figure 5-6. Present Weather

| CLOUD ABBREVIATION | $C_L$ | DESCRIPTION (Abridged from W.M.O. Code) | | $C_M$ | DESCRIPTION (Abridged from W.M.O. Code) | | $C_H$ | DESCRIPTION (Abridged from W.M.O. Code) |
|---|---|---|---|---|---|---|---|---|
| St or Fs - Stratus or Fractostratus | 1 |  Cu, fair weather, little vertical development & flattened | 1 | 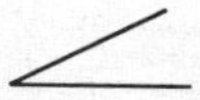 Thin As (most of cloud layer semitransparent) | 1 | Filaments of Cl, or "mares tails," scattered and not increasing |
| Cl - Cirrus | 2 |  Cu, considerable development, towering with or without other Cu or SC bases at same level | 2 |  Thick As, greater part sufficiently dense to hide sun (or moon), or Ns | 2 | 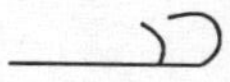 Dense Cl in Patches or twisted sheaves, usually not increasing, sometimes like remains of CB; or towers tufts |
| Cs - Cirrostratus | 3 |  Cb with tops lacking clear-cut outlines, but distinctly not cirriform or anvil shaped; with or without Cu, Sc, St | 3 |  Thin Ac, mostly semi-transparent; cloud elements not changing much at a single level | 3 | 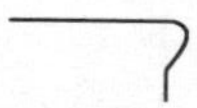 Dense Cl, often anvil-shaped derived from or associated Cb |
| Cc - Cirrocumulus | 4 |  Sc formed by spreading out of Cu; Cu often present also | 4 |  Thin Ac in patches; cloud elements continually changing and/or occurring at more than one level | 4 |  Cl, often hook-shaped gradually spreading over the sky and usually thickening as a whole |
| Ac - Altocumulus | 5 |  Sc not formed by spreading out of Cu | 5 |  Thin Ac in bands or in a layer gradually spreading over sky and usually thickening as a whole | 5 |  Cl and Cs, often in converging bands or Cs alone; generally overspreading and growing denser; the continuous layer not reaching 45 altitude |
| As - Altostratus | 6 |  St or Fs or both, but no Fs of bad weather | 6 | 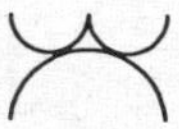 Ac formed by the spreading out of Cu | 6 |  Cl & Cs often in converging bands or Cs alone; generally overspreading and growing denser the continuous layer exceeding 45 altitude |
| Sc - Stratocumulus | 7 |  Fs and/or Fc of bad weather (scud) | 7 |  Double-layered Ac, or a thick layer of Ac, not increasing; or Ac with As and/or Ns | 7 | Veil of Cs covering the entire sky |
| Ns - Nimbostratus | 8 | 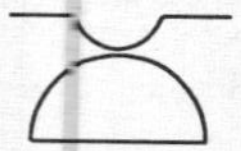 Cu and Sc (not formed by spreading out of Cu) with bases at different levels | 8 |  Ac in the form of Cu-shaped tufts or Ac with turrets | 8 | Cs not increasing and not covering entire sky |
| Cu or Fc - Cumulus or Fractocumulus | 9 |  Cb having a clearly fibrous (cirriform) top, often anvil-shaped, with or without Cu, Sc, ST or scud | 9 |  Ac of a chaotic sky, usually at different levels; patches of dense Cl are usually present | 9 |  Cc alone or Cc with some Cl or Cs but the Cc being the main cirriform cloud |
| Cb - Cumulonimbus | | | | | | |

Figure 5-7.  Cloud Symbols

# END OF CHAPTER

# CHAPTER SEVENTEEN
# WEATHER DEPICTION CHART

> Please take a few minutes to study each of the concepts listed above and anticipate/imagine what they are and how they relate to the other listed concepts.

## A. Introduction

1. The weather depiction chart, Figure 6-3 on page 305, is computer-prepared (with the frontal analysis from a forecaster) from aviation routine weather reports (METARs).

   a. The weather depiction chart gives a broad overview of the observed flying category conditions at the valid time of the chart.

   b. This chart begins at 01Z (0100Z) each day, is transmitted at 3-hr. intervals, and is valid at the time of the plotted data.

## B. Plotted Data.
Data for the chart come from the observations reported by both manual and automated observation stations. The automated stations are denoted by a bracket (]) plotted to the right of the station circle. If the stations on the chart are crowded together, the weather, visibility, and cloud height may be moved up to 90° around the station circle for better legibility. When reports are frequently updated, as at some automated stations (every 20 min.) or when weather changes significantly, the observation used is the latest METAR received instead of using one closest to the stated analysis time. The plotted data for each station includes the following:

1. **Total Sky Cover.** The amount of sky cover is shown by the station circle shaded as in Figure 6-1.

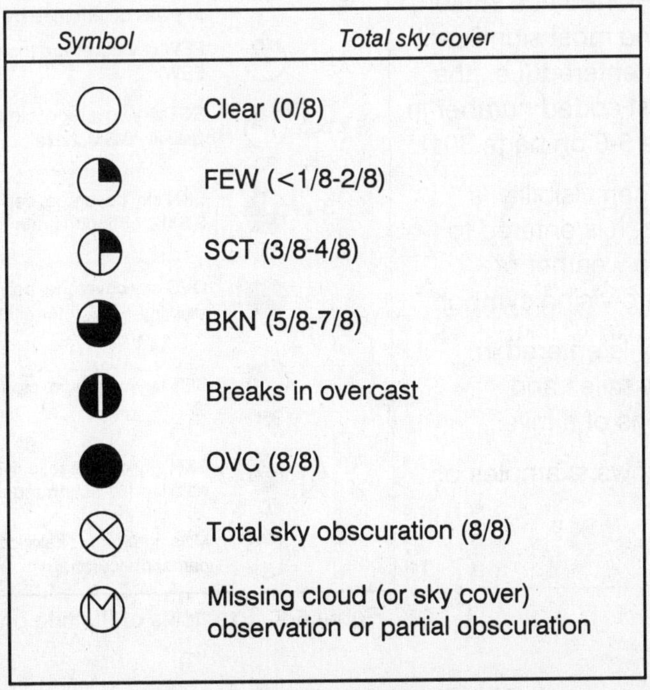

| Symbol | Total sky cover |
|---|---|
| | Clear (0/8) |
| | FEW (<1/8-2/8) |
| | SCT (3/8-4/8) |
| | BKN (5/8-7/8) |
| | Breaks in overcast |
| | OVC (8/8) |
| | Total sky obscuration (8/8) |
| | Missing cloud (or sky cover) observation or partial obscuration |

Figure 6-1. Total Sky Cover

2.  **Cloud Height.** Cloud height above ground level (AGL) is entered under the station circle in hundreds of feet, the same as coded in a METAR report.

    a.  If total sky cover is scattered, the cloud height entered is the base of the lowest layer.

    b.  If total sky cover is broken or greater, the cloud height entered is the ceiling.

    c.  A totally obscured sky is shown by the sky cover symbol "X" and is accompanied by the height entry of the obscuration (vertical visibility into the obscuration).

    d.  A partially obscured sky without a cloud layer above, however, is not recognized by the computer program reading the METAR report, and it is interpreted it as if it were a missing observation.

        1)  Thus, the sky cover symbol "M" is used for either a missing observation or a partial obscuration without a cloud layer above.

        2)  Consequently, to obtain the most accurate information, you must consult the METAR for that specific station.

    e.  A partially obscured sky with clouds above will have a cloud height entry for the cloud layer, but there will be no entry to indicate a partial obscuration at the surface.

        1)  Once again, the METAR would have the most accurate information.

3.  **Weather and Obstructions to Visibility**. Weather and obstructions to visibility symbols are entered to the left of the station circle. Figure 5-6 on page 301 explains most of the symbols used.

    a.  When clouds topping ridges are reported in the remarks section of a METAR, a symbol unique to the weather depiction chart is entered to the left of the station circle: ▲

    b.  When several types of weather and/or obstructions to visibility are reported at a station, only the most significant one is entered (i.e., the highest coded number in Figure 5-6 on page 301).

4.  **Visibility**. When visibility is 5 SM or less, it is entered to the left of the weather or obstruction-to-vision symbol.

    a.  Visibility is entered in statute miles and fractions of a mile.

5.  Figure 6-2 shows examples of plotted data.

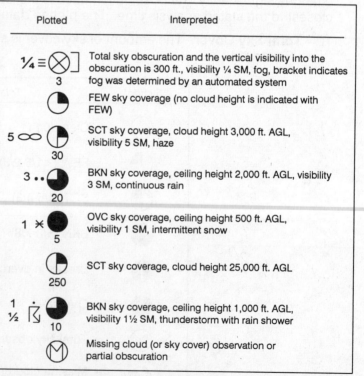

| Plotted | Interpreted |
| --- | --- |
| ¼ ≡ ⊗ ] / 3 | Total sky obscuration and the vertical visibility into the obscuration is 300 ft., visibility ¼ SM, fog, bracket indicates fog was determined by an automated system |
| ◕ | FEW sky coverage (no cloud height is indicated with FEW) |
| 5 ∞ ◔ / 30 | SCT sky coverage, cloud height 3,000 ft. AGL, visibility 5 SM, haze |
| 3 .. ◕ / 20 | BKN sky coverage, ceiling height 2,000 ft. AGL, visibility 3 SM, continuous rain |
| 1 ✳ ● / 5 | OVC sky coverage, ceiling height 500 ft. AGL, visibility 1 SM, intermittent snow |
| ◔ / 250 | SCT sky coverage, cloud height 25,000 ft. AGL |
| 1 / ½ ⚡ ◕ / 10 | BKN sky coverage, ceiling height 1,000 ft. AGL, visibility 1½ SM, thunderstorm with rain shower |
| Ⓜ | Missing cloud (or sky cover) observation or partial obscuration |

Figure 6-2. Examples of Plotting on the Weather Depiction Chart

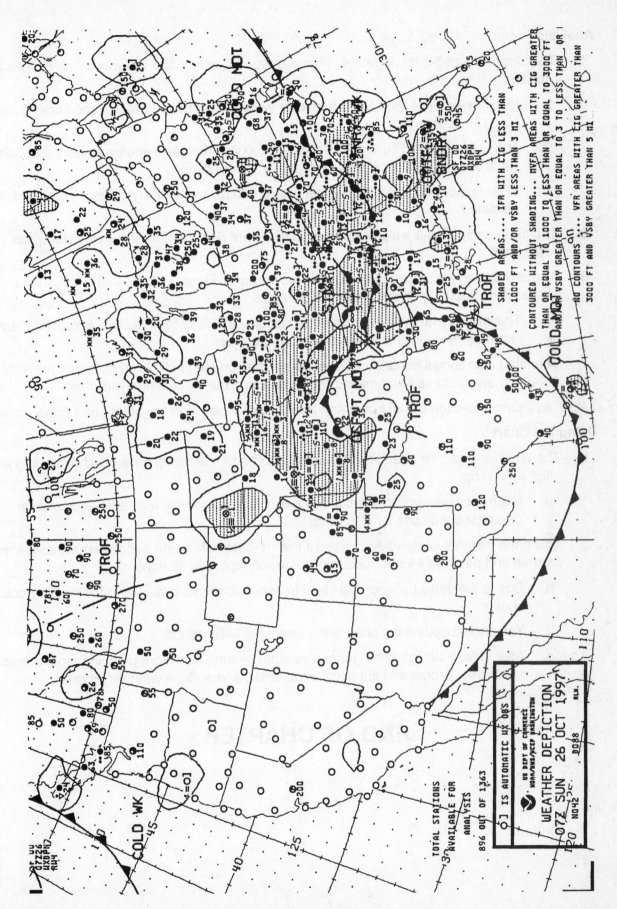

Figure 6-3. Weather Depiction Chart

C.  **Analysis**

1.  The chart (Figure 6-3 on page 305) shows observed ceiling and visibility by categories as follows:

    a.  IFR -- ceiling less than 1,000 ft. and/or visibility less than 3 SM; hatched area outlined by a smooth line

    b.  MVFR (marginal VFR) -- ceiling 1,000 to 3,000 ft. inclusive and/or visibility 3 to 5 SM inclusive; non-hatched area outlined by a smooth line

    c.  VFR -- no ceiling or ceiling greater than 3,000 ft., and visibility greater than 5 SM; not outlined

2.  The three categories are also explained in the lower right portion of the chart for quick reference.

3.  Above the chart identification box (lower left corner of Figure 6-3), a note states, "Total stations available for analysis 896 out of 1363."

    a.  This means that the total number of stations analyzed for this chart is greater than the number of stations actually plotted on the chart.

4.  The chart also shows fronts and troughs from the surface analysis chart for the preceding hour.  These features are depicted the same as on the surface analysis chart.

    a.  One exception is that fronts and troughs are omitted on the 1000Z and 2300Z charts.

D.  **Using the Chart**

1.  The weather depiction chart is an ideal place to begin preparing for a weather briefing and flight planning.

    a.  From this chart, you can get a bird's-eye view of areas of favorable and adverse weather conditions for chart time.

2.  This chart may not completely represent the en route conditions because of variations in terrain and possible weather occurring between reporting stations.

    a.  Due to the delay between data and transmission time, changes in the weather could occur.

    b.  You should update the chart with current METAR reports.

    c.  After initially sizing up the general weather picture, final flight planning must consider forecasts; progs; and the latest pilot, radar, and surface weather reports.

# END OF CHAPTER

# CHAPTER EIGHTEEN
# RADAR SUMMARY CHART

Please take a few minutes to study each of the concepts listed above and anticipate/imagine what they are and how they relate to the other listed concepts.

## A. Introduction

1. A radar summary chart, Figure 7-1 on page 308, is a computer-generated graphical display of a collection of automated radar weather reports (SDs).

2. The radar summary chart displays areas of precipitation as well as information about type, intensity, configuration, coverage, echo top, and cell movement of precipitation.

   a. Severe weather watches are plotted if they are in effect when the chart is valid.

3. The radar summary chart is available hourly with a valid time of H+35 (35 min. past each hour).

4. Figure 7-2, on page 309, depicts the WSR-88D radar network from which the radar summary chart is developed.

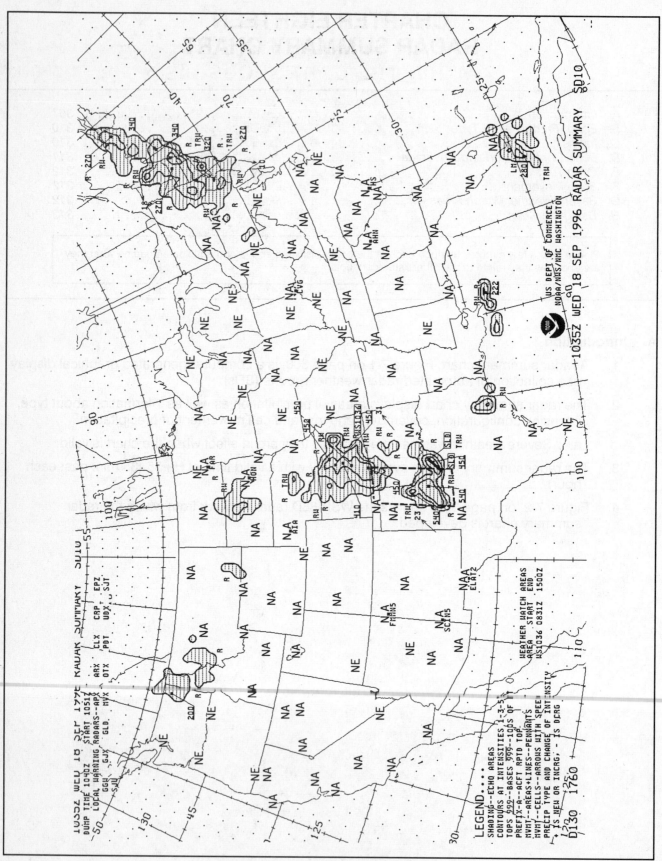

Figure 7-1.  Radar Summary Chart

# COMPLETED WSR-88D INSTALLATIONS WITHIN THE CONTIGUOUS U.S.

OPERATIONAL SUPPORT FACILITY
NORMAN, OKLAHOMA

# COMPLETED WSR-88D INSTALLATIONS

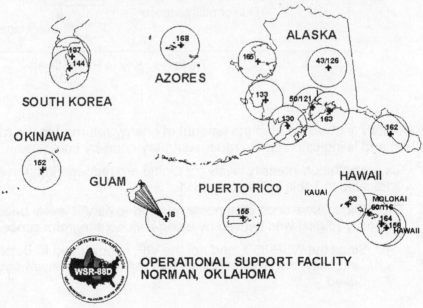

OPERATIONAL SUPPORT FACILITY
NORMAN, OKLAHOMA

Figure 7-2.  WSR-88D Radar Network

B. **Echo (Precipitation) Type**

1. The types of precipitation are indicated by symbols adjacent to the depicted precipitation areas on the radar summary chart.

2. The left column of Table 7-1 below lists the symbols used to denote types of precipitation. Note that these symbols (code) do not reflect the METAR code.

   a. Since the input data for the radar summary chart are the automated SDs, the type of precipitation is determined by computer models and is limited to the ones listed in Table 7-1.

---

## Symbols Used on Chart

| Symbol | Meaning | Symbol | Meaning |
|--------|---------|--------|---------|
| R | Rain | ↗ 35 | Cell movement to the northeast at 35 knots |
| RW | Rain shower | | |
| S | Snow | LM | Little movement |
| SW | Snow shower | WS999 | Severe thunderstorm watch number 999 |
| T | Thunderstorm | | |
| NA | Not available | WT210 | Tornado watch number 210 |
| NE | No echoes | SLD | 8/10 or greater coverage in a line |
| OM | Out for maintenance | ╱ | Line of echoes |

Table 7-1. Key to Radar Chart

---

C. **Intensity**

1. The intensity is obtained from the amount of energy returned to the radar from targets (DBZs) and is indicated on the radar summary chart by contours.

   a. Six precipitation intensity levels (or digits) are reduced into three contour intervals as indicated in Table 7-2 on page 311.

   b. These digits were once commonly referred to as VIP levels because precipitation intensity (digits) was derived by using a video integrator processor (VIP).

      1) Since the WSR-88D, and not the VIP, is now used to determine precipitation intensity, the NWS suggests that the term "VIP level" should no longer be used.

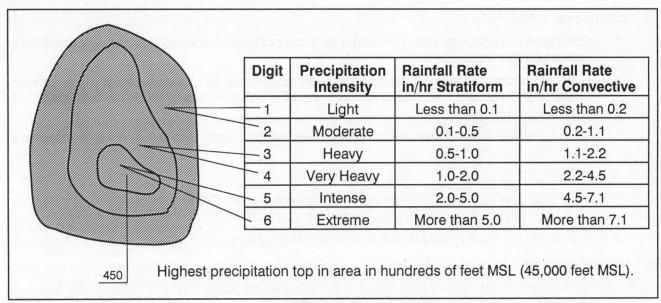

| Digit | Precipitation Intensity | Rainfall Rate in/hr Stratiform | Rainfall Rate in/hr Convective |
|---|---|---|---|
| 1 | Light | Less than 0.1 | Less than 0.2 |
| 2 | Moderate | 0.1-0.5 | 0.2-1.1 |
| 3 | Heavy | 0.5-1.0 | 1.1-2.2 |
| 4 | Very Heavy | 1.0-2.0 | 2.2-4.5 |
| 5 | Intense | 2.0-5.0 | 4.5-7.1 |
| 6 | Extreme | More than 5.0 | More than 7.1 |

450 | Highest precipitation top in area in hundreds of feet MSL (45,000 feet MSL).

Table 7-2. Precipitation Intensities

2. In Figure 7-1, on page 308, the area over southeast Montana and extreme north-central Wyoming has an area of precipitation depicted by one contour.

    a. The area would have light to possibly moderate precipitation.

    b. Whether there is moderate precipitation in the area cannot be determined.

        1) What can be said is that the maximum intensity is definitely below heavy.

3. When determining intensity levels from the radar summary chart, it is recommended that the maximum possible intensity for that contour be used.

4. Intensity can now be coded for snow and snow showers.

    a. This is possible because the WSR-88D is much more powerful and sensitive than previous radars.

5. Finally, it is very important to remember that the intensity trend (increasing or weakening) is no longer coded on either the radar summary chart or the radar weather report (SD).

## D. Echo Configuration and Coverage

1. The configuration is the arrangement of echoes. There are three designated arrangements: a **line** of echoes, an **area** of echoes, and an isolated **cell**.

    a. See Part III, Chapter 4, Radar Weather Report, on page 254 for the definitions of line, area, and cell.

2. Coverage is simply the area covered by echoes. All of the hatched area inside of the contours on the chart is considered to be covered by echoes.

3. When the echoes are reported as a line, a line will be drawn through them on the chart.

    a. If there is 8/10 coverage or more, the line is labeled as solid (SLD) at both ends.

        1) In the absence of this label, it can be assumed that there is less than 8/10 coverage.

    b. EXAMPLE: In Figure 7-1, on page 308, there is a solid line of thunderstorms with intense to extreme rain showers over north-central Texas.

E.   **Echo Tops**

1.   Echo tops are obtained from both radar and, on occasion, satellite data and displayed for **precipitation** tops.

2.   Echo tops are the maximum heights of the precipitation in hundreds of feet MSL, and these heights should be considered only as approximations because of radar wave propagation limitations.

   a.   Tops are entered above a short line, with the top height displayed being the highest in the indicated area.

3.   EXAMPLES (See Figure 7-1 on page 308):

   a.   200 over northeast Washington means the top is at 20,000 ft. MSL.
   b.   540 over north-central Texas means the top is at 54,000 ft. MSL.

4.   It is assumed that all precipitation is reaching the surface.

F.   **Echo Movement**

1.   Individual cell direction of movement is indicated by an arrow with the speed in knots entered as a number at the top of the arrowhead.

   a.   Little movement is identified by "LM."

2.   EXAMPLES (See Figure 7-1 on page 308):

   a.   The cells in the area of rain showers/rain over central Oklahoma are moving northeast at 31 kt.

   b.   The cells in the area of thunderstorms and rain showers over north-central Texas are moving northeast at 23 kt.

3.   Line or area movement is no longer indicated on the radar summary chart.

G.   **Severe Weather Watch Areas**

1.   Severe weather watch areas are outlined by heavy dashed lines, usually in the form of a large rectangular box.

   a.   There are two types, tornado watches and severe thunderstorm watches.

2.   The type of watch and the watch number are enclosed in a small rectangle (see Table 7-1 on page 310) and positioned as closely as possible to the northeast corner of the watch box.

   a.   Weather watch areas are also listed by type and number at the bottom of the chart (over Mexico) together with the issuance time and "valid until" time.

      1)   The absence of weather watch areas will be indicated by the word "NONE."

3.   EXAMPLE (see Figure 7-1 on page 308):  The boxed symbol WS1036 in southeast Kansas means that the watch area in northern Texas and western Oklahoma (marked by heavy dashed lines) is a severe thunderstorm watch and is the 1,036th severe thunderstorm watch issued so far in the year.

   a.   Over Mexico, the weather watch area is listed as WS1036.  The issuance (start) time was at 0831Z, and the severe thunderstorm watch is valid until (end) 1500Z.

## H.  **Using the Chart**

1.  The radar summary chart helps preflight planning by identifying general areas and movement of precipitation and/or thunderstorms.

    a.  Radar detects **only** drops or ice particles of precipitation size; it **does not** detect clouds and fog.

    b.  Thus, the absence of echoes does not guarantee clear weather, and cloud tops will most likely be higher than the precipitation tops detected by radar.

    c.  The chart must be used in conjunction with other charts, reports, and forecasts.

2.  Examine chart notations carefully.  Always determine location and movement of echoes.

    a.  If echoes are anticipated near your planned route, take special note of echo intensity.

    b.  Be sure to examine the chart for missing radar reports (e.g., NA, OM) before assuming that no echoes are present.

        1)  EXAMPLE:  In Figure 7-1, the Springfield (SGF) radar report in southwest Missouri is shown as not available (NA).  There could be echoes in southwest Missouri that are too far away to be detected by the other surrounding radars.

3.  Suppose your planned flight route goes through an area of widely scattered thunderstorms in which no increase is anticipated.  If these storms are separated by good VFR weather, they can be visually detected and circumnavigated.

    a.  However, widespread cloudiness may conceal the thunderstorms.

    b.  To avoid these embedded thunderstorms, you should either use airborne radar or detour the area.

    c.  Details on avoiding hazards of thunderstorms are given in Part I, Chapter 11, Thunderstorms, beginning on page 115.

4.  Remember that the radar summary chart is for preflight planning only and should be updated by current WSR-88D images.

    a.  Once airborne, you must avoid individual storms by in-flight observations either by visual detection, by using airborne radar, or by requesting radar echo information from FSS Flight Watch.

    b.  FSS Flight Watch has access to current WSR-88D imagery.

5.  There can be an interpretation problem concerning an area of precipitation that is reported by more than one radar site.

    a.  EXAMPLE:  Station A may report RW with cell movement toward the northeast at 10 kt.  For the same area, station B may be reporting TRW with cell movement toward the northeast at 30 kt.

    b.  This difference in reports may be due to a different perspective and distance of the radar site from the area of echoes.  The area may be moving away from station A and approaching station B.

        1)  The general rule is to use the plotted data associated with the area that presents the greatest hazard to aviation.  In this case, the station B report would be used.

# END OF CHAPTER

# CHAPTER NINETEEN
# CONSTANT PRESSURE ANALYSIS CHARTS

Please take a few minutes to study each of the concepts listed above and anticipate/imagine what they are and how they relate to the other listed concepts.

## A. Type and Time of Report

1. Weather information for computer-generated constant pressure charts is observed primarily by balloon-ascending radiosonde packages.

   a. Each package consists of weather instruments and a radio transmitter.

   b. During ascent, instrument data are continuously transmitted to the observation station. The data are used to prepare constant pressure charts twice a day.

   c. Radiosondes are released at selected observation sites across the U.S. at 0000Z and 1200Z.

2. Any surface of equal pressure in the atmosphere is a constant pressure surface. A constant pressure analysis chart is an upper-air weather map on which all the information depicted is at the specified pressure of the chart.

3. Constant pressure charts are prepared for the following selected pressure levels: 850 mb/hPa, 700 mb/hPa, 500 mb/hPa, 300 mb/hPa, 250 mb/hPa, and 200 mb/hPa.

   a. Charts with higher pressure present information at lower altitudes, and charts with lower pressure present information at higher altitudes.

   b. These charts are valid at the time of observations (0000Z and 1200Z).

4. Pressure altitude (height in the standard atmosphere) for each of the six pressure surfaces is shown in Table 8-1 on page 315.

   a. For example, 700 mb/hPa of pressure has a pressure altitude of 10,000 ft. in the standard atmosphere.

      1) In the real atmosphere, 700 mb/hPa of pressure only approximates 10,000 ft. because the real atmosphere is seldom standard.

   b. For direct use of a constant pressure chart, assume a flight is planned at 10,000 ft. MSL.

      1) The 700-mb/hPa chart is approximately 10,000 ft. MSL and is the best source for observed temperature, temperature-dew point spread, moisture, and wind for that flight level.

5. Figures 8-2 through 8-7, on pages 321 through 326, are the six standard charts that make up the constant pressure analysis charts.

| PRESSURE (millibars/hectoPascals) | PRESSURE ALTITUDE (feet) | PRESSURE ALTITUDE (meters) | TEMPERATURE-DEW POINT SPREAD | ISOTACHS | CONTOUR INTERVAL (meters) | DECODE STATION HEIGHT PLOT | | EXAMPLES OF STATION HEIGHT PLOTTING | |
|---|---|---|---|---|---|---|---|---|---|
| | | | | | | PREFIX TO PLOTTED VALUE | SUFFIX TO PLOTTED VALUE | PLOTTED | HEIGHT (meters) |
| 850 | 5,000 | 1,500 | yes | no | 30 | 1 | – | 530 | 1,530 |
| 700 | 10,000 | 3,000 | yes | no | 30 | 2 or 3* | – | 180 | 3,180 |
| 500 | 18,000 | 5,500 | yes | no | 60 | — | 0 | 582 | 5,820 |
| 300 | 30,000 | 9,000 | yes** | yes | 120 | — | 0 | 948 | 9,480 |
| 250 | 34,000 | 10,500 | yes** | yes | 120 | 1 | 0 | 063 | 10,630 |
| 200 | 39,000 | 12,000 | yes** | yes | 120 | 1 | 0 | 164 | 11,640 |

NOTE:
1. The pressure altitudes are rounded off to the nearest thousand for feet and to the nearest 500 for meters.
2. All heights are above mean sea level (MSL).
3. * Prefix a "2" or "3," whichever brings the height closer to 3,000 meters.
4. ** Omitted when the air is too cold (temperature less than −41°C).

Table 8-1. Features of the Constant Pressure Charts - U.S.

## B. Plotted Data

1. Figure 8-1 on page 316 illustrates and decodes the standard radiosonde data plot. Table 8-2 on page 316 gives a station data plot example for each chart level.

2. Data from each observation station are plotted around a station circle on each constant pressure chart. The station circle identifies the station location, and the following data are plotted on each chart:

   a. Temperature (°C)
   b. Temperature-dew point spread (°C)
   c. Wind (direction referenced to true north, wind speed in kt.)
   d. Height of the pressure surface above sea level in meters (m)
   e. Height change of the pressure surface over the previous 12-hr. period (meters)

3. The station circle is blackened when the temperature-dew point spread is 5° or less (moist atmosphere).

   a. The station circle is clear when the temperature-dew point spread is more than 5° (dry atmosphere).

4. Aircraft and satellite observations are also used as information sources for constant pressure charts. These observations are particularly useful over sparse radiosonde data areas. Data from aircraft and satellites are plotted on the constant pressure charts closest to their reporting altitudes.

a.    A square is used to identify an aircraft reporting position.  Data plotted include

1)    The flight level of the aircraft in hundreds of feet MSL
2)    Temperature
3)    Wind
4)    Time (to the nearest hour UTC)

b.    A star is used to identify a satellite reporting position.  Satellite information is determined by identifying cloud drift and height of cloud tops.  Data plotted include

1)    Flight level in hundreds of feet MSL
2)    Wind
3)    Time (to the nearest hour UTC)

| Code | Explanation |
|---|---|
| TT | Plotted temperature is the nearest whole degree Celsius.  A below-zero temperature is prefaced with a minus sign.  Position is left blank if data is missing.  A bracketed computer-generated temperature is plotted on the 850 mb/hPa chart in mountainous regions when stations have elevations above the 850 mb/hPa pressure level.  If two temperatures are plotted, one above the other, the top temperature is used in the analysis. |
| T-D | Plotted temperature-dew point spread to the nearest whole degree Celsius.  An "X" is plotted when the spread is greater than 29°.  The position is left blank when the temperature is at or below –41°C or the data is missing. |
| $H_c$ | Plot of constant pressure surface height change which occurred during the previous 12 hr. in tens of meters.  For example, a "+04" means the height of the surface rose 40 meters and a "–12" means the height fell by 120 meters.  $H_c$ data is superseded by "LV" or "M" when pertinent. |
| CIRCLE | Identifies station position.  Shaded black when T-D spread is 5° or less (moist).  Unshaded when spread is more than 5°. |

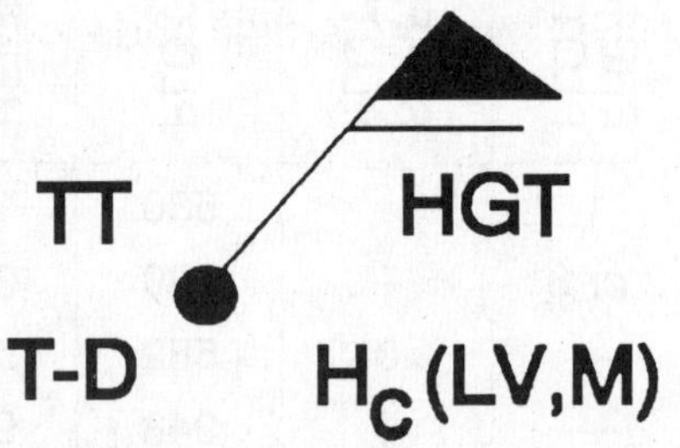

| Code | Explanation |
|---|---|
| WIND | Plotted wind direction and speed by symbol.  Direction is to the nearest 10° and speed is to the nearest 5 kt.  See Figure 5-3, on page 300, for the explanation of the symbols.  If the direction or speed is missing, the wind symbol is omitted and an "M" is plotted in the $H_c$ space.  If speed is less than 3 kt., the wind is light and variable, the wind symbol is omitted, and an "LV" is plotted in the $H_c$ space. |
| HGT | Plotted height of the constant pressure surface in meters above mean sea level.  (See Table 8-1, on page 317, for decoding.)  If data is missing, nothing is plotted in this position. |

Figure 8-1.  Radiosonde Data Station Plot

| | | | | | | |
|---|---|---|---|---|---|---|
| | 22 • 479  4 LV | 09 129  17 –03 | –19 558  X +03 | –46 919  +10 | –55 037  –01 | –60 191  M |
| | 850 mb/hPa | 700 mb/hPa | 500 mb/hPa | 300 mb/hPa | 250 mb/hPa | 200 mb/hPa |
| WIND | Light and variable | 010° true 20 kt. | 210° true 60 kt. | 270° true 25 kt. | 240° true 30 kt. | Missing |
| TT | 22°C | 9°C | –19°C | –46°C | –55°C | –60°C |
| T-D | 4°C | 17°C | >29°C | Not plotted | Not plotted | Not plotted |
| DEW POINT | 18°C | –8°C | Dry | Dry | Dry | Dry |
| HGT | 1,479 m | 3,129 m | 5,580 m | 9,190 m | 10,370 m | 11,910 m |
| $H_c$ | Not plotted | –30 m | +30 m | +100 m | –10 m | Not plotted |

Table 8-2.  Examples of Radiosonde Plotted Data

C.  **Analysis**

1.  All constant pressure charts contain analyses of height and temperature variations.  Also, selected charts have analyses of wind speed variations.

    a.  Contours are lines of equal heights, isotherms are lines of equal temperature, and isotachs are lines of equal wind speed.

2.  **Contours**

    a.  Contours are lines of constant height, in meters, above MSL.

        1)  Contours are used to map the height variations of the pressure surface, which fluctuates in altitude, and to identify and characterize pressure systems on the constant pressure charts.

    b.  Contours are drawn as solid lines and are identified by a three-digit code located on each contour.

        1)  See Table 8-1, on page 315, for information on how to decode the height plot and examples.

    c.  The **contour interval** is the height difference between each contour.  A standard contour interval is used for each chart.

        1)  The contour intervals are 30 meters (m) for the 850 and 700 mb/hPa charts; 60 m for the 500 mb/hPa chart; and 120 m for the 300, 250, and 200 mb/hPa charts.

    d.  The **contour gradient** is the distance between the contours.  Contour gradients identify slopes of the pressure surfaces.

        1)  Strong gradients are closely spaced contours and identify steep slopes.
        2)  Weak gradients are widely spaced contours and identify shallow slopes.

    e.  The contours display height patterns, with the common types being lows, highs, troughs, and ridges.  These contour patterns on constant pressure charts can be interpreted the same way as isobar patterns on the surface chart.  For example, an area of low height is the same as an area of low pressure.

    f.  The contour patterns on the constant pressure charts can be further characterized by size and intensity.

        1)  Sizes can range from large (more than 1,000 mi. across) to small (less than 1,000 mi. across).

        2)  Intensities can range from strong (closely spaced contours) to weak (widely spaced contours).

    g.  Since constant pressure charts are depicted levels above the surface friction layer, the winds are considered to flow parallel to the contours.

        1)  In the Northern Hemisphere, winds flow counterclockwise around lows and clockwise around highs.

        2)  Wind speeds are faster with stronger gradients (closely spaced contours) and slower with weaker gradients (widely spaced contours).

        3)  Contours have the effect of "channeling" the wind.

3.  **Isotherms**

    a.  Isotherms are lines of equal temperature and are used to map temperature variations over a pressure surface.

    b.  Isotherms are drawn as bold, dashed lines on the constant pressure charts.

        1)  Isotherm values are identified by a two-digit block on each line.

            a)  The two digits are prefaced by "+" for above freezing values as well as the zero isotherm and "–" for below freezing values.

        2)  Isotherms are drawn at 5° intervals on each chart.

    c.  Isotherm gradients identify the magnitude of temperature variations.

        1)  Strong gradients are closely spaced isotherms and identify large temperature variations.

        2)  Weak gradients are widely spaced isotherms and identify small temperature variations.

    d.  By inspecting the isotherm pattern, one can determine if a flight would be toward colder or warmer air.

        1)  Subfreezing temperatures and a temperature-dew point spread of 5°C or less would indicate the possibility of icing.

4.  **Isotachs**

    a.  Isotachs are lines of constant wind speed and are used to map wind speed variations over a pressure surface.

        1)  Isotachs are drawn on only the 300, 250, and 200 mb/hPa charts.

    b.  Isotachs are drawn as short, fine dashed lines.

        1)  Isotach values are identified by a two- or three-digit number followed by a "K" located on each line.

            a)  EXAMPLES:

                i)   "10K" means a wind speed of 10 kt.
                ii)  "130K" means a wind speed of 130 kt.

        2)  Isotachs are drawn at 20-kt. intervals beginning at 10 kt.

    c.  Isotach gradients identify the magnitude of wind speed variations.

        1)  Strong gradients are closely spaced isotachs and identify large wind speed variations.

        2)  Weak gradients are widely spaced isotachs and identify small wind speed variations.

    d.  Hatched and unhatched areas are alternated at 40-kt. intervals beginning at 70 kt. Areas between the 70-kt. and 110-kt. isotachs are hatched, and areas between the 110-kt. and 150-kt. isotachs are unhatched.

        1)  This alternating pattern is continued until the strongest winds on the chart are highlighted.

        2)  Highlighted isotachs assist in the identification of jet streams.

D. **Three-Dimensional Aspects**

    1.    It is important to assess weather in both the horizontal and the vertical dimensions. This advice applies not only to clouds, precipitation, and other significant conditions but also to pressure systems and winds.

        a.    The characteristics of pressure systems vary horizontally and vertically in the atmosphere.

    2.    The horizontal distribution of pressure systems is depicted by the constant pressure charts and the surface analysis chart. Pressure systems appear on each pressure chart as pressure patterns.

        a.    Pressure charts identify and characterize pressure systems by their location, type, size, and intensity.

        b.    For example, "a large, strong surface low over Illinois" could describe a pressure system as it is identified and characterized on a pressure chart.

    3.    The vertical distribution of pressure systems must be determined by comparing pressure patterns on vertically adjacent pressure charts.

        a.    For example, compare the surface analysis chart with the 850-mb/hPa chart, the 850-mb/hPa chart with the 700-mb/hPa chart, and so forth.

        b.    Changes of pressure patterns with height can be in the form of position, type, size, or intensity.

            1)    For example, a surface low over Illinois may change type and position with height and be a trough over Nebraska at the 500-mb/hPa level.

    4.    The three-dimensional assessment of pressure systems implies the assessment of the three-dimensional variations of wind.

E. **Using the Charts**

    1.    Constant pressure charts are used to provide an overview of selected observed en route flying conditions. Use all pressure charts for a general overview of conditions.

    2.    Select the chart closest to the desired flight altitude for assessment of en route conditions.

        a.    Review the winds along the route considering their direction and speed.

            1)    For high altitude flights, identify jet stream positions.

            2)    Note whether pressure patterns cause significant wind shifts or speed changes.

            3)    Determine if these winds will be favorable or unfavorable (tailwind, headwind, crosswind).

            4)    Consider vertically adjacent charts and determine if a higher or lower altitude would have a more desirable en route wind.

            5)    Interpolate winds between charts for flights between chart levels.

        b.    Review other conditions along the route.

            1)    Evaluate temperatures by identifying isotherm values and patterns.

            2)    Evaluate areas with moist air and cloud potential by identifying station circles shaded black.

3.    Consider the potential for hazardous flight conditions.

    a.    Evaluate the potential for icing.  Freezing temperatures and visible moisture produce icing.

    b.    Evaluate the potential for turbulence.  In addition to convective conditions and strong surface winds, turbulence is also associated with wind shear and mountain waves.

    c.    Wind shear occurs with strong curved flow and speed shear.  Strong lows and troughs and strong isotach gradients are indicators of strong shear.

        1)    Vertical wind shear can be identified by comparing winds on vertically adjacent charts.

    d.    Mountain waves are caused by strong perpendicular flow across mountain crests.  Use winds on the pressure charts near mountain crest level to evaluate mountain wave potential.

4.    Pressure patterns cause and characterize much of the weather.

    a.    As a general rule, lows and troughs are associated with clouds and precipitation, while highs and ridges are associated with good weather.

        1)    However, this rule is more complicated when pressure patterns change with height.

    b.    Compare pressure pattern features on the various pressure charts with other weather charts, such as the weather depiction and radar summary charts.

        1)    Note the association of pressure patterns on each chart with the weather.

5.    Pressure systems, winds, temperatures, and moisture change with time.  For example, pressure systems move, change size, and change intensity.  Forecast products predict these changes.

    a.    Compare observed conditions with forecast conditions and be aware of any changes.

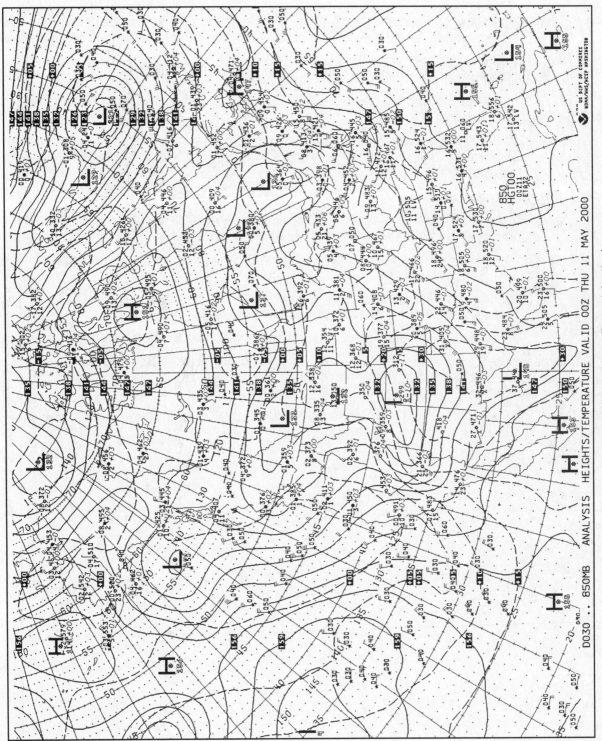

Figure 8-2. 850 Millibar/HectoPascal Analysis

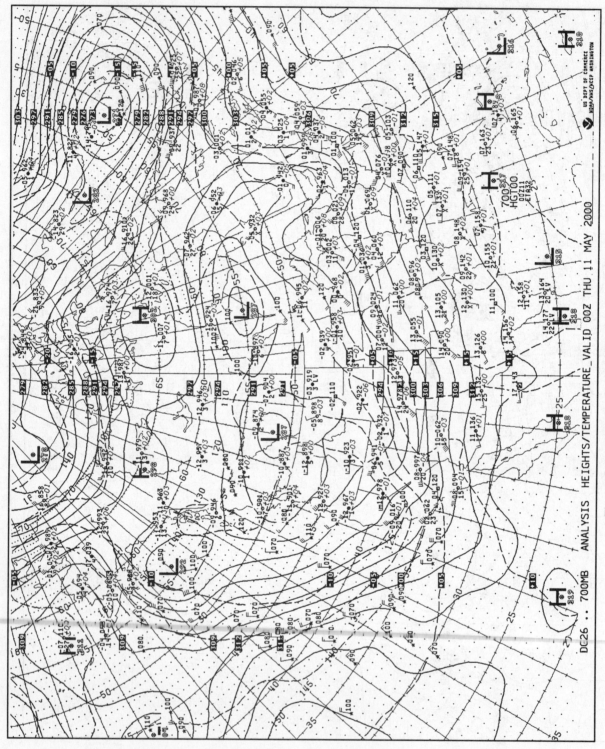

Figure 8-3.  700 Millibar/HectoPascal Analysis

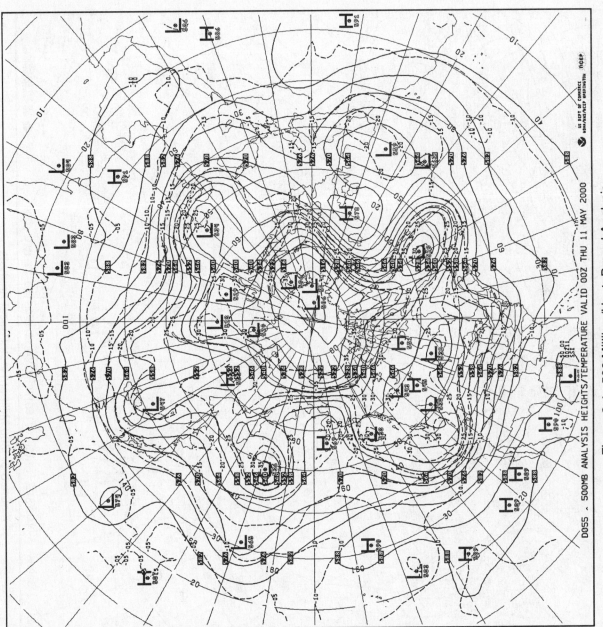

Figure 8-4. 500 Millibar/HectoPascal Analysis

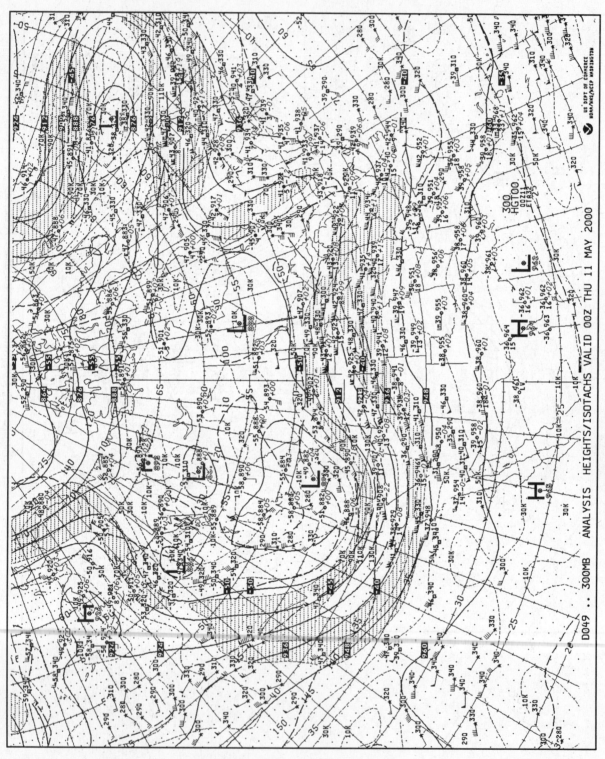

Figure 8-5.  300 Millibar/HectoPascal Analysis

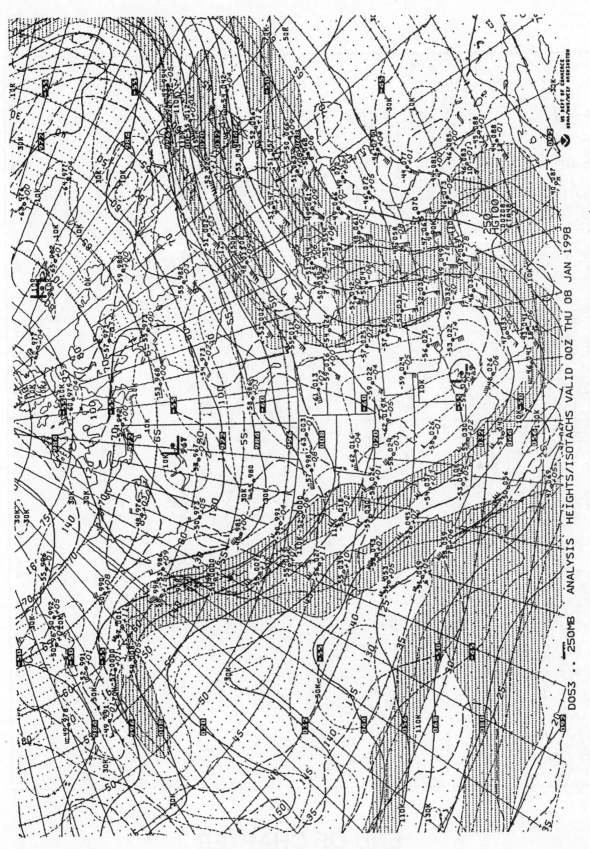

Figure 8-6. 250 Millibar/HectoPascal Analysis.

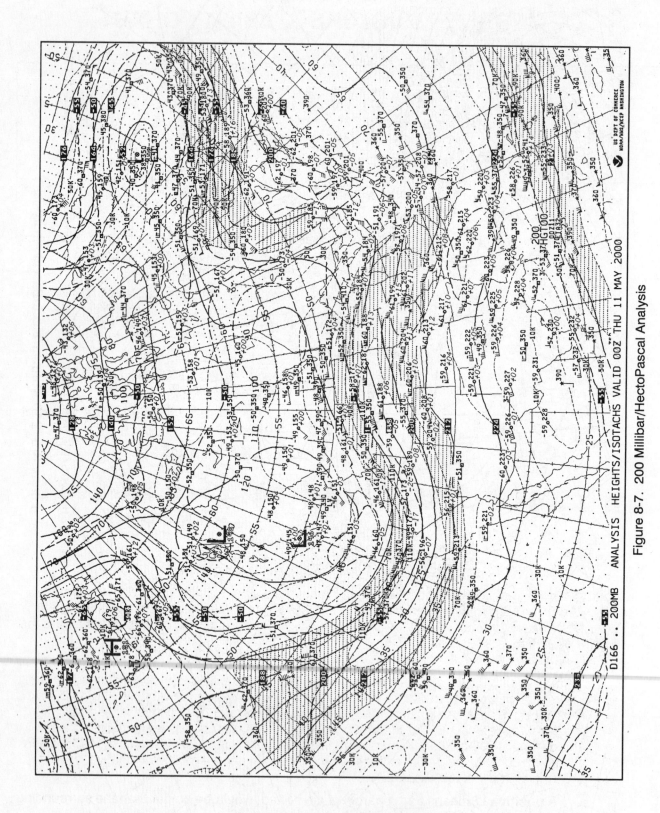

Figure 8-7.  200 Millibar/HectoPascal Analysis

**END OF CHAPTER**

# CHAPTER TWENTY
# COMPOSITE MOISTURE STABILITY CHART

## A.   Type and Time of Report

1.   The **composite moisture stability chart** (Figure 9-1 on page 328) is an analysis chart using observed upper-air data.  The chart is composed of four panels, including stability, freezing level, precipitable water, and average relative humidity.

   a.   This computer-generated chart is available twice daily with valid times of 1200Z and 0000Z.

2.   On the composite moisture stability chart, the mandatory levels used are the surface, 1000 mb/hPa, 850 mb/hPa, 700 mb/hPa, and 500 mb/hPa.

   a.   Significant levels are the levels where significant changes in temperature and/or moisture occur when compared to the levels below or above.

3.   The availability of upper-air data (on all the panels) for analysis is indicated by the shape of the station model.

   a.   The station model legend is shown on the lower left side of the precipitable water panel (see Figure 9-3 on page 332).

   b.   Station models are depicted four ways:

      1)   ○   No data missing
      2)   ✳   All significant levels missing
      3)   □   Some mandatory levels missing
      4)   ⓥ   Some mandatory and significant levels missing

## B.   Stability Panel.

The upper left panel of the composite moisture stability chart outlines areas of stable and unstable air.  (See Figure 9-2 on page 329 for an enlarged depiction of this panel.)  There are two stability indices that are computed for each upper-air station.  The top value is the lifted index, which is plotted above a short line, and the bottom value is the K index, which is plotted below the line.  An "M" indicates the value is missing.

1.   The **lifted index (LI)** is a common measure of atmospheric stability.  LI is obtained by lifting a parcel of air near the surface to 500 mb/hPa (approximately 18,000 ft. MSL).  As the air is lifted, it cools due to expansion.  This cooling rate is either 3°C/1,000 ft., if the air is unsaturated, or 2°C/1,000 ft., if the air is saturated.  The temperature the parcel of air would have at 500 mb/hPa is then subtracted from the actual (environmental) 500 mb/hPa temperature.  This difference is the lifted index, which can be positive, negative, or zero and indicates the stability of the parcel of air.

   a.   A **positive LI** means that a parcel of air, if lifted, would be colder than the surrounding air at 500 mb/hPa.

      1)   The air is stable and would resist vertical motion.
      2)   The more positive the LI, the more stable the air.
      3)   Large positive values (+8 or greater) would indicate very stable air.

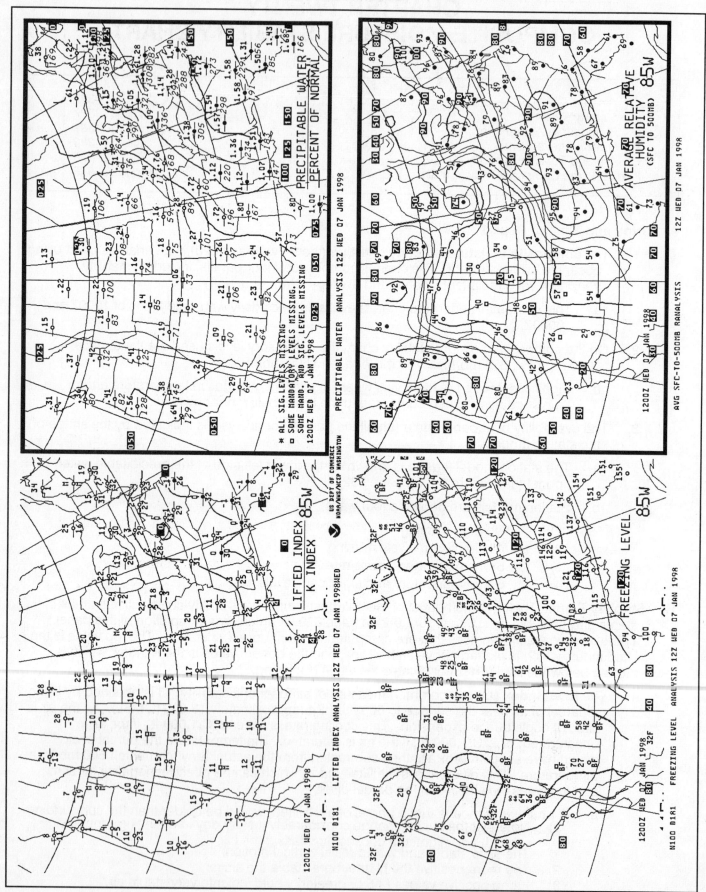

Figure 9-1.  Composite Moisture Stability Chart

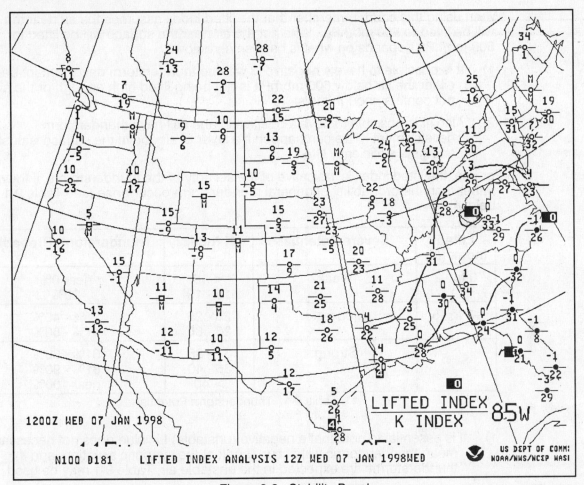

Figure 9-2. Stability Panel

b.  A **negative LI** means that a parcel of air, if lifted to 500 mb/hPa, would be warmer than the surrounding air.

   1)  The air is unstable and suggests the potential for thunderstorms.

   2)  The more negative the LI, the more unstable the air, and the stronger the updrafts are likely to be with any developing thunderstorms.

   3)  Large negative values (−6 or less) would indicate very unstable air.

c.  A **zero LI** means that a parcel of air, if lifted to 500 mb/hPa, would have the same temperature as the actual air at 500 mb/hPa.

   1)  The air is said to be neutrally stable (neither stable nor unstable).

d.  Lifted index values are decreased (made less stable) by increasing surface dew point (moisture), increasing surface temperature, and decreasing the 500 mb/hPa temperature.

   1)  Lifted index values are increased (made more stable) by decreasing surface dew point, decreasing surface temperature, and increasing the 500 mb/hPa temperature.

   2)  Note that the LI can change considerably just by daytime heating (decreasing the LI) and nighttime cooling (increasing the LI).

e.   When using this chart, remember that the lifted index assumes the air near the surface will be lifted to 500 mb/hPa.  Whether the air near the surface will be lifted to 500 mb/hPa depends on what is happening below.

1)   It is possible to have a negative LI with no thunderstorm development because either the air below 500 mb/hPa is not being lifted high enough or the air does not contain enough moisture.

2)   Also, it is possible to have a positive LI and still have thunderstorm development.  This can happen if a layer of air above the surface is unstable and gets sufficiently lifted.

f.   For use, the lifted index is indicative of the severity of the thunderstorms, if they occur, rather than the probability of general thunderstorm occurrence (see Table 9-1 below).

| Lifted Index (LI) | Severe Potential | K Index | Thunderstorm Probability |
|---|---|---|---|
| 0 to –2 | Weak | < 15<br>15 - 19 | near 0%<br>20% |
| –3 to –5 | Moderate | 20 - 25<br>26 - 30 | 21% - 40%<br>41% - 60% |
| ≤ –6 | Strong | 31 - 35<br>36 - 40<br>> 40 | 61% - 80%<br>81% - 90%<br>near 100% |

Table 9-1.  Thunderstorm Potential

1)   It is essential to note that a negative (unstable) LI value does not necessarily mean thunderstorms will occur.  Look at the synoptic situation, and if thunderstorms are expected in the unstable air, Table 9-1 may be used.

2.   The **K index** is primarily for the meteorologist.  It examines the temperature and moisture profile of the environment.  The K index is not really a stability index because the parcel of air is not lifted and compared to the environment.  The K index is computed using three terms:

K =  (850 mb/hPa temp – 500 mb/hPa temp) + (850 mb/hPa dew point) – (700 mb/hPa temp/dew point spread)

a.   The first term (850 mb/hPa temp – 500 mb/hPa temp) compares the temperature at 850 mb/hPa (approximately 5,000 ft. MSL) to the temperature at 500 mb/hPa (approximately 18,000 ft. MSL).

1)   The larger the temperature difference, the more unstable the air and the higher the K index.

b.   The second term (850 mb/hPa dew point) is a measure of low-level moisture.

1)   Note that, since the dew point variable is added, high moisture content at 850 mb/hPa increases the K index.

c.   The third term (700 mb/hPa temp/dew point spread) is a measure of saturation at 700 mb/hPa (approximately 10,000 ft. MSL).

1)   The greater the temperature/dew point spread, the drier the air; and since the term is subtracted, it lowers the K index.

2)   The greater the degree of saturation at 700 mb/hPa, the larger the K index.

d.   During the warm season, a large K index indicates conditions favorable for thunderstorms (see Table 9-1, Thunderstorm Potential, above).

1)   However, K index values and meanings can decrease significantly for thunderstorm development associated with steady state thunderstorms.

a)   Remember that steady state thunderstorms are usually associated with weather systems, and these systems will affect the variables used in computing the K index.

e.   In winter, because of cold temperatures and low moisture values, the temperature terms completely dominate the K index value computation.

1)   Because of the lack of moisture, even fairly large values do not mean conditions are favorable for thunderstorms.

2)   Be aware that the K index can change significantly over a short time period due to temperature and moisture advection.

### 3. Stability Analysis

a.   Stability is based on the lifted index only.

b.   Station circles are blackened for LI values of zero or below.

c.   Contour lines are drawn for values of +4 and below at intervals of 4 (+4, 0, −4, −8, etc.).

### 4. Using the Stability Panel

a.   When clouds and precipitation are forecast or are occurring, the LI is used to determine the type of clouds and precipitation.  That is, stratiform clouds and continuous precipitation occur with stable air, while convective clouds and showery precipitation occur with unstable air.

b.   Stability is also very important when considering the type, extent, and intensity of aviation weather hazards.  For example, a quick estimate of areas of probable convective turbulence can be made by associating the areas with unstable air.

1)   An area of extensive icing would be associated with stratiform clouds and steady precipitation, which are characterized by stable air.

C.   **Precipitable Water Panel**.  The upper right panel of the chart is an analysis of the water vapor content from the surface to the 500-mb/hPa level (see Figure 9-3 on page 332 for an enlarged depiction of this panel).  The amount of water vapor observed is shown as precipitation water, which is the amount of liquid precipitation that would result if all the water vapor were condensed.

### 1. Plotted Data

a.   At each station, precipitable water values to the nearest hundredth of an inch are plotted above a short line, and the percent of normal value for the month is plotted below the line.

1)   The percent of normal value is the amount of precipitable water actually present compared to what is normally expected.

2)   EXAMPLES (See Figure 9-3 on page 332):

a)   Glasgow, MT has a plot of ".22/100," indicating that 0.22 in. of precipitable water is present, which is 100% of normal (normal) for any day this month.

b)   Oklahoma City, OK has a plot of ".72/196," indicating that 0.72 in. of precipitable water is present, which is 196% of normal (above normal) for any day during this month.

b.   An "M" plotted above the line indicates missing data.

c.   At Aberdeen, SD, the percent of normal value is not plotted due to insufficient climatological data to compute this value.

2. **Analysis**

   a. Blackened circles indicate stations with precipitable water values of 1.00 in. or more.

   b. Isopleths (lines of equal values) of precipitable water are drawn and labeled for every 0.25 in., with heavier isopleths drawn at 0.50-in. intervals.

3. **Using the Precipitable Water Panel**

   a. This panel is used to determine water vapor content in the air between the surface and 500 mb/hPA (18,000 ft. MSL).

      1) It is especially useful to meteorologists concerned with flash flood events.

   b. By looking at the wind field upstream from a station, you can get an indication of changes that will occur in the moisture content; that is, you can determine whether the air is drying out or increasing in moisture with time.

   c. There are two constant factors that affect the values shown on the precipitable water panel.

      1) Warm air can hold higher quantities of water vapor. Thus, warm air masses generally have more precipitable water than cold air masses.

         a) EXAMPLE: Precipitable water values are generally higher during summer months than during winter months.

      2) High elevation stations have smaller vertical columns of air between the surface and the 500 mb/hPa level than low elevation stations. Thus, higher elevation stations tend to have lower precipitable water than lower stations.

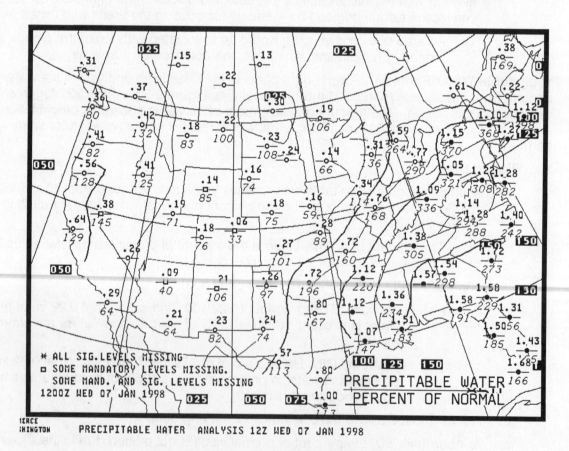

Figure 9-3. Precipitable Water Panel

D. **Freezing Level Panel**. The lower left panel of the composite moisture stability chart is an analysis of the observed freezing-level data from upper-air observations (see Figure 9-4, below).

1. **Analysis**

   a. Solid lines are contours of the lowest freezing level and are drawn for 4,000-ft. intervals and labeled in hundreds of feet MSL.

   b. When a station reports more than one crossing of the 0°C isotherm, the lowest crossing is used in the analysis.

      1) This is in contrast to the low-level significant weather prog on which the depicted forecast freezing level aloft is the highest freezing level.

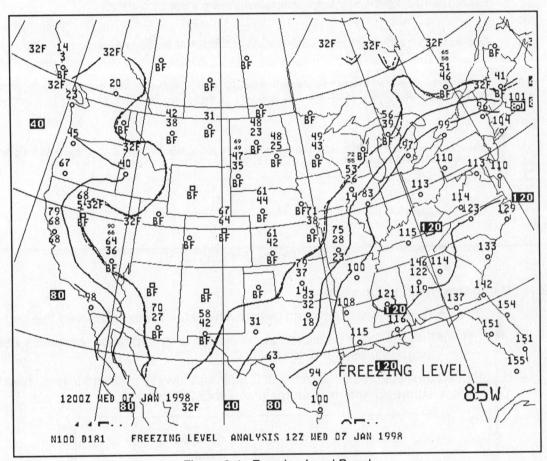

Figure 9-4. Freezing Level Panel

c.   When the freezing level is at the surface, a dashed line is drawn and labeled as the "32F" (0°C) isotherm and the dashed lines will outline an area of stations reporting "BF" (below freezing).

d.   See Table 9-3 below for interpretation of plotted freezing levels.

1)   All heights are above mean sea level (MSL).

| Plotted | Interpretation |
|---|---|
| O<br>BF | Entire observation is below freezing (0°C). |
| 28<br>✳ | Freezing level is at 2,800 ft. MSL; temperatures below freezing above 2,800 ft. MSL. |
| 120<br>□ | Freezing level at 12,000 ft. MSL; temperatures above 12,000 ft. MSL are below freezing. |
| 110<br>51<br>O<br>BF | Temperatures are below freezing from the surface to 5,100 ft. MSL; above freezing from 5,100 to 11,000 ft. MSL; and below freezing above 11,000 ft. MSL. |
| 90<br>34<br>O<br>3 | Lowest freezing level is at 300 ft. MSL; below freezing from 300 ft. MSL to 3,400 ft. MSL; above freezing from 3,400 to 9,000 ft. MSL; and below freezing above 9,000 ft. MSL (see Table 9-2 on page 336 for a vertical profile). |
| M<br>O | Data are missing. |

Table 9-3.  Plotting Freezing Levels

## 2.   **Using the Freezing Level Panel**

a.   The contour analysis shows an overall view of the lowest observed freezing level.

1)   Always plan for possible icing in clouds or precipitation, especially between the temperatures of 0°C and −10°C.

b.   Plotted multiple crossings of the 0°C isotherm always show an inversion with warm air above subfreezing temperatures (see Table 9-2 below).

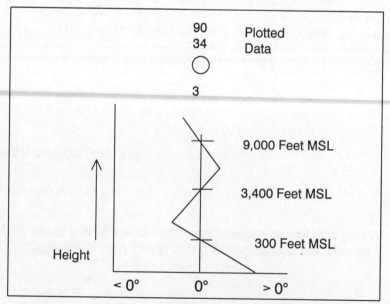

Table 9-2.  Vertical Temperature Profile with Freezing Levels

1)   This situation can produce very hazardous icing when precipitation is occurring.

a)   See AIRMET ZULU  (for icing and freezing level), which will state the areas of expected icing more specifically.

b)   The low-level significant weather prog shows anticipated changes in the freezing level.

c.   Freezing levels tend to lower behind a cold front and rise ahead of a warm front.

E.   **Average Relative Humidity Panel**.  The lower right panel of the chart is an analysis of the average relative humidity from the surface to 500 mb/hPa (see Figure 9-5 below for an enlarged depiction of this panel).  The values are plotted as a percentage for each reporting station.  An "M" indicates the value is missing.

1.   **Analysis**

a.   Blackened circles indicate stations with humidities of 50% and higher.

b.   Isopleths of relative humidity, called isohumes, are drawn and labeled every 10% with heavier isohumes drawn for values of 10%, 50%, and 90%.

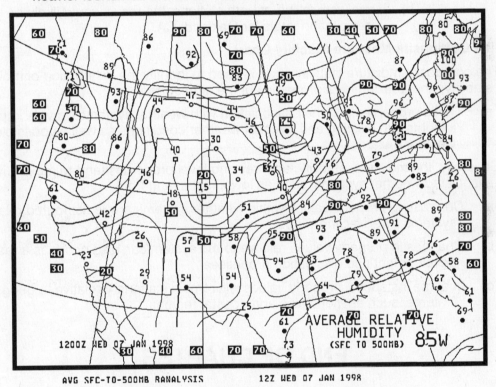

Figure 9-5.  Average Relative Humidity Panel

2.   **Using the Average Relative Humidity Panel**

a.   This panel is used to determine the average air saturation from the surface to 500 mb/hPa.

b.   Areas with high average relative humidities have a higher probability of thick clouds and possibly precipitation.

1)   Areas with low average relative humidities have a lower probability of thick clouds, although shallow cloud layers may be present.

    c.    Significant values of average relative humidities which support the possibility of developing clouds and precipitation are 50% and higher with unstable air and 70% and higher for stable air.

    d.    Weather-producing systems (such as lows and fronts) which are moving into areas with high relative humidities have a high probability of developing clouds and precipitation.

        1)    Weather-producing systems affecting areas with low average relative humidities (30% and less) may produce only a few clouds, if any.

    e.    It is important to remember that high values of relative humidity do not necessarily mean high values of water vapor content (precipitable water).

        1)    For example, in Figure 9-3 on page 332, Oakland, CA has less water vapor content than Miami, FL (.64 and 1.43, respectively).

            a)    However, in Figure 9-5 on page 335, the average relative humidities are the same for both stations (61%).

            b)    If rain were falling at both stations, the result would likely be lighter precipitation totals for Oakland, CA.

## F.   **Using the Composite Moisture Stability Chart**

    1.    The composite moisture stability chart is used to identify the distribution of moisture, stability, and freezing level properties of the atmosphere.

        a.    These properties and their association with weather systems provide important insights into existing and forecast weather conditions as well as possible aviation weather hazards.

    2.    Generally, the moisture, stability, and freezing level properties tend to move with the associated weather systems, such as lows, highs, and fronts.

        a.    Contrasting property values within weather systems are redistributed relative to the systems by advecting winds.

        b.    Changes in property values relative to the system can also occur as a result of development and dissipation processes.

        c.    In some instances, property values will remain stationary relative to geographical features, such as mountains and coastal regions.

# END OF CHAPTER

# CHAPTER TWENTY-ONE
# WINDS AND TEMPERATURES ALOFT CHARTS

Please take a few minutes to study each of the concepts listed above and anticipate/imagine what they are and how they relate to the other listed concepts.

This chapter contains two types of computer-generated charts for winds and temperatures aloft: forecast and observed. Also recall that winds and temperatures aloft forecasts are available in a table format as described and illustrated in Part III, Chapter 11, Winds and Temperatures Aloft Forecast (FD), beginning on page 286.

## A. Forecast Winds and Temperatures Aloft (FD)

1. Forecast winds and temperatures aloft are prepared for eight levels on eight separate panels.

    a. The levels are 6,000; 9,000; 12,000; 18,000; 24,000; 30,000; 34,000; 39,000 ft. MSL.

    b. They are available daily as 12-hr. progs valid at 1200Z and 0000Z.

        1) A legend on each panel shows the valid time and the level of the panel.

    c. Levels through 12,000 ft. are in true altitude, and levels at and above 18,000 ft. are in pressure altitude.

    d. Figure 10-1 on page 338 is an example of a forecast winds and temperatures aloft chart.

        1) Figure 10-2 on page 339 provides a closer view of the 34,000 ft. and 39,000 ft. MSL panels of the forecast winds and temperatures aloft chart.

2. Each station that prepares a winds and temperatures aloft forecast is represented on the panel by a station circle (see Table 10-1 on page 340).

    a. Temperature is in whole degrees Celsius for each forecast point and is entered above and to the right of the station circle.

    b. Arrows with pennants and barbs, similar to those used on the surface map, show wind direction and speed.

        1) Wind speed is indicated by the sum of three types of indicators.

            a) A half barb (ı) is 5 kt., a whole barb (ı) is 10 kt., and a pennant (▲) is 50 kt.

        2) See Figure 10-1 on page 338 for examples of these indicators on the chart.

    c. Wind direction is drawn to the nearest 10° with the second digit of the coded direction entered at the outer end of the arrow.

        1) To determine wind direction, obtain the general direction from the arrow and then use the digit to determine the direction to the nearest 10°.

        2) For example, wind in the northwest quadrant with a digit of "3," indicates 330°.

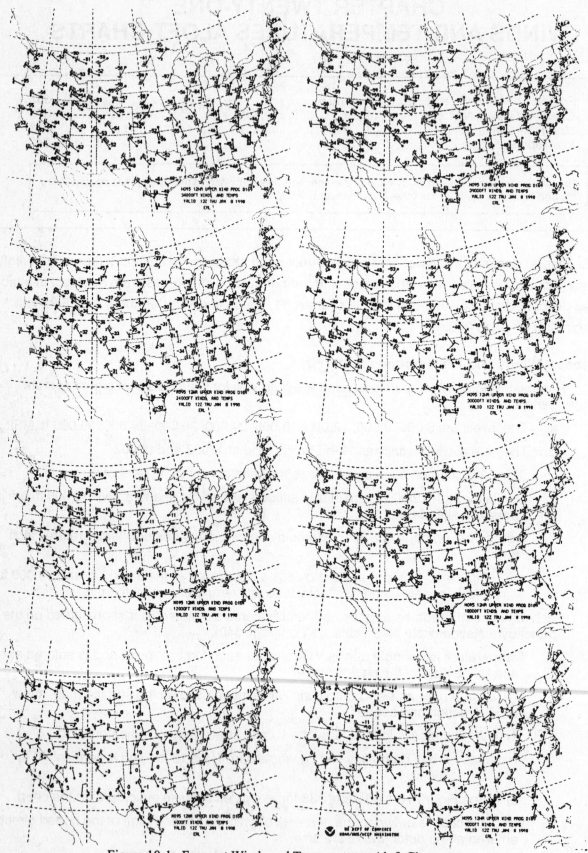

Figure 10-1.  Forecast Winds and Temperatures Aloft Chart.

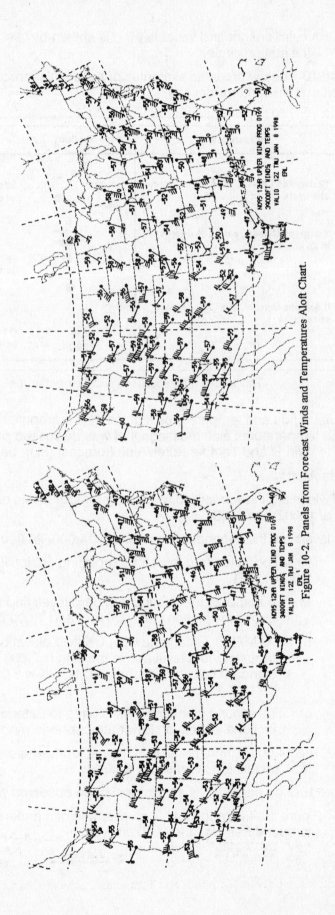

Figure 10-2. Panels from Forecast Winds and Temperatures Aloft Chart.

3)    A calm or light and variable wind is shown by "99" entered to the lower left of the station circle.

d.    Table 10-1 below presents examples of a station's forecast temperatures and winds aloft with their interpretations.

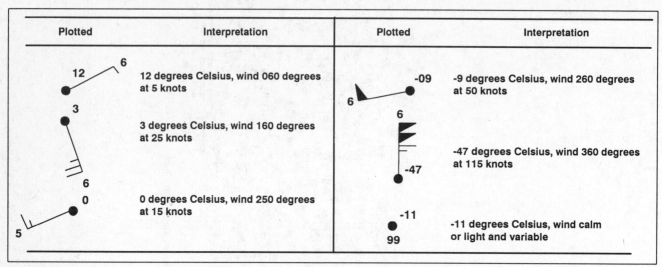

| Plotted | Interpretation | Plotted | Interpretation |
|---|---|---|---|
| | 12 degrees Celsius, wind 060 degrees at 5 knots | | -9 degrees Celsius, wind 260 degrees at 50 knots |
| | 3 degrees Celsius, wind 160 degrees at 25 knots | | -47 degrees Celsius, wind 360 degrees at 115 knots |
| | 0 degrees Celsius, wind 250 degrees at 15 knots | | -11 degrees Celsius, wind calm or light and variable |

Table 10-1.  Plotted Winds and Temperatures

3.    This forecast winds and temperatures aloft chart is a graphic representation of the forecast winds and temperatures aloft message that was discussed previously in Part III, Chapter 11, Winds and Temperatures Aloft Forecast (FD), beginning on page 286.

## B.    Observed Winds Aloft

1.    Observed winds aloft are prepared twice daily for four levels on four separate panels and are valid at 1200Z and 0000Z.

a.    The levels are the second standard level, 14,000, 24,000, and 34,000 ft. MSL.

1)    The second standard level for a reporting station is between 1,000 ft. and 2,000 ft. AGL.

a)    To compute the second standard level, find the next thousand-foot level above the station elevation and add 1,000 ft. to that level.

b)    EXAMPLE:  Oklahoma City, OK field elevation is 1,290 ft. MSL.  The next thousand-foot level above the field is 2,000 ft. MSL.  Thus, the second standard level for Oklahoma City, OK is 3,000 ft. MSL (2,000 + 1,000), or 1,710 ft. AGL.

c)    The second standard level is used to determine low-level windshear and frictional effects on lower atmospheric winds.

2)    The 14,000 ft. MSL panel is in true altitude, whereas the 24,000 and 34,000 ft. MSL panels are in pressure altitude.

b.    Figure 10-3 on page 341 is an example of an observed winds aloft chart.

1)    Figure 10-4 on page 342 is an example of a panel of observed winds and temperatures aloft for 24,000 ft. MSL.

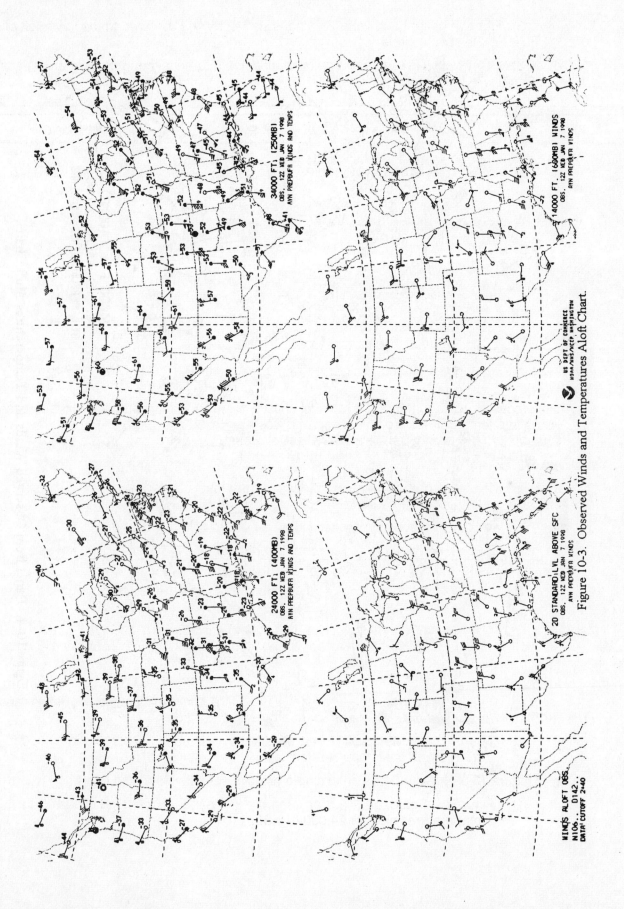

Figure 10-3.  Observed Winds and Temperatures Aloft Chart.

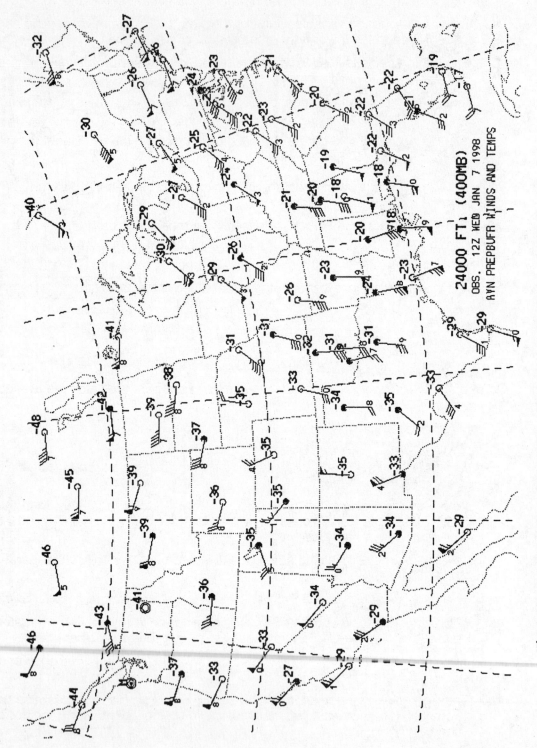

24000 FT₁ (400MB)
OBS. 12Z WED JAN 7 1998
AYN PREPBUFR WINDS AND TEMPS

Figure 10-4.  Panel from Observed Winds and Temperatures Aloft Chart.

2.   Wind direction and speed are shown by arrows with pennants and barbs, the same as shown on the forecast winds and temperatures aloft chart (see Table 10-1 on page 340).

   a.   A calm or light and variable wind is shown as "LV" and a missing wind as "M." Both are plotted to the lower right of the station circle.

   b.   The station circle is filled in when the reported temperature/dew point spread is 5°C or less.

   c.   Observed temperatures are included on the upper two panels of this chart (24,000 ft. and 34,000 ft.).

      1)   A dotted bracket around the temperature means a calculated temperature.

## C.   Using the Charts

1.   The winds aloft chart is used to determine winds at a proposed flight altitude or to select the best altitude for a proposed flight.

   a.   Temperatures can also be determined from the forecast charts.

   b.   Interpolation must be used to determine winds and temperatures at a level between charts and data when the time period is other than the valid time of the chart.

2.   Forecast winds are generally preferable to observed winds because they are more relevant to flight time.

   a.   Although observed winds are 5 to 8 hr. old when received, they can still be a useful reference to check for gross errors on the 12-hr. prog.

## D.   International Flights

1.   Computer-generated forecast charts of winds and temperatures aloft are available for international flights at specified levels.

2.   Figure 10-5, a polar stereographic forecast, on page 344 is a forecast winds and temperatures aloft chart for 34,000 ft. MSL.

   a.   This chart is part of a global winds and temperatures aloft forecast that is in a grid (latitude/longitude) format.

   b.   Polar stereographic is the type of projection used to produce this chart that allows the curvature of the Earth from the North Pole (upper right side of Figure 10-5) to be presented on a flat chart.

3.   Figure 10-6, a Mercator forecast, on page 345 is a forecast winds and temperatures aloft chart for 45,000 ft. MSL.

   a.   This chart is part of a global winds and temperatures aloft forecast in a grid format.

   b.   Mercator is the type of projection used to produce this chart (i.e., Mercator projection) that allows the curved Earth's surface to be presented on a flat chart.

4.   In Figures 10-5 and 10-6, the originating office (NCEP) is indicated in the lower right-hand corner.

   a.   The flight level of the chart, the valid time of the chart, and the database time (data from which the forecast was derived) make up the legend along the bottom of each chart.

5.   Forecast winds are expressed in knots for spot locations with direction and speed depicted in the same manner as on the U.S. forecast winds and temperatures aloft chart (Figure 10-1 on page 338).

   a.   Forecast temperatures, expressed in degrees Celsius, are depicted for spot locations inside circles.

   b.   For charts with flight levels at or below FL 180 (18,000 ft.), temperatures are depicted as negative (−) or positive (+). On charts for flight levels (FL) above FL 180, temperatures are always negative, and no sign is depicted.

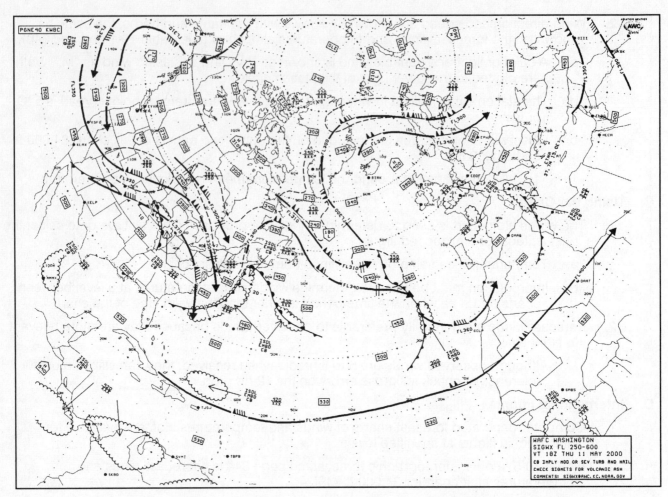

Figure 10-5.  A Polar Stereographic Forecast Winds and Temperatures Aloft Chart

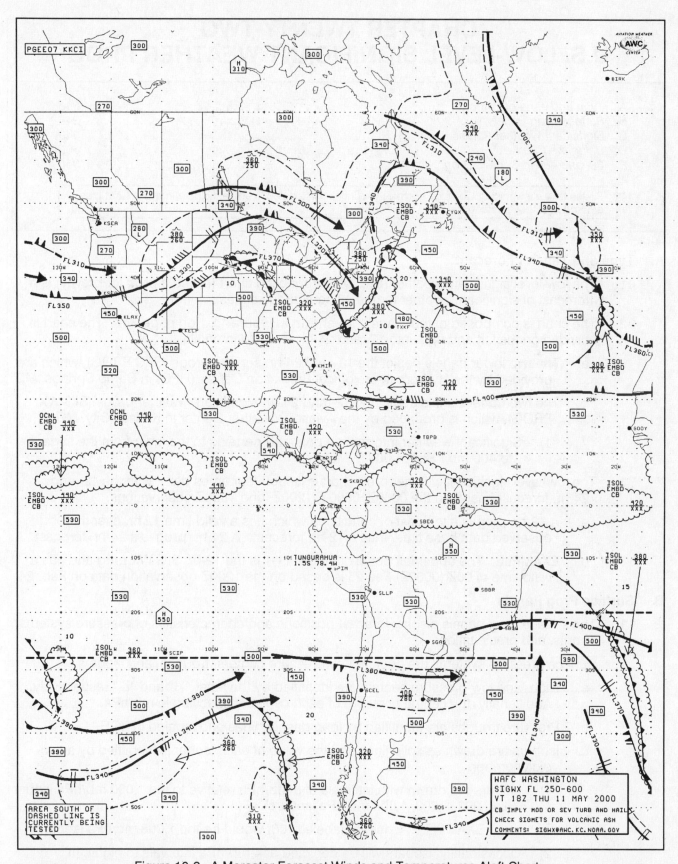

Figure 10-6.  A Mercator Forecast Winds and Temperatures Aloft Chart

# END OF CHAPTER

# CHAPTER TWENTY-TWO
# U.S. LOW-LEVEL SIGNIFICANT WEATHER PROG

Please take a few minutes to study each of the concepts listed above and anticipate/imagine what they are and how they relate to the other listed concepts.

## A.  Type and Time of Forecast

1.  The low-level significant weather (SIG WX) prognostic chart (called a prog for brevity) is a forecast of significant weather for the 48 contiguous states.

2.  The chart is composed of four panels, as shown in Figure 11-1 on page 347.  The chart is divided on the left and right into 12- and 24-hr. forecast intervals.

    a.  The two lower panels depict the 12- and 24-hr. surface progs (SFC PROG), which are provided by the Hydrometeorological Prediction Center (HPC) in Camp Springs, MD.

    b.  The two upper panels depict the 12- and 24-hr. significant weather progs (SIG WX PROG), which is produced by the Aviation Weather Center in Kansas City, MO.

        1)  Significant weather information provided pertains to the layer from the surface to 400 mb/hPa (24,000 ft. MSL).

3.  The low-level significant weather prog chart is issued four times daily, with the 12- and 24-hr. forecast based on the 0000Z, 0600Z, 1200Z, and 1800Z observations.

    a.  A 12-hr. prog is a forecast of conditions which has a valid time 12 hr. after the observed data base time, thus a 12-hr. forecast.  A 24-hr. prog is a 24-hr. forecast.

    b.  EXAMPLE:  The upper left panel of Figure 11-1 is the 12-hr. SIG WX prog and has a valid time of 00Z (0000Z) Feb 25 is based on the 1200Z observation data on Feb. 24.

## B.  Surface Prog Panels

1.  The surface prog panels display forecast positions and characteristics of pressure systems, fronts, and precipitation.

2.  Surface pressure systems are depicted by pressure centers, troughs, and isobars.

    a.  High- and low-pressure centers are identified by the letter "H" and "L," respectively.  Additionally, the central pressure of each center is specified in mb/hPa.

    b.  Pressure troughs are identified by long dashed lines and labeled "TROF."

    c.  Isobars are drawn as solid lines and the value of each isobar is identified by a two-digit number.

        1)  Isobars are drawn with intervals of 8 mb/hPa relative to the 1,000 mb/hPa isobar (i.e., 992, 1,000, 1,008 mb/hPa isobars).

            a)  EXAMPLE:  An isobar labeled "08" would be the 1,008 mb/hPa isobar.

        2)  Occasionally, isobars will be depicted at 4-mb/hPa intervals to highlight patterns with weak pressure gradients.  These isobars are depicted as dashed lines.

        3)  Isobars are shown on the 1200Z and 0000Z charts only.

Figure 11-1. U.S. Low-Level Significant Weather Prog

3.   Surface fronts and the three-digit characterization code are depicted on each panel.  See Part III, Chapter 16, Surface Analysis Chart, beginning on page 295 for the standard symbols and codes.

4.   Areas of forecast precipitation are enclosed by solid lines on each panel.

    a.   Symbols are used to specify the type and character of precipitation (see Tables 11-1 and 11-2 below).

    b.   A mix of precipitation is indicated by the use of two pertinent symbols separated by a slash.

    c.   Shading is used to provide information on the extent of coverage of precipitation in an area.

        1)   In areas of rain, drizzle, or snow (stable air), shading indicates continuous precipitation, which is a dominant and widespread event.

            a)   Absence of shading indicates intermittent precipitation, which is a periodic and patchy event.

        2)   In areas of showers or thunderstorms (unstable air), shading indicates coverage of more than half of the area.

            a)   Absence of shading indicates coverage of one-half or less.

    d.   In an area of forecast precipitation, a bold dashed line is used to separate precipitation of contrasting types, such as snow and rain.

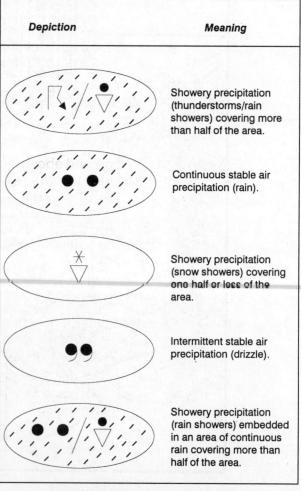

| Symbol | Meaning | Symbol | Meaning |
|---|---|---|---|
| $\wedge$ | Moderate Turbulence | ● ▽ | Rain Shower |
| $\triangle$ | Severe Turbulence | ⁎ ▽ | Snow Shower |
| ⊔ | Moderate Icing | | |
| ⊔ | Severe Icing | Ϟ | Thunderstorm |
| | | ∿ | Freezing Rain |
| ● ● | Rain | 6 | Tropical Storm |
| ✳ ✳ | Snow | 6 | Hurricane (typhoon) |
| ● ● | Drizzle | | |

Table 11-1.  Some standard weather symbols.

|  Depiction  |  Meaning  |
|---|---|
| | Showery precipitation (thunderstorms/rain showers) covering more than half of the area. |
| | Continuous stable air precipitation (rain). |
| | Showery precipitation (snow showers) covering one half or less of the area. |
| | Intermittent stable air precipitation (drizzle). |
| | Showery precipitation (rain showers) embedded in an area of continuous rain covering more than half of the area. |

Table 11-2.  Significant weather prognostic symbols.

e.   Look at the lower left panel of Figure 11-1.  At 0000Z, the forecast is for continuous precipitation (rain and snow) covering half or more of the shaded areas over the New England area, over Montana and Wyoming, and over Utah and Arizona.

1)   Within a shaded area, a dashed line represents the rain/snow line.

2)   The unshaded area over Pennsylvania is forecasting intermittent rain showers/snow showers, covering less than half of the area.

## C.   Significant Weather Panels

1.   The upper panels of Figure 11-1 depict IFR, MVFR, turbulence, and freezing levels from the surface to 24,000 ft. MSL.  Note that the legend near the center of the chart explains the methods of depiction.

2.   Smooth lines enclose areas of forecast IFR weather and scalloped lines enclose areas of marginal weather (MVFR).  VFR areas are not outlined.

a.   This is not the same manner of depiction used on the weather depiction chart (Part III, Chapter 17, beginning on page 303) to portray IFR and MVFR.

b.   Referring to the upper left panel of Figure 11-1, at 0000Z, an area of IFR is depicted over portions of Montana, Wyoming, and Idaho and is surrounded by an area of MVFR.

3.   Forecast areas of nonconvective turbulence of moderate or greater intensity are enclosed by bold, long dashed lines.

a.   Since thunderstorms always imply moderate or greater turbulence, areas of thunderstorm-related turbulence will not normally be outlined.

1)   However, for added emphasis, moderate to severe turbulence from the surface to above 24,000 ft. MSL is depicted for areas that have thunderstorms covering more than half of the area depicted on the surface prog.

2)   Other areas of thunderstorm-related turbulence can be implied by locating areas of forecast thunderstorms on the surface progs.

b.   A symbol entered within a general area of forecast turbulence denotes intensity (see Table 11-1).

c.   Numbers below and above a short line show expected bases and tops of the turbulent layer in hundreds of feet MSL.

1)   A top height of "240" indicates turbulence at or above 24,000 ft. MSL (the upper limit of the prog is 24,000 ft. MSL).

2)   The base height is omitted where turbulence is forecast from the surface upward.

a)   EXAMPLE: "080/" identifies a layer of turbulence from the surface to 8,000 ft. MSL.

d.   Referring to the upper right panel of Figure 11-1, at 1200Z, an area of moderate non-thunderstorm-related turbulence is forecast from the surface to 18,000 ft. MSL over the central U.S., from Texas to North Dakota.

1)   Within this area, a smaller area of moderate-to-severe turbulence is forecast from the surface to 14,000 ft. MSL over central Texas, Oklahoma, Kansas, and central Nebraska.

4.    Freezing-level height contours for the **highest** forecast freezing level are drawn as thin, short, dashed lines at 4,000-ft. intervals.

    a.    Contours are labeled in hundreds of feet MSL.

    b.    The zig-zag line shows where the freezing level is forecast to be at the surface and is labeled "SFC."

    c.    An upper freezing-level contour crossing the surface/32°F (0°C) line indicates multiple freezing levels.

        1)    Multiple freezing-levels indicate layers of warmer air aloft.

        2)    If clouds and precipitation are forecast in this area, icing hazards (freezing precipitation) should be considered.

    d.    Referring to the upper left panel of Figure 11-1, at 0000Z, the 12,000-ft. MSL freezing-level contour extends from northern Florida westward into central Oklahoma and then into western Texas.

## D.  Using the Chart

1.    The low-level significant weather prog chart provides an overview of selected flying weather conditions from the surface to 24,000 ft. MSL for a 24-hr. period (which is also called a day 1 forecast).

2.    Much insight can be gained by evaluating the individual fields of pressure patterns, fronts, precipitation, weather flying categories, freezing levels, and turbulence displayed on the chart.

    a.    Surface winds can be inferred from the surface pressure patterns.

    b.    Structural icing can be inferred in areas which have clouds and precipitation, at altitudes above the freezing level, and in areas of freezing precipitation.

3.    The low-level prog chart can also be used to obtain an overview of the progression of weather during day 1.

    a.    The progression of weather is the change in position, size, and intensity of weather with time.

    b.    Progression analysis is accomplished by comparing charts of observed conditions with the 12- and 24-hr. prog panels.

    c.    Progression analysis adds insight to the time continuity of weather from before flight time to after flight time.

4.    The low-level prog chart makes the comprehension of weather details easier to understand and more meaningful.

    a.    An effective overview of observed and prog charts allows the essential details of observed reports, forecast products, and weather advisories to fit into place and have continuity.

## E.  36- and 48-Hr. Surface Prog

1.    The 36- and 48-hr. surface prog chart (see Figure 11-2 on the next page) is a day 2 (up to 48-hr.) forecast of general weather for the 48 contiguous states issued by the HPC.

    a.    This chart is an extension of the low-level significant weather prog chart (day 1) issued from the same observation data base time.

    b.    This chart is issued twice daily at 0000Z and 1200Z.

        1)    A chart based on the 0000Z Tuesday observations has a 36-hr. valid time (VT) of 12Z (1200Z) Wednesday and a 48-hr. VT of 00Z (0000Z) Thursday.

2. The chart is composed of two panels (the 36- and 48-hr. surface progs) and a forecast discussion.

3. The surface prog panels display forecast positions and characteristics of pressure patterns, fronts, and precipitation.

   a. The format used is the same as the surface prog panels on the low-level prog chart.

4. The forecast discussion is a discussion of the day 1 and day 2 forecast package.

   a. The discussion will include identification and characterization of weather systems and associated weather conditions portrayed on the prog charts.

5. The 36- and 48-hr. surface prog chart provides an outlook of general weather conditions for day 2.

   a. The 36- and 48-hr. surface prog can also be used to assess the progression of weather through day 2.

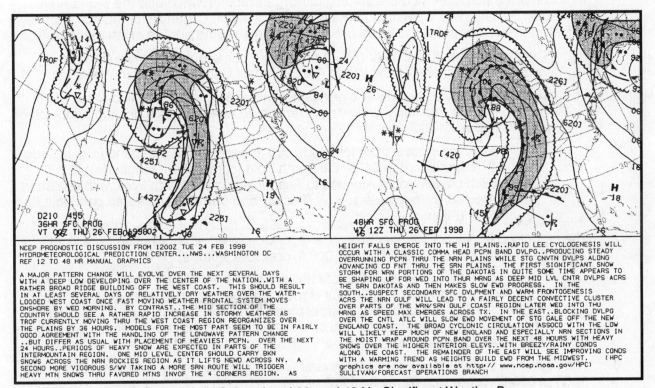

Figure 11-2.  U.S. Low-Level 36- and 48-Hr. Significant Weather Prog

# END OF CHAPTER

# CHAPTER TWENTY-THREE
# HIGH-LEVEL SIGNIFICANT WEATHER PROG

Please take a few minutes to study each of the concepts listed above and anticipate/imagine what they are and how they relate to the other listed concepts.

## A.  Introduction

1.  In accordance with the World Meteorological Organization (WMO) and the World Area Forecast System (WAFS) of the International Civil Aviation Organization (ICAO), high-level significant weather prog charts are provided for the en-route portion of international flights.

    a.  The Aviation Weather Center (AWC) provides a numerous high-level significant weather prog charts for the World Area Forecast Center (WAFC) in Washington, D.C.

    b.  The AWC produces high-level significant weather prog charts for two-thirds of the world, with the capability to cover the remaining one-third as a backup for the other WAFC in Bracknell, England.

        1)  These charts cover from longitude 100°E eastward to 38°E and from latitude 67°N southward to 54°S; the South Pole (centered on longitude 142°W), which covers South America to Australia; and the North Pole, with one chart centered on longitude 155°W (Pacific) and one chart centered on longitude 45°W (Atlantic).

2.  The high-level significant weather prog charts provide weather information from flight level (FL) 250 to FL 600 (i.e., 25,000 ft. to 60,000 ft. pressure altitude).

    a.  The significant weather elements are defined by the WMO and ICAO and include thunderstorms and cumulonimbus clouds, tropical cyclones, severe squall lines, moderate or severe turbulence, widespread sand storms and dust storms, well-defined surface convergence zones, surface fronts, tropopause heights, jet streams, and volcanic eruption sites.

3.  The high-level significant weather prog charts are issued four times daily with valid times (VT) of 0000Z, 0600Z, 1200Z, and 1800Z.

    a.  The charts are formatted on Mercator (see Figure 11-3 on page 353) or polar stereographic (see Figure 11-4 on page 354) projection background maps.

## B.  Content. This section discusses the content of high-level significant weather progs.

1.  **Thunderstorms.** The abbreviation or symbol "CB" (cumulonimbus clouds) is used to depict certain types of thunderstorm activity as described below and on page 355.

    a.  By definition, this symbol refers to either the occurrence or the expected occurrence of an area of widespread cumulonimbus clouds along a line with little or no space between individual clouds, or cumulonimbus clouds embedded in cloud layers or concealed by haze or dust.

        1)  It does not refer to isolated or scattered (occasional) cumulonimbus clouds that are not embedded in cloud layers or concealed by haze or dust.

    b.  The symbol "CB" automatically implies moderate or greater turbulence and icing, and these are not depicted separately.

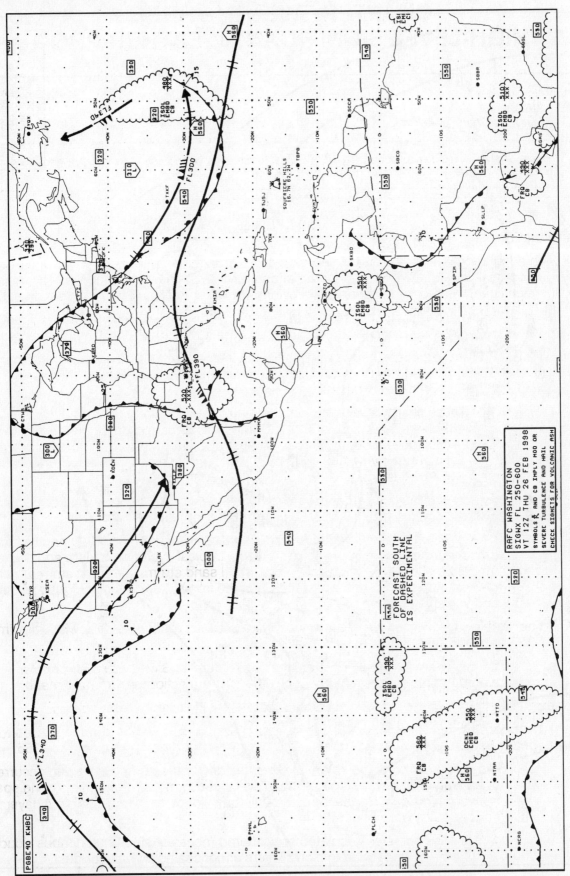

Figure 11-3. U.S. High-Level Significant Weather Prog

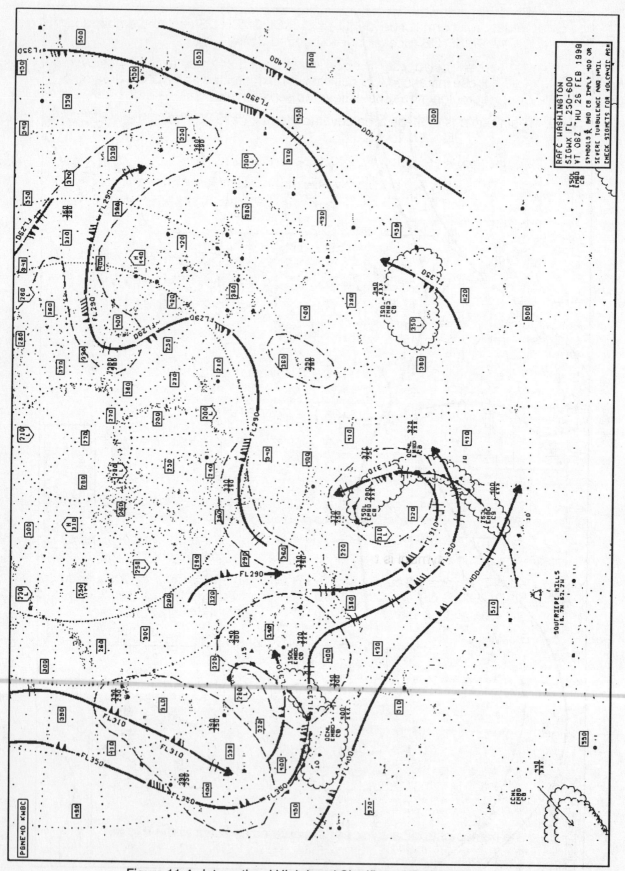

Figure 11-4. International High-Level Significant Weather Prog

c.   CB data will normally be identified as ISOL EMBD CB (isolated embedded CB) or OCNL EMBD CB (occasional embedded CB).

    1)   In rare instances, CB coverage above FL 240 may exceed 4/8 coverage; in these instances, CB activity will be described as FRQ CB (frequent cumulonimbus with little or no separation).

d.   The meanings of these area coverage terms are ISOL -- less than 1/8, OCNL -- 1/8 to 4/8, and FRQ -- 5/8 to 8/8.

e.   CB bases that are below FL 240 are shown as XXX.

    1)   CB tops are to be expressed in hundreds of feet MSL.
    2)   The area to which the forecast applies is shown by scalloped lines.

f.   EXAMPLE:

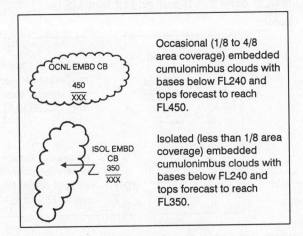

2.   **Tropical cyclones**

a.   The positions of hurricanes/typhoons and tropical storms are depicted by a symbol shown in the upper left-hand corner of the diagram below.

    1)   The only difference between the hurricane/typhoon symbol and the tropical storm symbol is that the circle of the hurricane/typhoon symbol is shaded in.

b.   Areas of associated cumulonimbus activity, if meeting the previously given criteria (ISOL EMBD CB or OCNL EMBD CB), are enclosed by scalloped lines and labeled with the vertical extent.

c.   EXAMPLE:

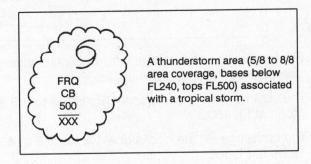

d.   The names of tropical cyclones are entered adjacent to the symbol.

3. **Moderate or Severe Turbulence** (in clouds or clear air)

   a. Areas of forecast moderate or greater turbulence are bounded by heavy (or bold), dashed lines.

      1) Areas of forecast turbulence are those associated with wind shear zones and mountain waves.

         a) Wind shear zones include speed shears associated with jet streams and areas with sharply curved flow.

      2) Turbulence associated with thunderstorms is not identified since thunderstorms imply moderate or severe turbulence.

   b. Areas are labeled with the appropriate turbulence symbol (Table 11-1 on page 348) and the vertical extent in hundreds of feet MSL.

      1) EXAMPLE:

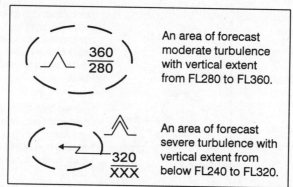

4. **Convergence zones** are areas of active thunderstorm activity associated with the inter-tropical convergence zone. These zones are enclosed by scalloped lines and labeled with the vertical extent.

   a. Note that the CBs must meet the previous given criteria (ISOL EMBD CB or OCNL EMBD CB, or FRQ CB).

   b. EXAMPLE:

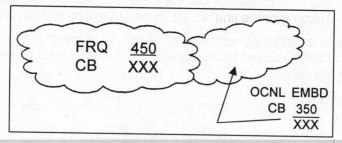

      1) The forecast position of the inter-tropical convergence zone is shown by the associated thunderstorm areas.

      2) The coverage for the frequent CBs is 5/8 to 8/8 with bases below FL 240 and tops at FL 450.

      3) The coverage for the occasional CBs is 1/8 to 4/8 with bases below FL 240 and tops at FL 350.

   c. Areas of CBs that are not associated with tropical storms may occur in the tropics.

5. **Surface Fronts**

   a. Surface fronts are added to the prog to provide added perspective.

      1) The standard symbols are used, as discussed in Part III, Chapter 16, Surface Analysis Chart, beginning on page 295.

b.  EXAMPLE:

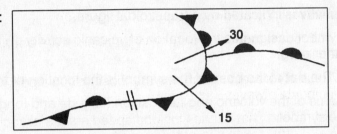

1)  A frontal system is forecast to be at the position and with the orientation indicated at the valid time of the prognostic chart.

2)  The forecast movement related to true north and the speed in knots are indicated by arrow shafts and adjacent numbers.

6.  **Tropopause heights** are depicted in hundreds of feet MSL.

a.  A five-sided polygon (as shown below) indicates areas of high and low tropopause heights.

1)  The example above indicates a low (L) tropopause height of 22,000 ft. MSL. Thus, the height of the tropopause is higher around this point.

b.  Other tropopause heights are used liberally to define the slope of the tropopause field. These tropopause heights are shown in small rectangular blocks.

1)  EXAMPLE: In Figure 11-3, the tropopause is sloping from 30,000 ft. MSL in North Dakota to 38,000 ft. MSL in Texas.

7.  **Jet Stream.** The height and maximum wind speed of jet streams having a core speed of 80 kt. or greater are shown by bold lines. The height is given as a flight level (FL).

a.  The beginning of the line indicates a core speed of 80 kt.

b.  Double hatch lines across the jet stream core either indicate a 20-kt. speed increase or decrease, or depict changes in the wind speed of the jet stream.

c.  The maximum core speed along the jet stream is depicted by arrow shafts, pennants, and barbs.

d.  EXAMPLE:

FL 420

1)  The jet stream depicted has a forecast maximum speed of 130 kt. (two pennants @ 50 kt. + three barbs @ 10 kt.) at a height of 42,000 ft. MSL, i.e., FL 420.

a)  The extreme left line starts at 80 kt.

b)  The first pair of hatch lines indicates a speed increase of 20 kt. to 100 kt., and the second pair shows an increase of 20 kt. to 120 kt.

c)  The double hatch lines to the right of the maximum speed indicate a decrease of 20 kt. to 120 kt.

2)  Wind directions are indicated by the orientation of arrow shafts in relation to true north.

8.  **Volcanic activity** is indicated by a trapezoidal figure.

    a.  The symbol designates the location of volcanic activity on the high-level significant weather prog.

        1)  The dot at the base of the symbol is the location of the volcano.

    b.  The name of the volcano and its position (latitude and longitude) are noted adjacent to the symbol.

        1)  A reminder to check SIGMETs for volcanic ash is included in the identifier box at the lower right-hand corner of the chart.

    c.  EXAMPLE:

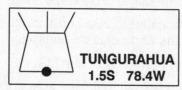

9.  **Severe squall lines** are identified by long dashed lines, and each dash is separated by the letter "V," as shown below.

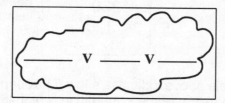

    a.  Severe squall lines are lines of thunderstorms (CB) with 5/8 coverage or greater.

    b.  CB activity meeting chart criteria is identified and characterized with each squall line.

10. **Sandstorms and Duststorms**

    a.  Areas of widespread sandstorms and duststorms are labeled as shown below:

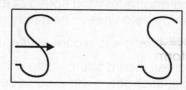

        1)  The symbol with the arrow depicts areas of widespread sandstorm or duststorm.

        2)  The symbol without the arrow depicts severe sandstorm or dust haze.

## C.  Using the Chart

1.  The high-level significant weather prog chart is used to obtain an overview of selected flying conditions above FL 240.

2.  This chart is useful in determining the progression of the weather above FL 240.

# END OF CHAPTER

# CHAPTER TWENTY-FOUR
# CONVECTIVE OUTLOOK CHART

> Please take a few minutes to study each of the concepts listed above and anticipate/imagine what they are and how they relate to the other listed concepts.

## A. Introduction

1. The convective outlook chart is to identify areas forecast to have thunderstorm activity. There are two charts:

   a. The first chart is the Day 1 Convective Outlook.
   b. The second chart is the Day 2 Convective Outlook.

2. See Part III, Chapter 14, Convective Outlook (AC), on page 292 for a discussion on the textual product.

3. The convective outlook chart is produced at the Storm Prediction Center (SPC) in Norman, OK.

## B. Day 1 Convective Outlook

1. The Day 1 Convective Outlook (see Figure 12-1 below) outlines areas in the continental U.S. where severe and general thunderstorms may develop during the next 6 to 30 hr.

   a. A line with an arrowhead delineates these areas. When facing in the direction of the arrow, the thunderstorm activity is expected to the right of the line.

2. The Day 1 Convective Outlook is issued five times daily with each chart valid until 1200Z the next day.

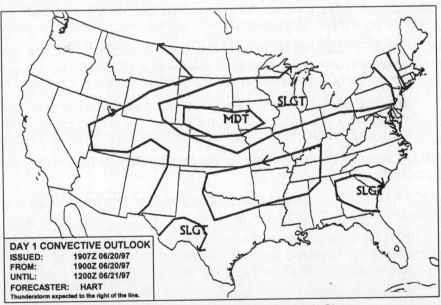

DAY 1 CONVECTIVE OUTLOOK
ISSUED:      1907Z 06/20/97
FROM:        1900Z 06/20/97
UNTIL:       1200Z 06/21/97
FORECASTER:  HART
Thunderstorm expected to the right of the line.

Figure 12-1. Day 1 Convective Outlook Chart

### C.  Day 2 Convective Outlook

1.  The Day 2 Convective Outlook is very similar to the Day 1 Convective Outlook, except it is issued only twice a day at 0830Z (standard time) or 0730Z (daylight savings time), and 1730Z.

2.  The Day 2 Convective Outlook chart covers a time period from 1200Z the following day to 1200Z the next day.

    a.  EXAMPLE:  If today is Monday, the Day 2 Convective Outlook will cover the period from 1200Z Tuesday until 1200Z Wednesday.

### D.  Levels of Risk

1.  Risk areas come in five varieties and are based on the forecaster's confidence and expected number and intensity of severe thunderstorm reports over an area.

    a.  The report criteria for each of those risks is valid for an area the size of Oklahoma without the panhandle, or about 50,000 square miles.  As the size of the risk area increases (or decreases) from 50,000 square miles, those expected severe weather numbers would increase (or decrease) proportionally.  Table 12-1 below indicates the labels that appear on both convective outlook charts.

| NOTATION | EXPLANATION |
|---|---|
| SEE TEXT | Used for those situations where a SLGT risk was considered but at the time of the forecast was not warranted. |
| SLGT (Slight risk) | A high probability of 5 to 29 reports of 1 inch or larger hail, and/or 3 to 5 tornadoes, and/or 5 to 29 wind events,... or... a low/moderate probability of moderate to high risk being issued later if some conditions come together. |
| MDT (Moderate risk) | A high probability of at least 30 reports of hail 1 inch or larger; or 6 to 19 tornadoes; or numerous wind events (30). |
| HIGH (High risk) | A high probability of at least 20 tornadoes with at least two of them rated F3 (or higher), or an extreme derecho causing widespread (50 or more) wind events with numerous higher-end wind (80 mph or higher) and structural damage reports. |

Table 12-1.  Notations of Risk

2.  The **SEE TEXT** will be used for those situations where a SLGT risk was considered, but at the time of the forecast, is not warranted.  Although there is no severe outlook for the labeled area, you should read the text of the outlook discussion to learn more about the potential for a threat to develop if some particular conditions do come together.

    a.  The SEE TEXT will be used in the Day 1 Outlooks to discuss areas where a few severe storms are possible or storms may approach severe levels, but the coverage or intensity is expected to be too small or marginal for a risk area.

    b.  The SEE TEXT is used on the Day 2 Outlooks for areas where severe weather is possible, but there is too much forecast uncertainty (questionable model data, capping, moisture return, or other such factors) to draw a risk area.

3.  A **SLGT** risk implies well-organized, severe thunderstorms are expected but in small numbers and/or low coverage.

4.  A **MDT** risk implies a greater concentration of severe thunderstorms and, in most situations, a greater magnitude of severe weather.

5.  The **HIGH** risk almost always means a major severe weather outbreak is expected, with great coverage of severe weather and enhanced likelihood of extreme events, such as violent tornadoes or unusually intense damaging wind.

6.  In addition to severe thunderstorm risk areas, general thunderstorm areas are also outlined but not labeled.  General thunderstorm areas are where nonsevere thunderstorms are expected.

7.  All of the areas in Figure 12-1 (except SEE TEXT) are enclosed by a line with an arrowhead.  When you are facing in the direction of the arrow, the expected convective activity is to the right of the line.

E.  **Explanation of Figure 12-1**

1.  The line through NM, CO, UT, WY, and MT indicates no convective activity west of this line because the arrow points to the north (convective activity is to the right of the arrow).

2.  Parts of the Central Plains and the upper Midwest eastward to PA and NY have a slight risk.

3.  NE and western IA have a moderate risk.

4.  Northeastern TX to western KY has no convective activity forecast (the area is to the left of the arrow).

5.  A portion of southwest TX and most of GA are forecast to have a slight risk.

6.  New England does not have any convective activity forecast (but the mid-Atlantic, Gulf Coast, etc., areas are forecast to have general thunderstorm activity).

F.  **Using the Chart**

1.  The convective outlook charts are for planning purposes only.

2.  Additional and more specific information is available from convective SIGMETs, the radar summary chart, METARs, and "live" radar images.

# END OF CHAPTER

# CHAPTER TWENTY-FIVE
# VOLCANIC ASH ADVISORY CENTER
# (VAAC) PRODUCTS

## A. Introduction

1. The International Civil Aviation Organization (ICAO) and other aviation concerns recognized the need to keep the aviation community informed of volcanic hazards.

   a. To meet this need, nine Volcanic Ash Advisory Centers (VAAC) were created around the world.

   b. The U.S. VAACs are the Anchorage VAAC located in Anchorage, AK and the Washington VAAC located in Camp Springs, MD.

2. The VAACs do not issue routine products, but will issue a Volcanic Ash Advisory Statement (VAAS) and Volcanic Ash Forecast Transport and Dispersion (VAFTAD) chart whenever a volcanic event occurs in their area of responsibility.

   a. These products are based on information from PIREPs, SIGMETs, satellite observations, and volcanic observatory reports.

3. Since the products issued by the VAACs are triggered by the occurrence of an eruption, PIREPs concerning volcanic activity are extremely important.

## B. Volcanic Ash Advisory Statement (VAAS)

1. The VAAS is normally the first product issued by a VAAC following a volcanic eruption.

   a. The VAAS is required to be issued within 6 hr. of an eruption and every 6 hr. after that.

      1) The VAAS can be issued more frequently if new information about the eruption is received.

   b. The VAAS summarizes the currently known information about a volcanic eruption.

2. The height of the ash cloud given in the VAAS is estimated by meteorologist analyzing satellite imagery and satellite cloud drift winds combined with any PIREPs, volcano observatory reports, and upper-air wind reports.

3. Example of a VAAS:

```
FVXX22 KWBC 210800
VOLCANIC ASH ADVISORY STATEMENT
ISSUED 0800 UTC 21 MAR 2000 BY THE WASHINGTON VAAC

SOUFRIERE HILLS 00-038    WEST INDIES 1643N 6211W

BACKGROUND:    SOUFRIERE HILLS  MONTSERRAT (1600-05)
               SUMMIT HEIGHT  3000 FT (915M)

SOURCES OF INFORMATION: GOES-8 INFRARED AND MULTISPECTRAL
IMAGERY. 21/0000 UTC UPPER AIR SOUNDING FROM GUADELOUPE.
MONTSERRAT VOLCANO OBSERVATORY REPORT.
```

.
ERUPTION DETAILS: MAJOR ERUPTION BEGINNING 20/2340 UTC WITH
ANOTHER ERUPTION STARTING AROUND 21/0530 UTC.

.
DETAILS OF ASH CLOUD: IN THE FIRST IMAGERY AFTER THE ECLIPSE
AT 0645 UTC NO VOLCANIC CLOUD CAN BE DETECTED FROM THE
REPORTED 0530 UTC ERUPTION. FROM THE 2340 UTC ERUPTION... THE
LOWER LEVEL ASH CLOUD (BELOW FL090) THAT WAS FANNING OUT TO
THE NORTH... WEST AND SOUTH CAN NO LONGER BE DETECTED IN NIGHT
TIME IMAGERY BUT MAY STILL EXIST AND BECOME DETECTABLE IN THE
FIRST VISIBLE IMAGERY LATER THIS MORNING. THE HIGHER LEVEL
ASH AND STEAM CLOUD HAS BECOME VERY FAINT WITH 2 AREAS OF
MORE DISTINCT ASH. THE FIRST ASH AREA IS ROUGHLY 95 KM (51 NMI)
WIDE BETWEEN POINTS 1601N 5544W AND 1440N 5844W AND THE SECOND
IS AROUND 57 KM WIDE BETWEEN POINTS 1626N 5346W AND 1545N
5343W. THE GENERAL VERY FAINT AREA OF ASH AND STEAM IS BOUNDED
BY POINTS 1646N 6143W  1347N 5936W  1557N 5254W  1634N 5310W
1636N 5738W  1730N 6027W. THE ESTIMATED HEIGHT IS AROUND
FL300.

.
TRAJECTORY: UPPER AIR PATTERNS SUGGEST ASH TO FL300 TO MOVE
EAST SOUTHEAST AT 25 TO 60 KNOTS. ASH BELOW FL090 WILL MOVE
MAINLY WEST AROUND 10 KNOTS.

.
OUTLOOK: SEE LATEST SIGMETS AND VAFTAD.

.
THE NEXT STATEMENT WILL BE ISSUED BY 21/1400 UTC.

.
PLEASE REFER TO SIGMETS FOR CURRENT WARNINGS.

.
NNNN

## C. Volcanic Ash Forecast Transport and Dispersion (VAFTAD) Chart

1.  The VAFTAD chart is generated by a three-dimensional time-dependent dispersion model developed by NOAA's Air Resources Laboratory (ARL).

    a.  The VAFTAD model focuses on hazards to aircraft flight operations caused by a volcanic eruption with emphasis on the ash cloud location in time and space.

2.  The VAFTAD model uses the National Centers for Environmental Prediction (NCEP) forecast data to determine the location of ash concentrations over 6-hr. and 12-hr. time intervals, with valid times beginning 6, 12, 24, and 36 hours following a volcanic eruption.

    a.  The VAFTAD Chart is computer-prepared and is made available on graphics distribution networks supported by the NWS and the Internet.

    b.  The VAFTAD Chart is not issued on a routine basis. Rather, it is issued as volcanic eruptions are reported.

3.  Since the VAFTAD Chart is triggered by the occurrence of a volcanic eruption, PIREPs concerning volcanic activity are very important. Initial input to the VAFTAD model and the resulting chart includes

    a.  Geographic region
    b.  Volcano name
    c.  Volcano location (latitude and longitude)
    d.  Eruption date and time
    e.  Initial ash cloud height

4.  NCEP meteorological forecast guidance is used to depict volcanic ash particle transport and dispersion horizontally and vertically through representative atmospheric layers.

    a.  The VAFTAD model does take into account that ash particles fall with the passage of time.

5.   VAFTAD charts from an actual volcanic eruption will be labeled with ALERT.

    a.   A VAFTAD labeled with WATCH (see Figure 13-1 on page 365) indicates the chart was prepared for a potential volcanic eruption.

## D.  VAFTAD Product

1.   The VAFTAD product presents the relative concentrations of ash following a volcanic eruption for three layers of the atmosphere in addition to a composite of ash concentration through the atmosphere.  Atmospheric layers depicted are

    a.   Surface to FL 200
    b.   FL 200 to FL 350
    c.   FL 350 to FL 550
    d.   Surface to FL 550 (composite layer)

2.   Figure 13-1 on page 365 is an example eight-panel VAFTAD chart of the forecast visual ash cloud at 12 hr. and 24 hr. after the eruption.

    a.   The left column is for 12 hr. after the eruption (ERUPTION+12H).
    b.   The right column is for 24 hr. after the eruption (ERUPTION+24H).

3.   Figure 13-2 on page 366 is an example eight-panel VAFTAD chart of the visual ash cloud at 18 hr. and 24 hr. after the eruption.

    a.   The left column is for 18 hr. after the eruption (ERUPTION+18H).
    b.   The right column is for 24 hr. after the eruption (ERUPTION+24H).

4.   VAFTAD charts are also available at 6 hr. and 12 hr., and for 36 hr. and 48 hr. after the eruption.

    a.   When all of the charts are placed in order, side by side, they give an easy to visualize time-dependent 3-D view of the forecast volcanic ash cloud.

5.   The four panels in any column are for a single valid time after the eruption.

    a.   Individual panels are for layers applicable to aviation operations and are identified at the side of the panel with the upper and lower flight levels (FL).

    b.   The bottom panel is the composite layer, from the SURFACE to FL 550, and is useful as an aid for issuing SIGMETs or for satellite imagery comparisons.

    c.   For each column, the forecast valid time separates the upper three panels from the composite panel.

6.   Additional information is provided at the bottom of the chart.

    a.   Volcanic eruption information is at the lower left and includes the

        1)   Volcano name (with location symbol)
        2)   Volcano location (latitude and longitude)
        3)   Volcano summit height
        4)   Eruption date and time
        5)   Ash column height

    b.   The visual ash cloud symbol (■) is shown at the lower center.
    c.   At the lower right is a message to see current SIGMET for warning area.

## E.  Using the Chart

1.   The VAFTAD chart is strictly for advance flight planning purposes.

    a.   It is not intended to take the place of SIGMETs issued regarding volcanic eruptions and ash.

2.   The actual presentation of the VAFTAD chart may change as international requirements are reviewed.

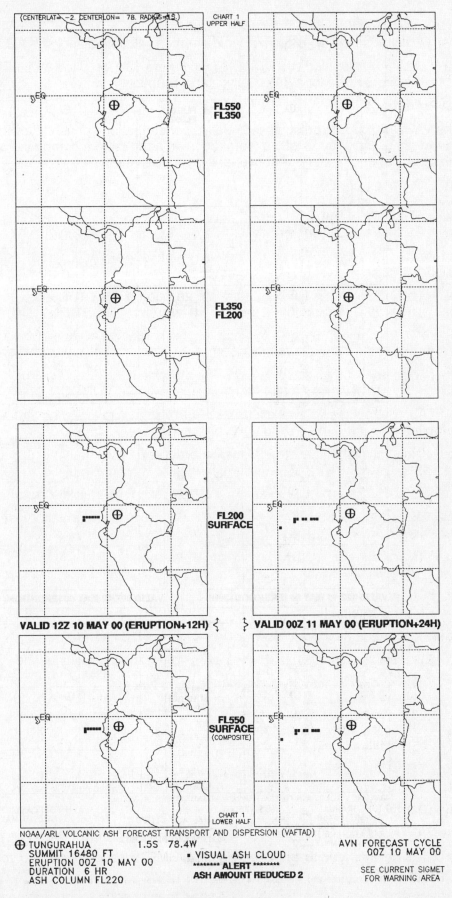

Figure 13-1.  Volcanic Ash Forecast Chart

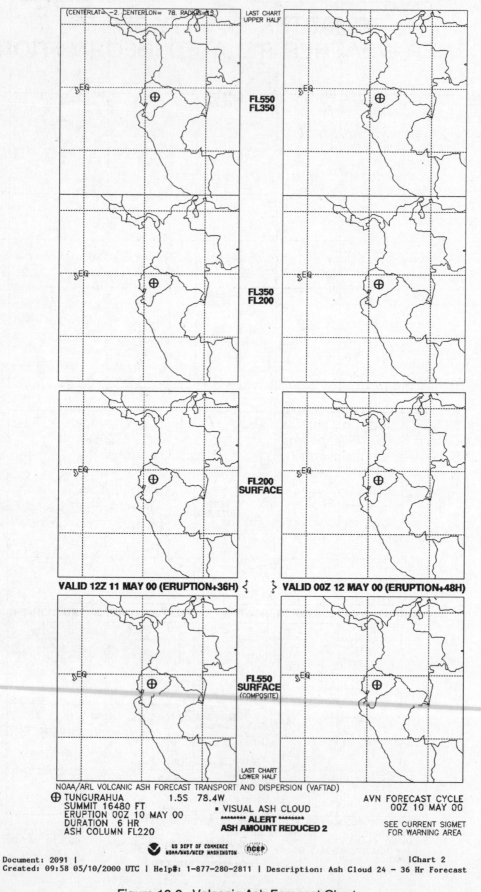

Figure 13-2.  Volcanic Ash Forecast Chart

# END OF CHAPTER

# CHAPTER TWENTY-SIX
# OTHER WEATHER-RELATED INFORMATION

Please take a few minutes to study each of the concepts listed above and anticipate/imagine what they are and how they relate to the other listed concepts.

## A. Introduction

1. This chapter provides other weather-related information that you can use to further understand the weather. Information included covers

   a. Standard conversion table
   b. Density altitude and chart
   c. Contractions and acronyms
   d. Scheduled issuance and valid times of forecast products
   e. National Weather Service station identifiers and WSR-88D sites
   f. Internet addresses

## B.  Standard Conversion Table

1.   This table can be used as a quick reference for conversion between metric and English units.

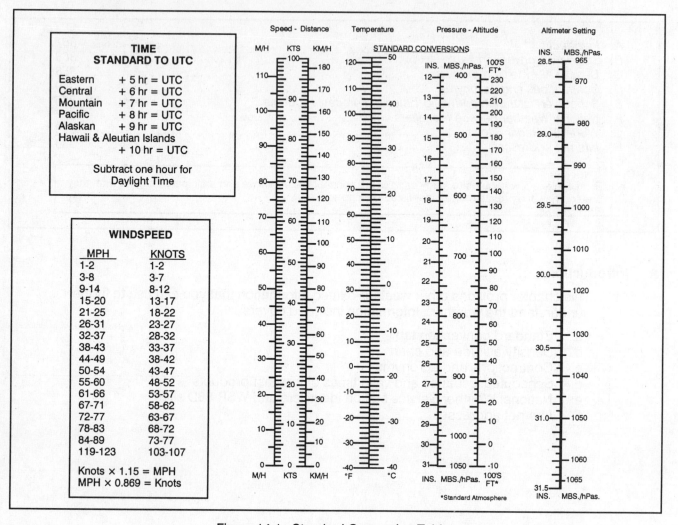

Figure 14-1.  Standard Conversion Table

## C.  Density Altitude Computation

1.   Use the graph, Figure 14-2, on page 369 to find density altitude either on the ground or aloft.

   a.   Set the aircraft's altimeter to 29.92 in.; it now indicates pressure altitude.
   b.   Read the outside air temperature.
   c.   Enter the graph at the pressure altitude and move horizontally to the temperature.
   d.   Read the density altitude from the sloping lines.

2.   **Examples of Density Altitude**

   a.   Density altitude in flight.  Pressure altitude is 9,500 ft., and the temperature is −8°C. Find 9,500 ft. on the left of the graph and move to −8°C.  Density altitude is 9,000 ft., as marked by the dot labeled 1 on Figure 14-2.

   b.   Density altitude for takeoff.  Pressure altitude is 4,950 ft., and the temperature is 97°F. Enter the graph at 4,950 ft. and move across to 97°F.  Density altitude is 8,200 ft. as marked by the dot labeled 2 on Figure 14-2.  Note that, in the warm air, density altitude is considerably higher than pressure altitude.

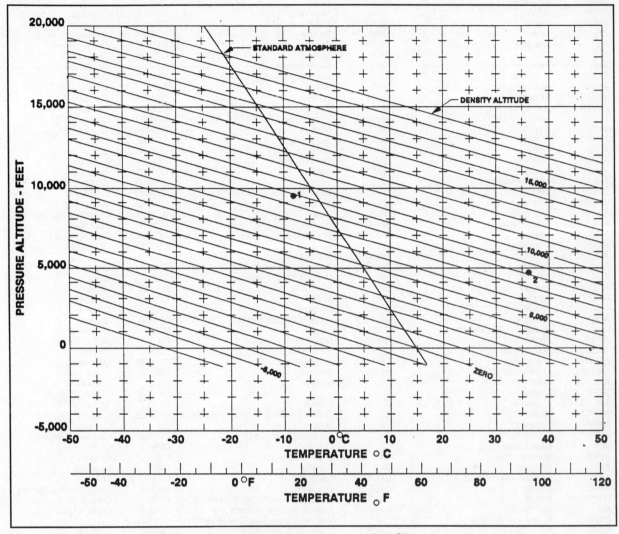

Figure 14-2. Density Altitude Computation Chart

## D. Contractions and Acronyms

1. Contractions and acronyms are used extensively in surface reports, pilot reports, and forecasts.

2. The following list of contractions and acronyms through page 375 is provided to assist you.

# Contractions and Acronyms

| | | | | | |
|---|---|---|---|---|---|
| A | Absolute (temperature) | ACFT | Aircraft | ADVCTG | Advecting |
| A | Alaskan Standard Time (time groups only) | ACLT | Accelerate | ADVCTN | Advection |
| | | ACLTD | Accelerated | ADVCTS | Advects |
| A | Arctic (air mass) | ACLTG | Accelerating | ADVN | Advance |
| A01 | Automated Observation without Precipitation Discriminator (rain/snow) (METAR) | ACLTS | Accelerates | ADVNG | Advancing |
| | | ACPY | Accompany | ADVY | Advisory |
| | | ACRS | Across | ADVYS | Advisories |
| A02 | Automated Observation with Precipitation Discriminator (rain/snow) (METAR) | ACSL | Altocumulus Standing Lenticular | AFC | Area Forecast Center |
| | | ACTV | Active | AFCT | Affect |
| | | ACTVTY | Activity | AFCTD | Affected |
| AAWF | Auxiliary Aviation Weather Facility | ACYC | Anticyclone | AFCTG | Affecting |
| AAWU | Alaska Aviation Weather Unit | ADJ | Adjacent | AFDK | After Dark |
| ABNDT | Abundant | ADL | Additional | AFOS | Automated Field Operations System |
| ABNML | Abnormal | ADQT | Adequate | AFSS | Aeronautical Flight Service Station |
| ABT | About | ADQTLY | Adequately | AFT | After |
| ABV | Above | ADRNDCK | Adirondack | AFTN | Afternoon |
| AC | Convective Outlook or Altocumulus | ADV | Advise | AGL | Above Ground Level |
| ACC | Altocumulus Castellanus | ADVCT | Advect | AGN | Again |
| ACCUM | Accumulate | ADVCTD | Advected | AGRD | Agreed |

|  |  |  |  |  |  |
|---|---|---|---|---|---|
| AGRMT | Agreement | BGN | Begin | CNTY | County |
| AGRS | Agrees | BGNG | Beginning | CNTYS | Counties |
| AHD | Ahead | BGNS | Begins | CNVG | Converge |
| AK | Alaska | BHND | Behind | CNVGG | Converging |
| AL | Alabama | BINOVC | Breaks in Overcast | CNVGNC | Convergence |
| ALF | Aloft | BKN | Broken | CNVTN | Convection |
| ALG | Along | BL | Between Layers | CNVTV | Convective |
| ALGHNY | Allegheny | BL | Blowing (METAR) | CNVTVLY | Convectively |
| ALQDS | All Quadrants | BLD | Build | CO | Colorado |
| ALSEC | All Sectors | BLDG | Building | COMPAR | Compare |
| ALSTG | Altimeter Setting | BLDUP | Buildup | COMPARD | Compared |
| ALT | Altitude | BLKHLS | Black Hills | COMPARG | Comparing |
| ALTA | Alberta | BLKT | Blanket | COMPARS | Compares |
| ALTHO | Although | BLKTG | Blanketing | COND | Condition |
| ALTM | Altimeter | BLKTS | Blankets | CONFDC | Confidence |
| ALUTN | Aleutian | BLO | Below clouds | CONT | Continue |
| ALWF | Actual Wind Factor | BLW | Below | CONTD | Continued |
| AM | Ante Meridiem | BLZD | Blizzard | CONTDVD | Continental Divide |
| AMD | Amend | BMS | Basic Meteorological Services | CONTG | Continuing |
| AMDD | Amended | BN | Blowing Sand | CONTLY | Continually |
| AMDG | Amending | BND | Bound | CONTRAILS | Condensation Trails |
| AMDT | Amendment | BNDRY | Boundary | CONTS | Continues |
| AMP | Amplify | BNDRYS | Boundaries | CONUS | Continental U.S. |
| AMPG | Amplifying | BNTH | Beneath | COORD | Coordinate |
| AMPLTD | Amplitude | BOOTHEEL | Bootheel | COR | Correction |
| AMS | Air Mass | BOVC | Base of Overcast | CPBL | Capable |
| AMT | Amount | BR | Mist (METAR) or Branch | CRC | Circle |
| ANLYS | Analysis | BRF | Brief | CRCLC | Circulate |
| ANS | Answer | BRK | Break | CRCLN | Circulation |
| AOA | At or Above | BRKG | Breaking | CRNR | Corner |
| AOB | At or Below | BRKHIC | Breaks in Higher Clouds | CRNRS | Corners |
| AP | Anomalous Propagation | BRKS | Breaks | CRS | Course |
| APCH | Approach | BRKSHR | Berkshire | CS | Cirrostratus |
| APCHG | Approaching | BRM | Barometer | CSDR | Consider |
| APCHS | Approaches | BS | Blowing Snow | CSDRBL | Considerable |
| APLCN | Appalachian | BTWN | Between | CST | Coast |
| APLCNS | Appalachians | BYD | Beyond | CSTL | Coastal |
| APPR | Appear | | | CT | Connecticut |
| APPRG | Appearing | C | Celsius | CTGY | Category |
| APPRS | Appears | C | Central Standard Time (time groups only) | CTSKLS | Catskills |
| APRNT | Apparent | | | CU | Cumulus |
| APRNTLY | Apparently | C | Continental (air mass) | CUFRA | Cumulus Fractus |
| APRX | Approximate | CA | California | CVR | Cover |
| APRXLY | Approximately | CAA | Cold Air Advection | CVRD | Covered |
| AR | Arkansas | CAN | Canada | CVRG | Covering |
| ARL | Air Resources Laboratory | CARIB | Caribbean | CVRS | Covers |
| ARND | Around | CASCDS | Cascades | CWSU | Center Weather Service Units |
| ARPT | Airport | CAVOK | Cloud and Visibility OK (METAR) | CYC | Cyclonic |
| AS | Altostratus | CAVU | Clear or Scattered Clouds and Visibility Greater Than Ten Miles | CYCLGN | Cyclogenesis |
| ASAP | As Soon As Possible | | | | |
| ASOS | Automated Surface Observing System | CAWS | Common Aviation Weather Sub-system | DABRK | Daybreak |
| | | | | DALGT | Daylight |
| ASSOCD | Associated | CB | Cumulonimbus | DBL | Double |
| ASSOCN | Association | CBMAM | Cumulonimbus Mamma | DC | District of Columbia |
| ATLC | Atlantic | CC | Cirrocumulus | DCAVU | Clear or Scattered Clouds and Visibility Greater Than Ten, Remainder of Report Missing (weather reports only) |
| ATTM | At This Time | CCLDS | Clear of Clouds | | |
| ATTN | Attention | CCLKWS | Counterclockwise | | |
| AURBO | Aurora Borealis | CCSL | Cirrocumulus Standing Lenticular | | |
| AVBL | Available | CDFNT | Cold Front | DCR | Decrease |
| AVG | Average | CFP | Cold Front Passage | DCRD | Decreased |
| AVN | Aviation Model | CHARC | Characteristic | DCRG | Decreasing |
| AWC | Aviation Weather Center | CHC | Chance | DCRGLY | Decreasingly |
| AWP | Aviation Weather Processors | CHCS | Chances | DCRS | Decreases |
| AWI | Awaiting | CHG | Change | DE | Delaware |
| AZ | Arizona | CHGD | Changed | DEG | Degree |
| AZM | Azimuth | CHGG | Changing | DEGS | Degrees |
| | | CHGS | Changes | DELMARVA | Delaware-Maryland-Virginia |
| B | Beginning of Precipitation (time in minutes) (weather reports only) | CHSPK | Chesapeake | DFCLT | Difficult |
| | | CI | Cirrus | DFCLTY | Difficulty |
| B | Bering Standard Time (time groups only) | CIG | Ceiling | DFNT | Definite |
| | | CIGS | Ceilings | DFNTLY | Definitely |
| BACLIN | Baroclinic | CLD | Cloud | DFRS | Differs |
| BAJA | Baja, California | CLDNS | Cloudiness | DFUS | Diffuse |
| BATROP | Barotropic | CLDS | Clouds | DGNL | Diagonal |
| BC | British Columbia | CLKWS | Clockwise | DGNLLY | Diagonally |
| BC | Patches (METAR) | CLR | Clear | DIGG | Digging |
| BCFG | Patchy Fog (METAR) | CLRG | Clearing | DIR | Direction |
| BCH | Beach | CLRS | Clears | DISC | Discontinue |
| BCKG | Backing | CMPLX | Complex | DISCD | Discontinued |
| BCM | Become | CNCL | Cancel | DISCG | Discontinuing |
| BCMG | Becoming | CNCLD | Canceled | DISRE | Disregard |
| BCMS | Becomes | CNCLG | Canceling | DISRED | Disregarded |
| BDA | Bermuda | CNCLS | Cancels | DISREG | Disregarding |
| BDRY | Boundary | CNDN | Canadian | DKTS | Dakotas |
| BECMG | Becoming | CNTR | Center | DLA | Delay |
| BFDK | Before Dark | CNTRD | Centered | DLAD | Delayed |
| BFR | Before | CNTRL | Central | DLT | Delete |

| | | | | | | | |
|---|---|---|---|---|---|
| DLTD | Deleted | ENCTR | Encounter | FQTLY | Frequently |
| DLTG | Deleting | ENDG | Ending | FRM | Form |
| DLY | Daily | ENE | East-Northeast | FRMG | Forming |
| DMG | Damage | ENELY | East-Northeasterly | FRMN | Formation |
| DMGD | Damaged | ENERN | East-Northeastern | FROPA | Frontal Passage |
| DMGG | Damaging | ENEWD | East-Northeastward | FROSFC | Frontal Surface |
| DMNT | Dominant | ENHNC | Enhance | FRST | Frost |
| DMSH | Diminish | ENHNCD | Enhanced | FRWF | Forecast Wind Factor |
| DMSHD | Diminished | ENHNCG | Enhancing | FRZ | Freeze |
| DMSHG | Diminishing | ENHNCMNT | Enhancement | FRZG | Freezing |
| DMSHS | Diminishes | ENHNCS | Enhances | FRZLVL | Freezing Level |
| DNDFTS | Downdrafts | ENTR | Entire | FRZN | Frozen |
| DNS | Dense | EOF | Expected Operations Forecast | FT | Feet |
| DNSLP | Downslope | ERN | Eastern | FTHR | Further |
| DNSTRM | Downstream | ERY | Early | FU | Smoke (METAR) |
| DNWND | Downwind | ERYR | Earlier | FULRY | Smoke Layer Aloft |
| DP | Deep | ESE | East-Southeast | FUOCTY | Smoke over City |
| DPND | Deepened | ESELY | East-Southeasterly | FVRBL | Favorable |
| DPNG | Deepening | ESERN | East-Southeastern | FWC | Fleet Weather Central |
| DPNS | Deepens | ESEWD | East-Southeastward | FWD | Forward |
| DPR | Deeper | ESNTL | Essential | FYI | For Your Information |
| DPTH | Depth | EST | Estimate | FZ | Supercooled/Freezing (METAR) |
| DR | Low Drifting (METAR) | ESTAB | Establish | | |
| DRFT | Drift | ETA | Estimated Time of Arrival or ETA | G | Gust |
| DRFTD | Drifted | | Model | GA | Georgia |
| DRFTG | Drifting | ETC | Et Cetera | GEN | General |
| DRFTS | Drifts | ETIM | Elapsed Time | GENLY | Generally |
| DRZL | Drizzle | EVE | Evening | GEO | Geographic |
| DS | Dust Storm (METAR) | EWD | Eastward | GEOREF | Geographical Reference |
| DSCNT | Descent | EXCLV | Exclusive | GF | Ground Fog |
| DSIPT | Dissipate | EXCLVLY | Exclusively | GICG | Glaze Icing |
| DSIPTD | Dissipated | EXCP | Except | GLFALSK | Gulf of Alaska |
| DSIPTG | Dissipating | EXPC | Expect | GLFCAL | Gulf of California |
| DSIPTN | Dissipation | EXPCD | Expected | GLFMEX | Gulf of Mexico |
| DSIPTS | Dissipates | EXPCG | Expecting | GLFSTLAWR | Gulf of St. Lawrence |
| DSND | Descend | EXTD | Extend | GND | Ground |
| DSNDG | Descending | EXTDD | Extended | GR | Hail (METAR) |
| DSNDS | Descends | EXTDG | Extending | GRAD | Gradient |
| DSNT | Distant | EXTDS | Extends | GRBNKS | Grand Banks |
| DSTBLZ | Destabilize | EXTN | Extension | GRDL | Gradual |
| DSTBLZD | Destabilized | EXTRAP | Extrapolate | GRDLY | Gradually |
| DSTBLZG | Destabilizing | EXTRAPD | Extrapolated | GRT | Great |
| DSTBLZN | Destabilization | EXTRM | Extreme | GRTLKS | Great Lakes |
| DSTBLZS | Destabilizes | EXTRMLY | Extremely | GRTLY | Greatly |
| DSTC | Distance | EXTSV | Extensive | GS | Small Hail/Snow Pellets (METAR) |
| DTLN | International Dateline | | | GSTS | Gusts |
| DTRT | Deteriorate | F | Fahrenheit | GSTY | Gusty |
| DTRTD | Deteriorated | FA | Aviation Area Forecast | GTS | Global Telecommunications System |
| DTRTG | Deteriorating | FAM | Familiar | | |
| DTRTS | Deteriorates | FC | Funnel Cloud (METAR) | HAZ | Hazard |
| DU | Widespread Dust (METAR) | +FC | Tornado/Water Spout (METAR) | HCVIS | High Clouds Visible |
| DURC | During Climb | FCST | Forecast | HDFRZ | Hard Freeze |
| DURD | During Descent | FCSTD | Forecasted | HDSVLY | Hudson Valley |
| DURG | During | FCSTG | Forecasting | HDWND | Head Wind |
| DURN | Duration | FCSTR | Forecaster | HGT | Height |
| DVLP | Develop | FCSTS | Forecasts | HI | Hawaii or High |
| DVLPD | Developed | FEW | 1 or 2 Octas (eighths) Cloud | HIEAT | Highest Temperature Equaled for All |
| DVLPG | Developing | | Coverage (METAR) | | Time |
| DVLPMT | Development | FG | Fog (METAR) | HIEFM | Highest Temperature Equaled for the |
| DVLPS | Develops | FIBI | Filed but Impractical to Transmit | | Month |
| DVRG | Diverge | FIG | Figure | HIER | Higher |
| DVRGG | Diverging | FILG | Filling | HIESE | Highest Temperature Equaled So |
| DVRGNC | Divergence | FINO | Weather Report Will Not Be Filed for | | Early |
| DVRGS | Diverges | | Transmission | HIESL | Highest Temperature Equaled So |
| DVV | Downward Vertical Velocity | FIR | Flight Information Region | | Late |
| DWNDFTS | Downdrafts | FIRAV | First Available | HIFOR | High Level Forecast |
| DWPNT | Dew Point | FL | Florida or Flight Level | HITMP | Highest Temperature |
| DWPNTS | Dew Points | FLDST | Flood Stage | HIXAT | Highest Temperature Exceeded for |
| DZ | Drizzle (METAR) | FLG | Falling | | All Time |
| | | FLRY | Flurry | HIXFM | Highest Temperature Exceeded for |
| E | East | FLRYS | Flurries | | the Month |
| E | Eastern Standard Time (time groups | FLT | Flight | HIXSE | Highest Temperature Exceeded So |
| | only) | FLW | Follow | | Early |
| E | Ending of Precipitation (time in | FLWG | Following | HIXSL | Highest Temperature Exceeded So |
| | minutes) (weather reports only) | FLWIS | Flood Warning Issued | | Late |
| E | Equatorial (air mass) | FM | From | HLF | Half |
| E | Estimated (weather reports only) | FMT | Format | HLSTO | Hailstones |
| EBND | Eastbound | FNCTN | Function | HLTP | Hilltop |
| EFCT | Effect | FNT | Front | HLYR | Haze Layer Aloft |
| ELNGT | Elongate | FNTGNS | Frontogenesis | HND | Hundred |
| ELNGTD | Elongated | FNTL | Frontal | HPC | Hydrometeorological Center |
| ELSW | Elsewhere | FNTLYS | Frontolysis | HR | Hour |
| EMBDD | Embedded | FNTS | Fronts | HRS | Hours |
| EMERG | Emergency | FORNN | Forenoon | HRZN | Horizon |
| EMSU | Environment Meteorological Support | FPM | Feet per Minute | HTG | Heating |
| | Unit | FQT | Frequent | HURCN | Hurricane |

| | | | | | |
|---|---|---|---|---|---|
| HUREP | Hurricane Report | LCTMP | Little Change in Temperature | MCD | Mesoscale Discussion |
| HV | Have | LCTN | Location | MD | Maryland |
| HVY | Heavy | LDG | Landing | MDFY | Modify |
| HVYR | Heavier | LEVEL | Level | MDFYD | Modified |
| HVYST | Heaviest | LFT | Lift | MDFYG | Modifying |
| HWVR | However | LFTG | Lifting | MDL | Model |
| HWY | Highway | LGRNG | Long Range | MDLS | Models |
| HX | High Index | LGT | Light | MDT | Moderate |
| HZ | Haze (METAR) | LGTR | Lighter | MDTLY | Moderately |
| | | LGWV | Long Wave | ME | Maine |
| IA | Iowa | LI | Lifted Index | MED | Medium |
| IC | Ice (in PIREPs only) | LIFR | Low IFR (weather reports only) | MEGG | Merging |
| ICAO | International Civil Aviation Organization | LIS | Lifted Indices | MESO | Mesoscale |
| | | LK | Lake | MET | Meteorological |
| ICG | Icing | LKLY | Likely | METAR | Aviation Routine Weather Report |
| ICGIC | Icing in Clouds | LKS | Lakes | METRO | Metropolitan |
| ICGICIP | Icing in Clouds and in Precipitation | LLJ | Low Level Jet | MEX | Mexico |
| ICGIP | Icing in Precipitation | LLWAS | Low-Level Wind Shear Alert System | MHKVLY | Mohawk Valley |
| ID | Idaho | LLWS | Low-Level Wind Shear | MI | Michigan |
| IFR | Instrument Flight Rules | LMTD | Limited | MI | Shallow (METAR) |
| IL | Illinois | LMTG | Limiting | MID | Middle |
| IMDT | Immediate | LMTS | Limits | MIDN | Midnight |
| IMDTLY | Immediately | LN | Line | MIFG | Patches of Shallow Fog Not Deeper Than Two Meters (METAR) |
| IMPL | Impulse | LO | Low | | |
| IMPLS | Impulses | LOEAT | Lowest Temperature Equaled for All Time | MIL | Military |
| IMPT | Important | | | MIN | Minimum |
| INCL | Include | LOEFM | Lowest Temperature Equaled for the Month | MISG | Missing |
| INCLD | Included | | | MLTLVL | Melting Level |
| INCLG | Including | LOESE | Lowest Temperature Equaled So Early | MMO | Main Meteorological Office |
| INCLS | Includes | | | MN | Minnesota |
| INCR | Increase | LOESL | Lowest Temperature Equaled So Late | MNLD | Mainland |
| INCRD | Increased | | | MNLY | Mainly |
| INCRG | Increasing | LONG | Longitude | MO | Missouri |
| INCRGLY | Increasingly | LONGL | Longitudinal | MOGR | Moderate or Greater |
| INCRS | Increases | LOTMP | Lowest Temperature | MONTR | Monitor |
| INDC | Indicate | LOXAT | Lowest Temperature Exceeded for All Time | MOV | Move |
| INDCD | Indicated | | | MOVD | Moved |
| INDCG | Indicating | LOXFM | Lowest Temperature Exceeded for the Month | MOVG | Moving |
| INDCS | Indicates | | | MOVMT | Movement |
| INDEF | Indefinite | LOXSE | Lowest Temperature Exceeded So Early | MOVS | Moves |
| INFO | Information | | | MPH | Miles per Hour |
| INLD | Inland | LOXSL | Lowest Temperature Exceeded So Late | MRGL | Marginal |
| INSTBY | Instability | | | MRGLLY | Marginally |
| INTCNTL | Intercontinental | LRG | Large | MRNG | Morning |
| INTL | International | LRGLY | Largely | MRTM | Maritime |
| INTMD | Intermediate | LRGR | Larger | MS | Minus |
| INTMT | Intermittent | LRGST | Largest | MS | Mississippi |
| INTMTLY | Intermittently | LSR | Loose Snow on Runway | MSG | Message |
| INTR | Interior | LST | Local Standard Time | MSL | Mean Sea Level |
| INTRMTRGN | Intermountain Region | LTD | Limited | MST | Most |
| INTS | Intense | LTG | Lightning | MSTLY | Mostly |
| INTSFCN | Intensification | LTGCC | Lightning Cloud-to-Cloud | MSTR | Moisture |
| INTSFY | Intensify | LTGCCCG | Lightning Cloud-to-Cloud Cloud-to-Ground | MT | Montana |
| INTSFYD | Intensified | | | MTN | Mountain |
| INTSFYG | Intensifying | LTGCG | Lightning Cloud-to-Ground | MTNS | Mountains |
| INTSFYS | Intensifies | LTGCW | Lightning Cloud-to-Water | MULT | Multiple |
| INTSTY | Intensity | LTGIC | Lightning in Cloud | MULTILVL | Multilevel |
| INTVL | Interval | LTL | Little | MVFR | Marginal VFR |
| INVOF | In Vicinity Of | LTLCG | Little Change | MWO | Meteorological Watch Office |
| INVRN | Inversion | LTR | Later | MXD | Mixed |
| IOVC | In Overcast | LTST | Latest | | |
| IP | Ice Pellets | LV | Leaving | N | North |
| IPV | Improve | LVL | Level | NAB | Not Above |
| IPVG | Improving | LVLS | Levels | NAT | North Atlantic |
| IR | Ice on Runway | LWR | Lower | NATL | National |
| ISOL | Isolate | LWRD | Lowered | NAV | Navigation |
| ISOLD | Isolated | LWRG | Lowering | NB | New Brunswick |
| | | LX | Low Index | NBND | Northbound |
| JCTN | Junction | LYR | Layer | NBRHD | Neighborhood |
| JTSTR | Jet Stream | LYRD | Layered | NC | North Carolina |
| | | LYRS | Layers | NCEP | National Centers for Environmental Prediction |
| KFRST | Killing Frost | | | | |
| KLYR | Smoke Layer Aloft | M | Maritime (air mass) | NCWX | No Change in Weather |
| KOCTY | Smoke over City | M | In Temperature Field Means "Minus" or below Zero (METAR) | ND | North Dakota |
| KS | Kansas | | | NE | Northeast |
| KT | Knots | M | In RVR Field, Indicates Visibility Less Than Lowest Reportable Sensor Value (e.g., M0600FT) | NEB | Nebraska |
| KY | Kentucky | | | NEC | Necessary |
| | | | | NEG | Negative |
| LA | Louisiana | M | Missing (weather reports only) | NEGLY | Negatively |
| LABRDR | Labrador | M | Mountain Standard Time (time groups only) | NELY | Northeasterly |
| LAT | Latitude | | | NERN | Northeastern |
| LAWRS | Limited Aviation Weather Reporting Station | MA | Map Analysis | NEW ENG | New England |
| | | MA | Massachusetts | NEWD | Northeastward |
| LCL | Local | MAN | Manitoba | NFLD | Newfoundland |
| LCLY | Locally | MAX | Maximum | NGM | Nested Grid Model |
| LCTD | Located | MB | Millibars | NGT | Night |

|  |  |
|---|---|
| NH | New Hampshire |
| NIL | None |
| NJ | New Jersey |
| NL | No Layers |
| NLT | Not Later Than |
| NLY | Northerly |
| NM | New Mexico |
| NMBR | Number |
| NMBRS | Numbers |
| NML | Normal |
| NMRS | Numerous |
| NNE | North-Northeast |
| NNELY | North-Northeasterly |
| NNERN | North-Northeastern |
| NNEWD | North-Northeastward |
| NNNN | End of Message |
| NNW | North-Northwest |
| NNWLY | North-Northwesterly |
| NNWRN | North-Northwestern |
| NNWWD | Northwestward |
| NO | Not Available (e.g., SLPNO, RVRNO) |
| NOAA | National Oceanic and Atmospheric Administration |
| NOPAC | Northern Pacific |
| NORPI | No Pilot Balloon Observation Will Be Filed Next Collection Unless Weather Changes Significantly |
| NPRS | Nonpersistent |
| NR | Near |
| NRLY | Nearly |
| NRN | Northern |
| NRW | Narrow |
| NS | Nimbostratus |
| NS | Nova Scotia |
| NSCSWD | No Small Craft or Storm Warning Being Displayed |
| NSW | No Significant Weather (TAF) |
| NTFY | Notify |
| NTFYD | Notified |
| NV | Nevada |
| NVA | Negative Vorticity Advection |
| NW | Northwest |
| NWD | Northward |
| NWLY | Northwesterly |
| NWRN | Northwestern |
| NWS | National Weather Service |
| NY | New York |
| NXT | Next |
| OAT | Outside Air Temperature |
| OBND | Outbound |
| OBS | Observation |
| OBSC | Obscure |
| OBSCD | Obscured |
| OBSCG | Obscuring |
| OCFNT | Occluded Front |
| OCLD | Occlude |
| OCLDD | Occluded |
| OCLDG | Occluding |
| OCLDS | Occludes |
| OCLN | Occlusion |
| OCNL | Occasional |
| OCNLY | Occasionally |
| OCR | Occur |
| OCRD | Occurred |
| OCRG | Occurring |
| OCRS | Occurs |
| OFC | Office |
| OFP | Occluded Frontal Passage |
| OFSHR | Offshore |
| OH | Ohio |
| OK | Oklahoma |
| OMTNS | Over Mountains |
| ONSHR | On Shore |
| ONT | Ontario |
| OR | Oregon |
| ORGPHC | Orographic |
| ORIG | Original |
| OSV | Ocean Station Vessel |
| OTAS | On Top and Smooth |
| OTLK | Outlook |
| OTP | On Top |
| OTR | Other |
| OTRW | Otherwise |
| OUTFLO | Outflow |
| OVC | Overcast |

|  |  |
|---|---|
| OVHD | Overhead |
| OVNGT | Overnight |
| OVR | Over |
| OVRN | Overrun |
| OVRNG | Overrunning |
| OVTK | Overtake |
| OVTKG | Overtaking |
| OVTKS | Overtakes |
| P | Pacific Standard Time (time group only) |
| P | Polar (air mass) |
| P | In RVR Field, Indicates Visibility Greater Than Highest Reportable Sensor Value (e.g., P6000FT) |
| P6SM | Visibility Greater Than 6 Statute Miles (TAF only) |
| PA | Pennsylvania |
| PAC | Pacific |
| PBL | Probable or Planetary Boundary Layer |
| PCPN | Precipitation |
| PD | Period |
| PDMT | Predominant |
| PDW | Priority Delayed Weather |
| PEN | Peninsula |
| PERM | Permanent |
| PGTSND | Puget Sound |
| PHYS | Physical |
| PIBAL | Pilot Balloon Observation |
| PIREP | Pilot Weather Report |
| PISE | No Pilot Balloon Observation Due to Unfavorable Sea Conditions |
| PISO | No Pilot Balloon Observation Due to Snow |
| PIWI | No Pilot Balloon Observation Due to High, or Gusty, Surface Wind |
| PL | Ice Pellets |
| PLNS | Plains |
| PLS | Please |
| PLTO | Plateau |
| PLW | Plow (snow) |
| PM | Postmeridian |
| PNHDL | Panhandle |
| PO | Dust/Sand Whirls (METAR) |
| POS | Positive |
| POSLY | Positively |
| PPINA | Radar Weather Report Not Available |
| PPINE | Radar Weather Report No Echoes Observed |
| PPINO | Radar Weather Report Equipment Inoperative Due to Breakdown |
| PPIOK | Radar Weather Report Equipment Operation Resumed |
| PPIOM | Radar Weather Report Equipment Inoperative Due to Maintenance |
| PPSN | Present Position |
| PR | Partial (METAR) |
| PRBL | Probable |
| PRBLTY | Probability |
| PRBLY | Probably |
| PRECD | Precede |
| PRECDD | Preceded |
| PRECDG | Preceding |
| PRECDS | Precedes |
| PRES | Pressure |
| PRESFR | Pressure Falling Rapidly |
| PRESRR | Pressure Rising Rapidly |
| PRIM | Primary |
| PRIN | Principal |
| PRIND | Present Indications Are |
| PRJMP | Pressure Jump |
| PROB30 | Probability 30 Percent (TAF) |
| PROB40 | Probability 40 Percent (TAF) |
| PROB | Probability |
| PROC | Procedure |
| PROD | Produce |
| PRODG | Producing |
| PROG | Prognosis or Prognostic or Forecast |
| PROGD | Forecasted |
| PROGS | Forecasts |
| PRSNT | Present |
| PRSNTLY | Presently |
| PRST | Persist |
| PRSTNC | Persistence |
| PRSTS | Persists |

|  |  |
|---|---|
| PRSTNT | Persistent |
| PRVD | Provide |
| PRVDD | Provided |
| PRVDG | Providing |
| PRVDS | Provides |
| PS | Plus |
| PSBL | Possible |
| PSBLTY | Possibility |
| PSBLY | Possibly |
| PSG | Passage |
| PSG | Passing |
| PSN | Position |
| PSND | Positioned |
| PTCHY | Patchy |
| PTLY | Partly |
| PTNL | Potential |
| PTNLY | Potentially |
| PTNS | Portions |
| PUGET | Puget Sound |
| PY | Spray (METAR) |
| PVA | Positive Vorticity Advection |
| PVL | Prevail |
| PVLD | Prevailed |
| PVLG | Prevailing |
| PVLS | Prevails |
| PVLT | Prevalent |
| PWB | Pilot Weather Briefing |
| PWR | Power |
| QN | Question |
| QSTNRY | Quasistationary |
| QTR | Quarter |
| QUAD | Quadrant |
| QUE | Quebec |
| R | Rain (Radar Weather Report) |
| R | Runway (used in RVR measurement) |
| RA | Rain (METAR) |
| RABA | No RAWIN Obs., No Balloons Available |
| RABAL | Radiosonde Balloon Wind Data |
| RABAR | Radiosonde Balloon Release |
| RACO | No RAWIN Obs., Communications Out |
| RADAT | Radiosonde Additional Data |
| RADNO | Report Missing Account Radio Failure |
| RAFI | Radiosonde Observation Not Filed |
| RAFRZ | Radiosonde Observation Freezing Levels |
| RAHE | No RAWIN Obs., No Gas Available |
| RAICG | Radiosonde Observation Icing |
| RAOB | Radiosonde Observation |
| RAVU | Radiosonde Analysis and Verification Unit |
| RAWE | No RAWIN Obs., Unfavorable Weather |
| RAWI | No RAWIN Obs., High and Gusty Winds |
| RAWIN | Upper Winds Obs. (by radio methods) |
| RCD | Radar Cloud Detection Report |
| RCDNA | Radar Cloud Detection Report Not Available |
| RCDNE | Radar Cloud Detection Report No Echoes Observed |
| RCDNO | Radar Cloud Detection Inoperative Due to Breakdown Until |
| RCDOM | Radar Cloud Detection Inoperative Due to Maintenance Until |
| RCH | Reach |
| RCHD | Reached |
| RCHG | Reaching |
| RCHS | Reaches |
| RCKY | Rocky |
| RCKYS | Rockies |
| RCMD | Recommend |
| RCMDD | Recommended |
| RCMDG | Recommending |
| RCMDS | Recommends |
| RCRD | Record |
| RCRDS | Records |
| RCV | Receive |
| RCVD | Received |
| RCVG | Receiving |
| RCVS | Receives |

| | | | | | |
|---|---|---|---|---|---|
| RDC | Reduce | S | South | SNWFL | Snowfall |
| RDG | Ridge | SA | Sand (METAR) | SOP | Standard Operating Procedure |
| RDGG | Ridging | SAB | Satellite Analysis Branch | SP | Station Pressure |
| RDVLP | Redevelop | SASK | Saskatchewan | SPC | Storm Prediction Center |
| RDVLPG | Redeveloping | SATFY | Satisfactory | SPCLY | Especially |
| RDVLPMT | Redevelopment | SBND | Southbound | SPD | Speed |
| RDWND | Radar Dome Wind | SBSD | Subside | SPECI | Special Report (METAR) |
| RE | Regard | SBSDD | Subsided | SPKL | Sprinkle |
| RECON | Reconnaissance | SBSDNC | Subsidence | SPLNS | Southern Plains |
| REF | Reference | SBSDS | Subsides | SPRD | Spread |
| REPL | Replace | SC | South Carolina or Stratocumulus | SPRDG | Spreading |
| REPLD | Replaced | SCND | Second | SPRDS | Spreads |
| REPLG | Replacing | SCNDRY | Secondary | SPRL | Spiral |
| REPLS | Replaces | SCSL | Stratocumulus Standing Lenticular | SQ | Squall (METAR) |
| REQ | Request | SCT | Scatter | SQAL | Squall |
| REQS | Requests | SCTD | Scattered | SQLN | Squall Line |
| REQSTD | Requested | SCTR | Sector | SR | Sunrise |
| RES | Reserve | SD | Radar Weather Report | SRN | Southern |
| RESP | Response | SD | South Dakota | SRND | Surround |
| RESTR | Restrict | SE | Southeast | SRNDD | Surrounded |
| RGD | Ragged | SEC | Second | SRNDG | Surrounding |
| RGL | Regional Model | SELS | Severe Local Storms | SRNDS | Surrounds |
| RGLR | Regular | SELY | Southeasterly | SS | Sandstorm (METAR) |
| RGN | Region | SEPN | Separation | SS | Sunset |
| RGNS | Regions | SEQ | Sequence | SSE | South-Southeast |
| RGT | Right | SERN | Southeastern | SSELY | South-Southeasterly |
| RH | Relative Humidity | SEV | Severe | SSERN | South-Southeastern |
| RHINO | Radar Echo Height Information Not Available | SEWD | Southeastward | SSEWD | South-Southeastward |
| | | SFC | Surface | SSW | South-Southwest |
| RHINO | Radar Range Height Indicator Not Operating on Scan | SFERICS | Atmospherics | SSWLY | South-Southwesterly |
| | | SG | Snow Grains (METAR) | SSWRN | South-Southwestern |
| RI | Rhode Island | SGFNT | Significant | SSWWD | South-Southwestward |
| RIOGD | Rio Grande | SGFNTLY | Significantly | ST | Stratus |
| RLBL | Reliable | SGD | Solar-Geophysical Data | STAGN | Stagnation |
| RLTV | Relative | SH | Showers (METAR) | STBL | Stable |
| RLTVLY | Relatively | SHFT | Shift | STBLTY | Stability |
| RMK | Remark | SHFTD | Shifted | STD | Standard |
| RMN | Remain | SHFTG | Shifting | STDY | Steady |
| RMND | Remained | SHFTS | Shifts | STFR | Stratus Fractus |
| RMNDR | Remainder | SHLD | Shield | STFRM | Stratiform |
| RMNG | Remaining | SHLW | Shallow | STG | Strong |
| RMNS | Remains | SHRT | Short | STGLY | Strongly |
| RNFL | Rainfall | SHRTLY | Shortly | STGR | Stronger |
| ROBEPS | Radar Operating Below Prescribed Standard | SHRTWV | Shortwave | STGST | Strongest |
| | | SHUD | Should | STM | Storm |
| ROT | Rotate | SHWR | Shower | STMS | Storms |
| ROTD | Rotated | SIERNEV | Sierra Nevada | STN | Station |
| ROTG | Rotating | SIG | Signature | STNRY | Stationary |
| ROTS | Rotates | SIGMET | Significant Meteorological Information | SUB | Substitute |
| RPD | Rapid | | | SUBTRPCL | Subtropical |
| RPDLY | Rapidly | SIMUL | Simultaneous | SUF | Sufficient |
| RPLC | Replace | SIR | Snow and Ice on Runway | SUFLY | Sufficiently |
| RPLCD | Replaced | SKC | Sky Clear | SUG | Suggest |
| RPLCG | Replacing | SKED | Schedule | SUGG | Suggesting |
| RPLCS | Replaces | SLD | Solid | SUGS | Suggests |
| RPRT | Report | SLGT | Slight | SUP | Supply |
| RPRTD | Reported | SLGTLY | Slightly | SUPG | Supplying |
| RPRTG | Reporting | SLO | Slow | SUPR | Superior |
| RPRTS | Reports | SLOLY | Slowly | SUPSD | Supersede |
| RPT | Repeat | SLOR | Slower | SUPSDG | Superseding |
| RPTG | Repeating | SLP | Slope | SUPSDS | Supersedes |
| RPTS | Repeats | SLP | Sea Level Pressure (e.g., 1013.2 reported as 132) | SVG | Serving |
| RQR | Require | | | SVR | Severe |
| RQRD | Required | SLPG | Sloping | SVRL | Several |
| RQRG | Requiring | SLR | Slush on Runway | SW | Southwest |
| RQRS | Requires | SLT | Sleet | SWD | Southward |
| RSG | Rising | SLW | Slow | SWLG | Swelling |
| RSN | Reason | SLY | Southerly | SWLY | Southwesterly |
| RSNG | Reasoning | SM | Statute Mile(s) | SWRN | Southwestern |
| RSNS | Reasons | SMK | Smoke | SWWD | Southwestward |
| RSTR | Restrict | SML | Small | SX | Stability Index |
| RSTRD | Restricted | SMLR | Smaller | SXN | Section |
| RSTRG | Restricting | SMRY | Summary | SYNOP | Synoptic |
| RSTRS | Restricts | SMTH | Smooth | SYNS | Synopsis |
| RTRN | Return | SMTHR | Smoother | SYS | System |
| RTRND | Returned | SMTHST | Smoothest | | |
| RTRNG | Returning | SMTM | Sometime | T | Trace |
| RTRNS | Returns | SMWHT | Somewhat | T | Tropical (air mass) |
| RUF | Rough | SN | Snow (METAR) | TAF | Aviation Terminal Forecast |
| RUFLY | Roughly | SNBNK | Snowbank | TCNTL | Transcontinental |
| RVS | Revise | SNFLK | Snowflake | TCU | Towering Cumulus |
| RVSD | Revised | SNGL | Single | TDA | Today |
| RVSG | Revising | SNOINCR | Snow Increase | TEMP | Temperature |
| RVSS | Revises | SNOINCRG | Snow Increasing | TEMPO | Temporary Changes Expected (between 2-digit beginning hour and 2-digit ending hour) |
| RWY | Runway | SNST | Sunset | | |
| | | SNW | Snow | | |

| | | | | | |
|---|---|---|---|---|---|
| THD | Thunderhead (non METAR) | UNUSBL | Unusable | WLY | Westerly |
| THDR | Thunder (non METAR) | UP | Unknown Precipitation (automated | WND | Wind |
| THK | Thick | | observations) | WNDS | Winds |
| THKNG | Thickening | UPDFTS | Updrafts | WNW | West-Northwest |
| THKNS | Thickness | UPR | Upper | WNWLY | West-Northwesterly |
| THKR | Thicker | UPSLP | Upslope | WNWRN | West-Northwestern |
| THKST | Thickest | UPSTRM | Upstream | WNWWD | West-Northwestward |
| THN | Thin | URG | Urgent | WO | Without |
| THNG | Thinning | USBL | Usable | WPLTO | Western Plateau |
| THNR | Thinner | UT | Utah | WR | Wet Runway |
| THNST | Thinnest | UTC | Coordinated Universal Time | WRM | Warm |
| THR | Threshold | UVV | Upward Vertical Velocity | WRMFNT | Warm Front |
| THRFTR | Thereafter | UWNDS | Upper Winds | WRMFNTL | Warm Frontal |
| THRU | Through | | | WRMG | Warming |
| THRUT | Throughout | V | Varies (wind direction and RVR) | WRMR | Warmer |
| THSD | Thousand | V | Variable (weather reports only) | WRMST | Warmest |
| THTN | Threaten | VA | Virginia | WRN | Western |
| THTND | Threatened | VA | Volcanic Ash (METAR) | WRNG | Warning |
| THTNG | Threatening | VAAC | Volcanic Ash Advisory Center | WRS | Worse |
| THTNS | Threatens | VAAS | Volcanic Ash Advisory Statement | WS | Wind Shear |
| TIL | Until | VAL | Valley | WS | SIGMET |
| TKOF | Takeoff | VARN | Variation | WSFO | Weather Service Forecast Office |
| TMPRY | Temporary | VC | Vicinity | WSHFT | Wind Shift |
| TMPRYLY | Temporarily | VCNTY | Vicinity | WSOM | Weather Service Operations Manual |
| TMW | Tomorrow | VCOT | VFR Conditions on Top | WSR | Wet Snow on Runway |
| TN | Tennessee | VCTR | Vector | WSTCH | Wasatch Range |
| TNDCY | Tendency | VFR | Visual Flight Rules | WSW | West-Southwest |
| TNDCYS | Tendencies | VFY | Verify | WSWLY | West-Southwesterly |
| TNGT | Tonight | VFYD | Verified | WSWRN | West-Southwestern |
| TNTV | Tentative | VFYG | Verifying | WSWWD | West-Southwestward |
| TNTVLY | Tentatively | VFYS | Verifies | WTR | Water |
| TOP | Cloud Top | VLCTY | Velocity | WTSPT | Waterspout |
| TOPS | Tops | VLCTYS | Velocities | WUD | Would |
| TOVC | Top of Overcast | VLNT | Violent | WV | Wave |
| TPG | Topping | VLNTLY | Violently | WV | West Virginia |
| TRBL | Trouble | VLY | Valley | WVS | Waves |
| TRIB | Tributary | VMC | Visual Meteorological Conditions | WW | Severe Weather Watch |
| TRKG | Tracking | VOL | Volume | WWD | Westward |
| TRML | Terminal | VORT | Vorticity | WX | Weather |
| TRMT | Terminate | VR | Veer | WXCON | Weather Reconnaissance Flight Pilot |
| TRMTD | Terminated | VRB | Variable Wind Direction When Speed | | Report |
| TRMTG | Terminating | | Is Less Than or Equal to 6 Knots | WY | Wyoming |
| TRMTS | Terminates | VRBL | Variable | | |
| TRNSP | Transport | VRG | Veering | XCP | Except |
| TRNSPG | Transporting | VRISL | Vancouver Island, BC | XPC | Expect |
| TROF | Trough | VRS | Veers | XPCD | Expected |
| TROFS | Troughs | VRT MOTN | Vertical Motion | XPCG | Expecting |
| TROP | Tropopause | VRY | Very | XPCS | Expects |
| TRPCD | Tropical Continental Air Mass | VSB | Visible | XPLOS | Explosive |
| TRPCL | Tropical | VSBY | Visibility | XTND | Extend |
| TRPLYR | Trapping Layer | VSBYDR | Visibility Decreasing Rapidly | XTNDD | Extended |
| TRRN | Terrain | VSBYIR | Visibility Increasing Rapidly | XTNDG | Extending |
| TRSN | Transition | VT | Vermont | XTRM | Extreme |
| TS | Thunderstorm | VV | Vertical Velocity | XTRMLY | Extremely |
| TSFR | Transfer | VV | Vertical Visibility (indefinite ceiling) | | |
| TSFRD | Transferred | | (METAR) | Y | Yukon Standard Time (time groups |
| TSFRG | Transferring | | | | only) |
| TSFRS | Transfers | W | Warm (air mass) | YDA | Yesterday |
| TSHWR | Thundershower (non METAR) | W | West | YKN | Yukon |
| TSNT | Transient | WA | AIRMET | YLSTN | Yellowstone |
| TSQLS | Thundersqualls (non METAR) | WA | Washington | | |
| TSTM | Thunderstorm (non METAR) | WAA | Warm Air Advection | ZI | Zonal Index |
| TURBC | Turbulence | WAFS | World Area Forecast System | ZI | Zone of Interior |
| TURBT | Turbulent | WBND | Westbound | ZN | Zone |
| TWD | Toward | WDC-1 | World Data Centers in Western | ZNS | Zones |
| TWDS | Towards | | Europe | | |
| TWI | Twilight | WDC-2 | World Data Centers Throughout Rest | | |
| TWRG | Towering | | of World | | |
| TX | Texas | WDLY | Widely | | |
| | | WDSPRD | Widespread | | |
| UA | Pilot Weather Reports | WEA | Weather | | |
| UAG | Upper Atmosphere Geophysics | WFO | Weather Forecast Office | | |
| UDDF | Up and Down Drafts | WFSO | Weather Forecast Service Office | | |
| UN | Unable | WFP | Warm Front Passage | | |
| UNAVBL | Unavailable | WI | Wisconsin | | |
| UNEC | Unnecessary | WIBIS | Will Be Issued | | |
| UNKN | Unknown | WINT | Winter | | |
| UNL | Unlimited | WK | Weak | | |
| UNRELBL | Unreliable | WKDAY | Weekday | | |
| UNRSTD | Unrestricted | WKEND | Weekend | | |
| UNSATFY | Unsatisfactory | WKN | Weaken | | |
| UNSBL | Unseasonable | WKNG | Weakening | | |
| UNSTBL | Unstable | WKNS | Weakens | | |
| UNSTDY | Unsteady | WKR | Weaker | | |
| UNSTL | Unsettle | WKST | Weakest | | |
| UNSTLD | Unsettled | WL | Will | | |

E.  **Scheduled Issuance and Valid Times of Forecast Products**

1.  Table 14-1 below shows scheduled issuance and valid times of the TAFs. All times are UTC.

| Scheduled Issuance Times | Valid Period | Transmission Period |
|---|---|---|
| 0000 | 0000-0000 | 2320-2340 |
| 0600 | 0600-0600 | 0520-0540 |
| 1200 | 1200-1200 | 1120-1140 |
| 1800 | 1800-1800 | 1720-1740 |

Table 14-1.  Scheduled Issuance and Valid Times of TAFs

2.  Table 14-2 below has scheduled issuance and valid times of TWEBs. All times are UTC.

| Scheduled Issuance Times | Valid Period | Transmission Period |
|---|---|---|
| 0200 | 0200-1400 | 0130-0140 |
| 0800 | 0800-2000 | 0730-0740 |
| 1400 | 1400-0200 | 1330-1340 |
| 2000 | 2000-0800 | 1930-1940 |

Table 14-2.  Scheduled Issuance and Valid Times of TWEBs

3.  Table 14-3 below shows the scheduled issuance times of the FAs for their respective areas. The FA is valid 1 hr. after issuance time. All times are UTC. The times the FA is issued depends on whether the FA area is in local daylight time (LDT) or local standard time (LST).

| Area Forecast (FA) | Boston and Miami (LDT/LST) | Chicago and Ft. Worth (LDT/LST) | San Francisco and Salt Lake City (LDT/LST) | Alaska (LDT/LST) | Hawaii |
|---|---|---|---|---|---|
| 1st issuance | 0845/0945 | 0945/1045 | 1045/1145 | 0145/0245 | 0345 |
| 2nd issuance | 1745/1845 | 1845/1945 | 1945/2045 | 0745/0845 | 0945 |
| 3rd issuance | 0045/0145 | 0145/0245 | 0245/0345 | 1345/1445 | 1545 |
| 4th issuance | | | | 1945/2045 | 2145 |

Table 14-3.  Scheduled Issuance Times of FAs

4.  Table 14-4 below shows the scheduled issuance times of the Gulf of Mexico FA. All times are UTC.

| Gulf of Mexico FA | Issuance Times (LDT/LST) |
|---|---|
| 1st issuance | 1040/1140 |
| 2nd issuance | 1740/1840 |

Table 14-4.  Scheduled Issuance Times of the Gulf of Mexico FA

## F. National Weather Service Identifiers

### 1. Northeast Region

| | | | |
|---|---|---|---|
| AKQ | Norfolk/Wakefield, VA | GYX | Portland/Gray, ME |
| ALY | Albany/East Berne, NY | ILN | Cincinnati/Wilmington, OH |
| BGM | Binghamton, NY | LWX | Washington, DC/Sterling, VA |
| BOX | Boston/Taunton, MA | OKX | New York City/Brookhaven, NY |
| BTV | Burlington, VT | PBZ | Pittsburgh/Coraopolis, PA |
| BUF | Buffalo, NY | PHI | Philadelphia, PA/Mount Holly, NJ |
| CLE | Cleveland, OH | RLX | Charleston/Ruthdale, WV |
| CTP | State College, PA | RNK | Roanoke/Blacksburg, VA |

### 2. South Central Region

| | | | |
|---|---|---|---|
| AMA | Amarillo, TX | LUB | Lubbock, TX |
| BMX | Birmingham, AL | LZK | North Little Rock, AR |
| BRO | Brownsville, TX | MAF | Midland, TX |
| CRP | Corpus Christi, TX | MEG | Memphis/Germantown, TN |
| EPZ | El Paso, TX/Santa Theresa, NM | MOB | Mobile, MS |
| EWX | Austin/San Antonio, TX | MRX | Knoxville/Tri Cities, TN |
| FWD | Dallas/Forth Worth, TX | OHX | Nashville/Old Hickory, TN |
| HGX | Houston/Dickinson, TX | OUN | Oklahoma City/Norman, OK |
| JAN | Jackson, MS | SHV | Shreveport, LA |
| LCH | Lake Charles, LA | SJT | San Angelo, TX |
| LIX | New Orleans/Slidell, LA | TSA | Tulsa, OK |

### 3. Southeast Region

| | | | |
|---|---|---|---|
| CAE | Columbia, SC | MHX | Morehead City/Newport, NC |
| CHS | Charleston, SC | MLB | Melbourne, FL |
| FFC | Atlanta/Peachtree City, GA | RAH | Raleigh/Durham, NC |
| GSP | Greenville-Spartanburg/Greer, SC | TAE | Tallahassee, FL |
| ILM | Wilmington, NC | TBW | Tampa/Ruskin, FL |
| JAX | Jacksonville, FL | TJSJ | San Juan, PR |
| MFL | Miami, FL | | |

### 4. Mountain Region

| | | | |
|---|---|---|---|
| ABQ | Albuquerque, NM | PIH | Pocatello, ID |
| BIL | Billings, MT | PSR | Phoenix, AZ |
| BOI | Boise, ID | PUB | Pueblo, CO |
| BOU | Denver/Boulder, CO | REV | Reno, NV |
| CYS | Cheyenne, WY | RIW | Riverton, WY |
| FGZ | Flagstaff/Bellemont, AZ | SLC | Salt Lake City, UT |
| GGW | Glasgow, MT | TFX | Great Falls, MT |
| GJT | Grand Junction, CO | TWC | Tucson, AZ |
| LKN | Elko, NV | VEF | Las Vegas, NV |
| MSO | Missoula, MT | | |

### 5. West Coast Region

| | | | |
|---|---|---|---|
| EKA | Eureka, CA | PDT | Pendleton, OR |
| HNX | Hanford, CA | PQR | Portland, OR |
| LOX | Los Angeles/Oxnard, CA | SEW | Seattle, WA |
| MFR | Medford, OR | SGX | San Diego, CA |
| MTR | San Francisco/Monterey, CA | STO | Sacramento, CA |
| OTX | Spokane, WA | | |

6. **North Central Region**

| | | | | |
|---|---|---|---|---|
| ABR | Aberdeen, SD | | ICT | Wichita, KS |
| APX | Alpena/Gaylord, MI | | ILX | Lincoln, IL |
| ARX | La Crosse, WI | | IND | Indianapolis, IN |
| BIS | Bismarck, ND | | JKL | Jackson/Noctor, KY |
| DDC | Dodge City, KS | | LBF | North Platte, NE |
| DLH | Duluth, MN | | LMK | Louisville, KY |
| DMX | Des Moines/Johnston, IA | | LOT | Chicago/Romeoville, IL |
| DTX | Detroit/Pontiac, MI | | LSX | St. Louis, MO |
| DVN | Quad Cities/Davenport, IA | | MPX | Minneapolis/Chanhassen, MN |
| FGF | Fargo/Grand Forks, ND | | MKX | Milwaukee/Dousman, WI |
| EAX | Kansas City/Pleasant Hill, MO | | MQT | Marquette, MI |
| FSD | Sioux Falls, SD | | OAX | Omaha/Valley, NE |
| GID | Hastings, NE | | PAH | Paducah, KY |
| GLD | Goodland, KS | | SGF | Springfield, MO |
| GRB | Green Bay, WI | | TOP | Topeka, KS |
| GRR | Grand Rapids, MI | | UNR | Rapid City, SD |

7. **Alaskan Region**

| | |
|---|---|
| PAFC | Anchorage, AK |
| PAFG | Fairbanks, AK |
| PAJK | Juneau, AK |

8. **Pacific Region**

| | |
|---|---|
| PGUA | Tiyan, GU |
| PHFO | Honolulu, HI |

G. **WSR-88D Radar Sites**

| | | | | |
|---|---|---|---|---|
| ABC | Bethel, AK | | BUF | Buffalo/Cheektowaga, NY |
| ABR | Aberdeen, SD | | BYX | Key West/Boca Chica Key, FL |
| ABX | Albuquerque, NM | | CAE | Columbia, SC |
| ACG | Sitka/Biorka Island, AK | | CBW | Caribou/Hodgdon, ME |
| AEC | Nome, AK | | CBX | Boise/Ada County, ID |
| AHG | Anchorage/Nikiski, AK | | CCX | State College/Rush, PA |
| AIH | Middleton Island, AK | | CLE | Cleveland, OH |
| AKC | King Salmon, AK | | CLX | Charleston/Grays, SC |
| AKQ | Norfolk/Wakefield, VA | | CRP | Corpus Christi, TX |
| AMA | Amarillo, TX | | CXX | Burlington/Colchester, VT |
| AMX | Miami, FL | | CYS | Cheyenne, WY |
| APD | Fairbanks, AK | | DAX | Sacramento, CA |
| APX | Gaylord, MI | | DDC | Dodge City, KS |
| ARX | La Crosse, WI | | DFX | Del Rio/Laughlin AFB, TX |
| ATX | Seattle-Tacoma/Camano Island, WA | | DIX | Philadelphia, PA/Fort Dix, NJ |
| BBX | Marysville/Beale AFB, CA | | DLH | Duluth, MN |
| BGM | Binghamton, NY | | DMX | Des Moines/Johnston, IA |
| BHX | Eureka/Bunker Hill, CA | | DOX | Dover AFB, DE |
| BIS | Bismarck, ND | | DTX | Detroit-Pontiac/White Lake, MI |
| BIX | Keesler AFB, MS | | DVN | Quad Cities/Davenport, IA |
| BLX | Billings/Yellowstone County, MT | | DYX | Abilene/Dyess AFB, TX |
| BMX | Birmingham/Alabaster, AL | | EAX | Kansas City/Pleasant Hill, MO |
| BOX | Boston/Taunton, MA | | EMX | Tucson/Pima County, AZ |
| BRO | Brownsville, TX | | ENX | Albany/East Berne, NY |

| | | | |
|---|---|---|---|
| EOX | Fort Rucker, AL | LVX | Louisville/Fort Knox, KY |
| EPZ | El Paso, TX/Santa Teresa, NM | LWX | Baltimore, MD-Washington, DC/Sterling, VA |
| ESX | Las Vegas/Nelson, NV | | |
| EVX | Red Bay/Eglin AFB, FL | LZK | North Little Rock, AR |
| EWX | Austin-San Antonio/New Braunfels, TX | MAF | Midland/Odessa, TX |
| EYX | Edwards AFB, CA | MAX | Medford/Mount Ashland, OR |
| FCX | Roanoke/Coles Knob, VA | MBX | Minot AFB, ND |
| FDR | Frederick/Altus AFB, OK | MHX | Morehead City/Newport, NC |
| FDX | Clovis/Cannon AFB, NM | MKX | Milwaukee/Dousman, WI |
| FFC | Atlanta/Peachtree City, GA | MLB | Melbourne, FL |
| FSD | Sioux Falls, SD | MOB | Mobile, AL |
| FSX | Flagstaff/Coconino, AZ | MPX | Minneapolis/Chanhassen, MN |
| FTG | Denver/Boulder, CO | MQT | Marquette/Negaunee, MI |
| FWS | Dallas/Fort Worth, TX | MRX | Knoxville-Cities/Morristown, TN |
| GGW | Glasgow, MT | MSX | Missoula/Point Six Mountain, MT |
| GJX | Grand Junction/Mesa, CO | MTX | Salt Lake City/Promontory Point, UT |
| GLD | Goodland, KS | MUX | San Francisco/Mount Umunhum, CA |
| GRB | Green Bay/Ashwaubenon, WI | MVX | Fargo-Grand Forks/Mayville, ND |
| GRK | Killeen/Fort Hood, TX | MXX | Carrville/Maxwell AFB, AL |
| GRR | Grand Rapids, MI | NKX | San Diego/Miramar Nas, CA |
| GSP | Greenville-Spartanburg/Greer, SC | NQA | Memphis/Millington, TN |
| GUA | Agana, GU | OAX | Omaha/Valley, NE |
| GWX | Columbus AFB, MS | OHX | Nashville/Old Hickory, TN |
| GYX | Portland/Gray, ME | OKX | New York City/Upton, NY |
| HDX | Alamogordo/Holloman AFB, NM | OTX | Spokane, WA |
| HGX | Houston-Galveston/Dickinson, TX | PAH | Paducah, KY |
| HKI | South Kauai/Numila, HI | PBZ | Pittsburgh/Coraopolis, PA |
| HKM | Kamuela/Puu Mala, HI | PDT | Pendleton, OR |
| HMO | Molokai/Kukui, HI | POE | Fort Polk, LA |
| HNX | San Joaquin Valley/Hanford, CA | PUX | Pueblo, CO |
| HPX | Fort Campbell, KY | RAX | Raleigh-Durham/Clayton, NC |
| HTX | Hytop, AL | RGX | Reno/Virginia Peak, NV |
| HWA | South Hawaii/Naalehu, HI | RIW | Riverton, WY |
| ICT | Wichita, KS | RLX | Charleston/Ruthdale, WV |
| ICX | Cedar City, UT | RMX | Rome/Griffiss AFB, NY |
| ILN | Cincinnati/Wilmington, OH | RTX | Portland/Scappoose, OR |
| ILX | Lincoln, IL | SFX | Pocatello-Idaho Falls/Springfield, ID |
| IND | Indianapolis, IN | SGF | Springfield, MO |
| INX | Tulsa/Inola, OK | SHV | Shreveport, LA |
| IWA | Phoenix/Mesa, AZ | SJT | San Angelo, TX |
| IWX | North Webster, IN | SOX | Santa Ana Mountains/Orange County, CA |
| JAN | Jackson, MS | SRX | Slatington Mountain, AR |
| JAX | Jacksonville, FL | TBW | Tampa/Ruskin, FL |
| JGX | Warner Robins/Robins AFB, GA | TFX | Great Falls, MT |
| JKL | Jackson/Noctor, KY | TLH | Tallahassee, FL |
| JUA | San Juan/Cayey, PR | TLX | Oklahoma City/Norman, OK |
| LBB | Lubbock, TX | TWX | Topeka/Alma, KS |
| LCH | Lake Charles, LA | TYX | Fort Drum, NY |
| LIX | New Orleans-Baton Rouge/Slidell, LA | UDX | Rapid City/New Underwood, SD |
| LNX | North Platte/Thedford, NE | UEX | Hastings/Blue Hill, NE |
| LOT | Chicago/Romeoville, IL | VAX | Valdosta/Moody AFB, GA |
| LRX | Elko/Sheep Creek Mountain, NV | VBX | Lompoc/Vandenberg AFB, CA |
| LSX | St. Louis/Research Park, MO | VNX | Enid/Vance AFB, OK |
| LTX | Wilmington/Shallotte, NC | VTX | Los Angeles/Sulphur Mountain, CA |
| | | YUX | Yuma, AZ |

H.  **Internet Addresses**

1.  National Weather Service Home Page
    http://www.nws.noaa.gov

2.  Interactive Weather Information Network (IWIN)
    http://iwin.nws.noaa.gov/iwin/main.html

3.  Weather Graphics
    http://weather.noaa.gov/fax/graph.shtml

4.  Aviation Digital Data Service
    http://www.nws.noaa.gov/adds/

5.  National Centers for Environmental Prediction
    http://www.ncep.noaa.gov

6.  Aviation Weather Center
    http://www.awc-kc.noaa.gov

7.  NWS Links
    http://nimbo.wrh.noaa.gov/wrhq/nwspage.html

# END  OF CHAPTER

# ADDITIONAL USEFUL INFORMATION

The following five appendices contain information that will facilitate your access to additional weather forecasts and briefings.

A.  FAA Glossary of Weather Terms
B.  The Weather Channel
C.  Automated Flight Service Stations
D.  Direct User Access Terminal System (DUATS)
E.  Aviation Weather Resources on the Internet

Appendix A is a glossary of weather terms reproduced from the FAA's *Aviation Weather* (AC 00-6A).  Appendices B, C, and D provide details of weather forecasts available on television, preflight and in-flight briefings available through Automated Flight Service Stations, and weather briefings and flight planning that are accessible anywhere by a personal computer with a modem. Appendix E contains information on various web sites that provide weather information on the Internet.

# APPENDIX A
# FAA GLOSSARY OF WEATHER TERMS

The following glossary of weather terms is taken from AC 00-6A *Aviation Weather*.

## A

**absolute instability**—A state of a layer within the atmosphere in which the vertical distribution of temperature is such that an air parcel, if given an upward or downward push, will move away from its initial level without further outside force being applied.

**absolute temperature scale**—*See* Kelvin Temperature Scale.

**absolute vorticity**—*See* vorticity.

**adiabatic process**—The process by which fixed relationships are maintained during changes in temperature, volume, and pressure in a body of air without heat being added or removed from the body.

**advection**—The horizontal transport of air or atmospheric properties. In meteorology, sometimes referred to as the horizontal component of *convection*.

**advection fog**—Fog resulting from the transport of warm, humid air over a cold surface.

**air density**—The mass density of the air in terms of weight per unit volume.

**air mass**—In meteorology, an extensive body of air within which the conditions of temperature and moisture in a horizontal plane are essentially uniform.

**air mass classification**—A system used to identify and to characterize the different *air masses* according to a basic scheme. The system most commonly used classifies air masses primarily according to the thermal properties of their *source regions*: "tropical" (T); "polar" (P); and "Arctic" or "Antarctic" (A). They are further classified according to moisture characteristics as "continental" (c) or "maritime" (m).

**air parcel**—*See* parcel.

**albedo**—The ratio of the amount of electromagnetic *radiation* reflected by a body to the amount incident upon it, commonly expressed in percentage; in meteorology, usually used in reference to *insolation* (solar radiation); i.e., the albedo of wet sand is 9, meaning that about 9% of the incident insolation is reflected; albedoes of other surfaces range upward to 80–85 for fresh snow cover; average albedo for the earth and its atmosphere has been calculated to range from 35 to 43.

**altimeter**—An instrument which determines the altitude of an object with respect to a fixed level. *See* pressure altimeter.

**altimeter setting**—The value to which the scale of a *pressure altimeter* is set so as to read true altitude at field elevation.

**altimeter setting indicator**—A precision *aneroid barometer* calibrated to indicate directly the altimeter setting.

**altitude**—Height expressed in units of distance above a reference plane, usually above mean sea level or above ground.

(1) **corrected altitude**—Indicated altitude of an aircraft altimeter corrected for the temperature of the column of air below the aircraft, the correction being based on the estimated departure of existing temperature from standard atmospheric temperature; an approximation of true altitude.

(2) **density altitude**—The altitude in the standard atmosphere at which the air has the same density as the air at the point in question. An aircraft will have the same performance characteristics as it would have in a standard atmosphere at this altitude.

(3) **indicated altitude**—The altitude above mean sea level indicated on a *pressure altimeter* set at current local *altimeter setting*.

(4) **pressure altitude**—The altitude in the standard atmosphere at which the pressure is the same as at the point in question. Since an altimeter operates solely on pressure, this is the uncorrected altitude indicated by an altimeter set at standard sea level pressure of 29.92 inches or 1013 millibars.

(5) **radar altitude**—The altitude of an aircraft determined by radar-type radio altimeter; thus the actual distance from the nearest terrain or water feature encompassed by the downward directed radar beam. For all practical purposes, it is the "actual" distance above a ground or inland water surface or the true altitude above an ocean surface.

(6) **true altitude**—The exact distance above mean sea level.

**altocumulus**—White or gray layers or patches of cloud, often with a waved appearance; cloud elements appear as rounded masses or rolls; composed mostly of liquid water droplets which may be supercooled; may contain ice crystals at subfreezing temperatures.

**altocumulus castellanus**—A species of middle cloud of which at least a fraction of its upper part presents some vertically developed, cumuliform protuberances (some of which are taller than they are wide, as castles) and which give the cloud a crenelated or turreted appearance; especially evident when seen from the side; elements usually have a common base arranged in lines. This cloud indicates instability and turbulence at the altitudes of occurrence.

**anemometer**—An instrument for measuring *wind speed*.

**aneroid barometer**—A *barometer* which operates on the principle of having changing atmospheric pressure bend a metallic surface which, in turn, moves a pointer across a scale graduated in units of pressure.

**angel**—In radar meteorology, an *echo* caused by physical phenomena not discernible to the eye; they have been observed when abnormally strong temperature and/or moisture *gradients* were known to exist; sometimes attributed to insects or birds flying in the radar beam.

**anomalous propagation (sometimes called AP)**—In radar meteorology, the greater than normal bending of the radar beam such that *echoes* are received from ground *targets* at distances greater than normal *ground clutter*.

**anticyclone**—An area of high atmospheric pressure which has a closed circulation that is anticyclonic, i.e., as viewed from above, the circulation is clockwise in the Northern Hemisphere, counterclockwise in the Southern Hemisphere, undefined at the Equator.

**anvil cloud**—Popular name given to the top portion of a *cumulonimbus* cloud having an anvil-like form.

**APOB**—A *sounding* made by an aircraft.

**Arctic air**—An air mass with characteristics developed mostly in winter over Arctic surfaces of ice and snow. Arctic air extends to great heights, and the surface temperatures are basically, but not always, lower than those of *polar air*.

**Arctic front**—The surface of discontinuity between very cold (Arctic) air flowing directly from the Arctic region and another less cold and, consequently, less dense air mass.

**astronomical twilight**—*See* twilight.

**atmosphere**—The mass of air surrounding the Earth.

**atmospheric pressure (also called barometric pressure)**— The pressure exerted by the atmosphere as a consequence of gravitational attraction exerted upon the "column" of air lying directly above the point in question.

**atmospherics**—Disturbing effects produced in radio receiving apparatus by atmospheric electrical phenomena such as an electrical storm. Static.

**aurora**—A luminous, radiant emission over middle and high latitudes confined to the thin air of high altitudes and centered over the earth's magnetic poles. Called "aurora borealis" (northern lights) or "aurora australis" according to its occurrence in the Northern or Southern Hemisphere, respectively.

**attenuation**—In radar meteorology, any process which reduces power density in radar signals.

(1) **precipitation attenuation**—Reduction of power density because of absorption or reflection of energy by precipitation.

(2) **range attenuation**—Reduction of radar power density because of distance from the antenna. It occurs in the outgoing beam at a rate proportional to $1/range^2$. The return signal is also attenuated at the same rate.

# B

**backing**—Shifting of the wind in a counterclockwise direction with respect to either space or time; opposite of *veering*. Commonly used by meteorologists to refer to a cyclonic shift (counterclockwise in the Northern Hemisphere and clockwise in the Southern Hemisphere).

**backscatter**—Pertaining to radar, the energy reflected or scattered by a *target*; an *echo*.

**banner cloud (also called cloud banner)**—A banner-like cloud streaming off from a mountain peak.

**barograph**—A continuous-recording *barometer*.

**barometer**—An instrument for measuring the pressure of the atmosphere; the two principle types are *mercurial* and *aneroid*.

**barometric altimeter**—*See* pressure altimeter.

**barometric pressure**—Same as *atmospheric pressure*.

**barometric tendency**—The change of barometric pressure within a specified period of time. In aviation weather observations, routinely determined periodically, usually for a 3-hour period.

**beam resolution**—*See* resolution.

**Beaufort scale**—A scale of wind speeds.

**black blizzard**—Same as *duststorm*.

**blizzard**—A severe weather condition characterized by low temperatures and strong winds bearing a great amount of snow, either falling or picked up from the ground.

**blowing dust**—A type of *lithometeor* composed of dust particles picked up locally from the surface and blown about in clouds or sheets.

**blowing sand**—A type of *lithometeor* composed of sand picked up locally from the surface and blown about in clouds or sheets.

**blowing snow**—A type of *hydrometeor* composed of snow picked up from the surface by the wind and carried to a height of 6 feet or more.

**blowing spray**—A type of *hydrometeor* composed of water particles picked up by the wind from the surface of a large body of water.

**bright band**—In radar meteorology, a narrow, intense *echo* on the *range-height indicator* scope resulting from water-covered ice particles of high reflectivity at the melting level.

**Buys Ballot's law**—If an observer in the Northern Hemisphere stands with his back to the wind, lower pressure is to his left.

# C

**calm**—The absence of wind or of apparent motion of the air.

**cap cloud (also called cloud cap)**—A standing or stationary cap-like cloud crowning a mountain summit.

**ceiling**—In meteorology in the U.S., (1) the height above the surface of the base of the lowest layer of clouds or *obscuring phenomena* aloft that hides more than half of the sky, or (2) the *vertical visibility* into an *obscuration*. See summation principle.

**ceiling balloon**—A small balloon used to determine the height of a cloud base or the extent of vertical visibility.

**ceiling light**—An instrument which projects a vertical light beam onto the base of a cloud or into surface-based obscuring phenomena; used at night in conjunction with a *clinometer* to determine the height of the cloud base or as an aid in estimating the vertical visibility.

**ceilometer**—A cloud-height measuring system. It projects light on the cloud, detects the reflection by a photoelectric cell, and determines height by triangulation.

**Celsius temperature scale (abbreviated C)**—A temperature scale with zero degrees as the melting point of pure ice and 100 degrees as the boiling point of pure water at standard sea level atmospheric pressure.

**Centigrade temperature scale**—Same as *Celsius temperature scale*.

**chaff**—Pertaining to radar, (1) short, fine strips of metallic foil dropped from aircraft, usually by military forces, specifically for the purpose of jamming radar; (2) applied loosely to *echoes* resulting from chaff.

**change of state**—In meteorology, the transformation of water from one form, i.e., solid (ice), liquid, or gaseous (water vapor), to any other form. There are six possible transformations designated by the five terms following:

(1) **condensation**—The change of water vapor to liquid water.

(2) **evaporation**—The change of liquid water to water vapor.

(3) **freezing**—The change of liquid water to ice.

(4) **melting**—The change of ice to liquid water.

(5) **sublimation**—The change of (a) ice to water vapor or (b) water vapor to ice. See latent heat.

**Chinook**—A warm, dry *foehn* wind blowing down the eastern slopes of the Rocky Mountains over the adjacent plains in the U.S. and Canada.

**cirriform**—All species and varieties of *cirrus, cirrocumulus,* and *cirrostratus* clouds; descriptive of clouds composed mostly or entirely of small ice crystals, usually transparent and white; often producing *halo* phenomena not observed with other cloud forms. Average height ranges upward from 20,000 feet in middle latitudes.

**cirrocumulus**—A *cirriform* cloud appearing as a thin sheet of small white puffs resembling flakes or patches of cotton without shadows; sometimes confused with *altocumulus*.

**cirrostratus**—A *cirriform* cloud appearing as a whitish veil, usually fibrous, sometimes smooth; often produces *halo* phenomena; may totally cover the sky.

**cirrus**—A *cirriform* cloud in the form of thin, white feather-like clouds in patches or narrow bands; have a fibrous and/or silky sheen; large ice crystals often trail downward a considerable vertical distance in fibrous, slanted, or irregularly curved wisps called mares' tails.

**civil twilight**—*See* twilight.

**clear air turbulence (abbreviated CAT)**—Turbulence encountered in air where no clouds are present; more popularly applied to high level turbulence associated with *wind shear*.

**clear icing (or clear ice)**—Generally, the formation of a layer or mass of ice which is relatively transparent because of its homogeneous structure and small number and size of air spaces; used commonly as synonymous with *glaze*, particularly with respect to aircraft icing. Compare with *rime icing*. Factors which favor clear icing are large drop size, such as those found in *cumuliform* clouds, rapid accretion of supercooled water, and slow dissipation of *latent heat* of fusion.

**climate**—The statistical collective of the weather conditions of a point or area during a specified interval of time (usually several decades); may be expressed in a variety of ways.

**climatology**—The study of *climate*.

**clinometer**—An instrument used in weather observing for measuring angles of inclination; it is used in conjunction with a *ceiling light* to determine cloud height at night.

**cloud bank**—Generally, a fairly well-defined mass of cloud observed at a distance; it covers an appreciable portion of the horizon sky, but does not extend overhead.

**cloudburst**—In popular teminology, any sudden and heavy fall of *rain*, almost always of the *shower* type.

**cloud cap**—*See* cap cloud.

**cloud detection radar**—A vertically directed radar to detect cloud bases and tops.

**cold front**—Any non-occluded *front* which moves in such a way that colder air replaces warmer air.

**condensation**—*See* change of state.

**condensation level**—The height at which a rising *parcel* or layer of air would become saturated if lifted adiabatically.

**condensation nuclei**—Small particles in the air on which water vapor condenses or sublimates.

**condensation trail (or contrail) (also called vapor trail)**—A cloud-like streamer frequently observed to form behind aircraft flying in clear, cold, humid air.

**conditionally unstable air**—Unsaturated air that will become unstable on the condition it becomes saturated. *See* instability.

**conduction**—The transfer of heat by molecular action through a substance or from one substance in contact with another; transfer is always from warmer to colder temperature.

**constant pressure chart**—A chart of a constant pressure surface; may contain analyses of height, wind, temperature, humidity, and/or other elements.

**continental polar air**—*See* polar air.

**continental tropical air**—*See* tropical air.

**contour**—In meteorology, (1) a line of equal height on a constant pressure chart; analogous to contours on a relief map; (2) in radar meteorology, a line on a radar scope of equal *echo* intensity.

**contouring circuit**—On weather radar, a circuit which displays multiple contours of *echo* intensity simultaneously on the *plan position indicator* or *range-height indicator* scope. *See* contour (2).

**contrail**—Contraction for *condensation trail*.

**convection**—(1) In general, mass motions within a fluid resulting in transport and mixing of the properties of that fluid. (2) In meteorology, atmospheric motions that are predominantly vertical, resulting in vertical transport and mixing of atmospheric properties; distinguished from *advection*.

**convective cloud**—*See* cumuliform.

**convective condensation level (abbreviated CCL)**—The lowest level at which condensation will occur as a result of *convection* due to surface heating. When condensation occurs at this level, the layer between the surface and the CCL will be thoroughly mixed, temperature *lapse rate* will be dry adiabatic, and *mixing ratio* will be constant.

**convective instability**—The state of an unsaturated layer of air whose *lapse rates* of temperature and moisture are such that when lifted adiabatically until the layer becomes saturated, convection is spontaneous.

**convergence**—The condition that exists when the distribution of winds within a given area is such that there is a net horizontal inflow of air into the area. In convergence at lower levels, the removal of the resulting excess is accomplished by an upward movement of air; consequently, areas of low-level convergent winds are regions favorable to the occurrence of clouds and precipitation. Compare with *divergence*.

**Coriolis force**—A deflective force resulting from earth's rotation; it acts to the right of wind direction in the Northern Hemisphere and to the left in the Southern Hemisphere.

**corona**—A prismatically colored circle or arcs of a circle with the sun or moon at its center; coloration is from blue inside to red outside (opposite that of a *halo*); varies in size (much smaller) as opposed to the fixed diameter of the halo; characteristic of clouds composed of water droplets and valuable in differentiating between middle and cirriform clouds.

**corposant**—*See* St. Elmo's Fire.

**corrected altitude (approximation of true altitude)**—*See* altitude.

**cumuliform**—A term descriptive of all convective clouds exhibiting vertical development in contrast to the horizontally extended *stratiform* types.

**cumulonimbus**—A cumuliform cloud type; it is heavy and dense, with considerable vertical extent in the form of massive towers; often with tops in the shape of an *anvil* or massive plume; under the base of cumulonimbus, which often is very dark, there frequently exists *virga*, precipitation and low ragged clouds (*scud*), either merged with it or not; frequently accompanied by lightning, thunder, and sometimes hail; occasionally produces a tornado or a waterspout; the ultimate manifestation of the growth of a cumulus cloud, occasionally extending well into the stratosphere.

**cumulonimbus mamma**—A *cumulonimbus* cloud having hanging protuberances, like pouches, festoons, or udders, on the under side of the cloud; usually indicative of severe turbulence.

**cumulus**—A cloud in the form of individual detached domes or towers which are usually dense and well defined; develops vertically in the form of rising mounds of which the bulging upper part often resembles a cauliflower; the sunlit parts of these clouds are mostly brilliant white; their bases are relatively dark and nearly horizontal.

**cumulus fractus**—*See* fractus.

**cyclogenesis**—Any development or strengthening of cyclonic circulation in the atmosphere.

**cyclone**—(1) An area of low atmospheric pressure which has a closed circulation that is cyclonic, i.e., as viewed from above, the circulation is counterclockwise in the Northern Hemisphere, clockwise in the Southern Hemisphere, undefined at the Equator. Because cyclonic circulation and relatively low atmospheric pressure usually coexist, in common practice the terms cyclone and low are used interchangeably. Also, because cyclones often are accompanied by inclement (sometimes destructive) weather, they are frequently referred to simply as storms. (2) Frequently misused to denote a *tornado*. (3) In the Indian Ocean, a *tropical cyclone* of hurricane or typhoon force.

## D

**deepening**—A decrease in the central pressure of a pressure system; usually applied to a *low* rather than to a *high*, although technically, it is acceptable in either sense.

**density**—(1) The ratio of the mass of any substance to the volume it occupies—weight per unit volume. (2) The ratio of any quantity to the volume or area it occupies, i.e., population per unit area, *power density*.

**density altitude**—*See* altitude.

**depression**—In meteorology, an area of low pressure; a *low* or *trough*. This is usually applied to a certain stage in the development of a *tropical cyclone*, to migratory lows and troughs, and to upper-level lows and troughs that are only weakly developed.

**dew**—Water condensed onto grass and other objects near the ground, the temperatures of which have fallen below the initial dew point temperature of the surface air, but is still above freezing. Compare with *frost*.

**dew point (or dew-point temperature)**—The temperature to which a sample of air must be cooled, while the *mixing ratio* and barometric pressure remain constant, in order to attain saturation with respect to water.

**discontinuity**—A zone with comparatively rapid transition of one or more meteorological elements.

**disturbance**—In meteorology, applied rather loosely: (1) any low pressure or cyclone, but usually one that is relatively small in size; (2) an area where weather, wind, pressure, etc., show signs of cyclonic development; (3) any deviation in flow or pressure that is associated with a disturbed state of the weather, i.e., cloudiness and precipitation; and (4) any individual circulatory system within the primary circulation of the atmosphere.

**diurnal**—Daily, especially pertaining to a cycle completed within a 24-hour period, and which recurs every 24 hours.

**divergence**—The condition that exists when the distribution of winds within a given area is such that there is a net horizontal flow of air outward from the region. In divergence at lower levels, the resulting deficit is compensated for by subsidence of air from aloft; consequently the air is heated and the relative humidity lowered making divergence a warming and drying process. Low-level divergent regions are areas unfavorable to the occurrence of clouds and precipitation. The opposite of *convergence*.

**doldrums**—The equatorial belt of calm or light and variable winds between the two tradewind belts. Compare *intertropical convergence zone*.

**downdraft**—A relative small scale downward current of air; often observed on the lee side of large objects restricting the smooth flow of the air or in precipitation areas in or near *cumuliform* clouds.

**drifting snow**—A type of *hydrometeor* composed of snow particles picked up from the surface, but carried to a height of less than 6 feet.

**drizzle**—A form of *precipitation*. Very small water drops that appear to float with the air currents while falling in an irregular path (unlike *rain*, which falls in a comparatively straight path, and unlike *fog* droplets which remain suspended in the air).

**dropsonde**—A *radiosonde* dropped by parachute from an aircraft to obtain *soundings* (measurements) of the atmosphere below.

**dry adiabatic lapse rate**—The rate of decrease of temperature with height when unsaturated air is lifted adiabatically (due to expansion as it is lifted to lower pressure). *See* adiabatic process.

**dry bulb**—A name given to an ordinary thermometer used to determine temperature of the air; also used as a contraction for *dry-bulb temperature*. Compare *wet bulb*.

**dry-bulb temperature**—The temperature of the air.

**dust**—A type of *lithometeor* composed of small earthen particles suspended in the atmosphere.

**dust devil**—A small, vigorous *whirlwind*, usually of short duration, rendered visible by dust, sand, and debris picked up from the ground.

**duster**—Same as *duststorm*.

**duststorm (also called duster, black blizzard)**—An unusual, frequently severe weather condition characterized by strong winds and dust-filled air over an extensive area.

**D-value**—Departure of true altitude from pressure altitude (*see* altitude); obtained by algebraically subtracting true altitude from pressure altitude; thus it may be plus or minus. On a constant pressure chart, the difference between actual height and *standard atmospheric* height of a constant pressure surface.

**E**

**echo**—In radar, (1) the energy reflected or scattered by a *target;* (2) the radar scope presentation of the return from a target.

**eddy**—A local irregularity of wind in a larger scale wind flow. Small scale eddies produce turbulent conditions.

**estimated ceiling**—A ceiling classification applied when the ceiling height has been estimated by the observer or has been determined by some other method; but, because of the specified limits of time, distance, or precipitation conditions, a more descriptive classification cannot be applied.

**evaporation**—*See* change of state.

**extratropical low (sometimes called extratropical cyclone, extratropical storm)**—Any *cyclone* that is not a *tropical cyclone*, usually referring to the migratory frontal cyclones of middle and high latitudes.

**eye**—The roughly circular area of calm or relatively light winds and comparatively fair weather at the center of a well-developed *tropical cyclone*. A *wall cloud* marks the outer boundary of the eye.

**F**

**Fahrenheit temperature scale (abbreviated F)**—A temperature scale with 32 degrees as the melting point of pure ice and 212 degrees as the boiling point of pure water at standard sea level atmospheric pressure (29.92 inches or 1013.2 millibars).

**Fall wind**—A cold wind blowing downslope. Fall wind differs from *foehn* in that the air is initially cold enough to remain relatively cold despite compressional heating during descent.

**filling**—An increase in the central pressure of a pressure system; opposite of *deepening;* more commonly applied to a low rather than a high.

**first gust**—The leading edge of the spreading downdraft, *plow wind*, from an approaching thunderstorm.

**flow line**—A *streamline*.

**foehn**—A warm, dry downslope wind; the warmness and dryness being due to adiabatic compression upon descent; characteristic of mountainous regions. *See* adiabatic process, Chinook, Santa Ana.

**fog**—A *hydrometeor* consisting of numerous minute water droplets and based at the surface; droplets are small enough to be suspended in the earth's atmosphere indefinitely. (Unlike *drizzle*, it does not fall to the surface; differs from cloud only in that a cloud is not based at the surface; distinguished from haze by its wetness and gray color.)

**fractus**—Clouds in the form of irregular shreds, appearing as if torn; have a clearly ragged appearance; applies only to stratus and cumulus, i.e., *cumulus* fractus and *stratus* fractus.

**freezing**—*See* change of state.

**freezing level**—A level in the atmosphere at which the temperature is 0° C (32° F).

**front**—A surface, interface, or transition zone of discontinuity between two adjacent *air masses* of different densities; more simply the boundary between two different air masses. *See* frontal zone.

**frontal zone**—A *front* or zone with a marked increase of density gradient; used to denote that fronts are not truly a "surface" of discontinuity but rather a "zone" of rapid transition of meteorological elements.

**frontogenesis**—The initial formation of a *front* or *frontal* zone.

**frontolysis**—The dissipation of a *front*.

**frost (also hoarfrost)**—Ice crystal deposits formed by sublimation when temperature and dew point are below freezing.

**funnel cloud**—A *tornado* cloud or *vortex* cloud extending downward from the parent cloud but not reaching the ground.

## G

**glaze**—A coating of ice, generally clear and smooth, formed by freezing of supercooled water on a surface. *See* clear icing.

**gradient**—In meteorology, a horizontal decrease in value per unit distance of a parameter in the direction of maximum decrease; most commonly used with pressure, temperature, and moisture.

**ground clutter**—Pertaining to radar, a cluster of *echoes*, generally at short range, reflected from ground *targets*.

**ground fog**—In the United States, a *fog* that conceals less than 0.6 of the sky and is not contiguous with the base of clouds.

**gust**—A sudden brief increase in wind; according to U.S. weather observing practice, gusts are reported when the variation in wind speed between peaks and lulls is at least 10 knots.

## H

**hail**—A form of *precipitation* composed of balls or irregular lumps of ice, always produced by convective clouds which are nearly always *cumulonimbus*.

**halo**—A prismatically colored or whitish circle or arcs of a circle with the sun or moon at its center; coloration, if not white, is from red inside to blue outside (opposite

that of a *corona*); fixed in size with an angular diameter of 22° (common) or 46° (rare); characteristic of clouds composed of ice crystals; valuable in differentiating between *cirriform* and forms of lower clouds.

**haze**—A type of *lithometeor* composed of fine dust or salt particles dispersed through a portion of the atmosphere; particles are so small they cannot be felt or individually seen with the naked eye (as compared with the larger particles of *dust*), but diminish the visibility; distinguished from *fog* by its bluish or yellowish tinge.

**high**—An area of high barometric pressure, with its attendant system of winds; an *anticyclone*. Also high pressure system.

**hoar frost**—*See* frost.

**humidity**—Water vapor content of the air; may be expressed as *specific humidity, relative humidity,* or *mixing ratio*.

**hurricane**—A *tropical cyclone* in the Western Hemisphere with winds in excess of 65 knots or 120 km/h.

**hydrometeor**—A general term for particles of liquid water or ice such as rain, fog, frost, etc., formed by modification of water vapor in the atmosphere; also water or ice particles lifted from the earth by the wind such as sea spray or blowing snow.

**hygrograph**—The record produced by a continuous-recording *hygrometer*.

**hygrometer**—An instrument for measuring the water vapor content of the air.

## I

**ice crystals**—A type of *precipitation* composed of unbranched crystals in the form of needles, columns, or plates; usually having a very slight downward motion, may fall from a cloudless sky.

**ice fog**—A type of fog composed of minute suspended particles of ice; occurs at very low temperatures and may cause *halo* phenomena.

**ice needles**—A form of *ice crystals*.

**ice pellets**—Small, transparent or translucent, round or irregularly shaped pellets of ice. They may be (1) hard grains that rebound on striking a hard surface or (2) pellets of snow encased in ice.

**icing**—In general, any deposit of ice forming on an object. *See* clear icing, rime icing, glaze.

**indefinite ceiling**—A ceiling classification denoting *vertical visibility* into a surface based obscuration.

**indicated altitude**—*See* altitude.

**insolation**—Incoming solar *radiation* falling upon the earth and its atmosphere.

**instability**—A general term to indicate various states of the atmosphere in which spontaneous *convection* will occur when prescribed criteria are met; indicative of turbulence. *See* absolute instability, conditionally unstable air, convective instability.

**intertropical convergence zone**—The boundary zone between the trade wind system of the Northern and Southern Hemispheres; it is characterized in maritime climates by showery precipitation with cumulonimbus clouds sometimes extending to great heights.

**inversion**—An increase in temperature with height—a reversal of the normal decrease with height in the *troposphere;* may also be applied to other meteorological properties.

**isobar**—A line of equal or constant barometric pressure.

**iso echo**—In radar circuitry, a circuit that reverses signal strength above a specified intensity level, thus causing a void on the scope in the most intense portion of an echo when maximum intensity is greater than the specified level.

**isoheight**—On a weather chart, a line of equal height; same as *contour* (1).

**isoline**—A line of equal value of a variable quantity, i.e., an isoline of temperature is an *isotherm*, etc. *See* isobar, isotach, etc.

**isoshear**—A line of equal *wind shear.*

**isotach**—A line of equal or constant wind speed.

**isotherm**—A line of equal or constant temperature.

**isothermal**—Of equal or constant temperature, with respect to either space or time; more commonly, temperature with height; a zero *lapse rate.*

## J

**jet stream**—A quasi-horizontal stream of winds 50 knots or more concentrated within a narrow band embedded in the westerlies in the high *troposphere.*

## K

**katabatic wind**—Any wind blowing downslope. *See* fall wind, foehn.

**Kelvin temperature scale (abbreviated K)**—A temperature scale with zero degrees equal to the temperature at which all molecular motion ceases, i.e., absolute zero (0° K = −273° C); the Kelvin degree is identical to the Celsius degree; hence at standard sea level pressure, the melting point is 273° K and the boiling point 373° K.

**knot**—A unit of speed equal to one nautical mile per hour.

## L

**land breeze**—A coastal breeze blowing from land to sea, caused by temperature difference when the sea surface is warmer than the adjacent land. Therefore, it usually blows at night and alternates with a *sea breeze*, which blows in the opposite direction by day.

**lapse rate**—The rate of decrease of an atmospheric variable with height; commonly refers to decrease of temperature with height.

**latent heat**—The amount of heat absorbed (converted to kinetic energy) during the processes of change of liquid water to water vapor, ice to water vapor, or ice to liquid water; or the amount released during the reverse processes. Four basic classifications are:

(1) **latent heat of condensation**—Heat released during change of water vapor to water.

(2) **latent heat of fusion**—Heat released during change of water to ice or the amount absorbed in change of ice to water.

(3) **latent heat of sublimation**—Heat released during change of water vapor to ice or the amount absorbed in the change of ice to water vapor.

(4) **latent heat of vaporization**—Heat absorbed in the change of water to water vapor; the negative of latent heat of condensation.

**layer**—In reference to sky cover, clouds or other obscuring phenomena whose bases are approximately at the same level. The layer may be continuous or composed of detached elements. The term "layer" does not imply that a clear space exists between the layers or that the clouds or *obscuring phenomena* composing them are of the same type.

**lee wave**—Any stationary wave disturbance caused by a barrier in a fluid flow. In the atmosphere when sufficient moisture is present, this wave will be evidenced by *lenticular clouds* to the lee of mountain barriers; also called *mountain wave* or *standing wave.*

**lenticular cloud (or lenticularis)**—A species of cloud whose elements have the form of more or less isolated, generally smooth lenses or almonds. These clouds appear most often in formations of orographic origin, the result of *lee waves*, in which case they remain nearly stationary with respect to the terrain (standing cloud), but they also occur in regions without marked orography.

**level of free convection (abbreviated LFC)**—The level at which a *parcel* of air lifted dry-adiabatically until saturated and moist-adiabatically thereafter would become warmer than its surroundings in a conditionally unstable atmosphere. *See* conditional instability and adiabatic process.

**lifting condensation level (abbreviated LCL)**—The level at which a *parcel* of unsaturated air lifted dry-adiabatically would become saturated. Compare *level of free convection* and *convective condensation level.*

**lightning**—Generally, any and all forms of visible electrical discharge produced by a *thunderstorm.*

**lithometeor**—The general term for dry particles suspended in the atmosphere such as dust, haze, smoke, and sand.

**low**—An area of low barometric pressure, with its attendant system of winds. Also called a barometric depression or *cyclone.*

## M

**mammato cumulus**—Obsolete. *See* cumulonimbus mamma.

**mare's tail**—*See* cirrus.

**maritime polar air (abbreviated mP)**—*See* polar air.

**maritime tropical air (abbreviated mT)**—*See* tropical air.

**maximum wind axis**—On a constant pressure chart, a line denoting the axis of maximum wind speeds at that constant pressure surface.

**mean sea level**—The average height of the surface of the sea for all stages of tide; used as reference for elevations throughout the U.S.

**measured ceiling**—A ceiling classification applied when the ceiling value has been determined by instruments or the known heights of unobscured portions of objects, other than natural landmarks.

**melting**—*See* change of state.

**mercurial barometer**—A *barometer* in which pressure is determined by balancing air pressure against the weight of a column of mercury in an evacuated glass tube.

**meteorological visibility**—In U.S. observing practice, a main category of *visibility* which includes the subcategories of *prevailing visibility* and *runway visibility*. Meteorological visibility is a measure of horizontal visibility near the earth's surface, based on sighting of objects in the daytime or unfocused lights of moderate intensity at night. Compare *slant visibility, runway visual range, vertical visibility. See* surface visibility, tower visibility, and sector visibility.

**meteorology**—The science of the *atmosphere.*

**microbarograph**—An aneroid *barograph* designed to record atmospheric pressure changes of very small magnitudes.

**millibar (abbreviated mb.)**—An internationally used unit of pressure equal to 1,000 dynes per square centimeter. It is convenient for reporting *atmospheric pressure.*

**mist**—A popular expression for drizzle or heavy fog.

**mixing ratio**—The ratio by weight of the amount of water vapor in a volume of air to the amount of dry air; usually expressed as grams per kilogram (g/kg).

**moist-adiabatic lapse rate**—*See* saturated-adiabatic lapse rate.

**moisture**—An all-inclusive term denoting water in any or all of its three states.

**monsoon**—A wind that in summer blows from sea to a continental interior, bringing copious rain, and in winter blows from the interior to the sea, resulting in sustained dry weather.

**mountain wave**—A *standing wave* or *lee wave* to the lee of a mountain barrier.

### N

**nautical twilight**—*See* twilight.

**negative vorticity**—*See* vorticity.

**nimbostratus**—A principal cloud type, gray colored, often dark, the appearance of which is rendered diffuse by more or less continuously falling rain or snow, which in most cases reaches the ground. It is thick enough throughout to blot out the sun.

**noctilucent clouds**—Clouds of unknown composition which occur at great heights, probably around 75 to 90 kilometers. They resemble thin *cirrus*, but usually with a bluish or silverish color, although sometimes orange to red, standing out against a dark night sky. Rarely observed.

**normal**—In meteorology, the value of an element averaged for a given location over a period of years and recognized as a standard.

**numerical forecasting**—*See* numerical weather prediction.

**numerical weather prediction**—Forecasting by digital computers solving mathematical equations; used extensively in weather services throughout the world.

### O

**obscuration**—Denotes sky hidden by surface-based *obscuring phenomena* and *vertical visibility* restricted overhead.

**obscuring phenomena**—Any *hydrometeor* or *lithometeor* other than clouds; may be surface based or aloft.

**occlusion**—Same as *occluded front.*

**occluded front (commonly called occlusion, also called frontal occlusion)**—A composite of two fronts as a *cold front* overtakes a *warm front* or *quasi-stationary front.*

**orographic**—Of, pertaining to, or caused by mountains as in orographic clouds, orographic lift, or orographic precipitation.

**ozone**—An unstable form of oxygen; heaviest concentrations are in the stratosphere; corrosive to some metals; absorbs most ultraviolet solar radiation.

### P

**parcel**—A small volume of air, small enough to contain uniform distribution of its meteorological properties, and large enough to remain relatively self-contained and respond to all meteorological processes. No specific dimensions have been defined, however, the order of magnitude of 1 cubic foot has been suggested.

**partial obscuration**—A designation of sky cover when part of the sky is hidden by surface based *obscuring phenomena.*

**pilot balloon**—A small free-lift balloon used to determine the speed and direction of winds in the upper air.

**pilot balloon observation (commonly called PIBAL)**—A method of winds-aloft observation by visually tracking a *pilot balloon.*

**plan position indicator (PPI) scope**—A radar indicator scope displaying range and azimuth of *targets* in polar coordinates.

**plow wind**—The spreading downdraft of a *thunderstorm;* a strong, straight-line wind in advance of the storm. *See* first gust.

**polar air**—An air mass with characteristics developed over high latitudes, especially within the subpolar highs. Continental polar air (cP) has cold surface temperatures, low moisture content, and, especially in its source regions, has great stability in the lower layers. It is shallow in comparison with *Arctic air.* Maritime polar (mP) initially possesses similar properties to those of continental polar air, but in passing over warmer water 'it becomes unstable with a higher moisture content. Compare *tropical air.*

**polar front**—The semipermanent, semicontinuous *front* separating air masses of tropical and polar origins.

**positive vorticity**—*See* vorticity.

**power density**—In radar meteorology the amount of radiated energy per unit cross sectional area in the radar beam.

**precipitation**—Any or all forms of water particles, whether liquid or solid, that fall from the atmosphere and reach the surface. It is a major class of *hydrometeor*, distinguished from cloud and *virga* in that it must reach the surface.

**precipitation attenuation**—*See* attenuation.

**pressure**—*See* atmospheric pressure.

**pressure altimeter**—An *aneroid barometer* with a scale graduated in altitude instead of pressure using *standard atmospheric* pressure-height relationships; shows indicated altitude (not necessarily true altitude); may be set to measure altitude (indicated) from any arbitrarily chosen level. *See* altimeter setting, altitude.

**pressure altitude**—*See* altitude.

**pressure gradient**—The rate of decrease of pressure per unit distance at a fixed time.

**pressure jump**—A sudden, significant increase in *station pressure.*

**pressure tendency**—*See* barometric tendency.

**prevailing easterlies**—The broad current or pattern of persistent easterly winds in the Tropics and in polar regions.

**prevailing visibility**—In the U.S., the greatest horizontal visibility which is equaled or exceeded throughout half of the horizon circle; it need not be a continuous half.

**prevailing westerlies**—The dominant west-to-east motion of the atmosphere, centered over middle latitudes of both hemispheres.

**prevailing wind**—Direction from which the wind blows most frequently.

**prognostic chart (contracted PROG)**—A chart of expected or forecast conditions.

**pseudo-adiabatic lapse rate**—*See* saturated-adiabatic lapse rate.

**psychrometer**—An instrument consisting of a *wet-bulb* and a *dry-bulb* thermometer for measuring wet-bulb and dry-bulb temperature; used to determine water vapor content of the air.

**pulse**—Pertaining to radar, a brief burst of electromagnetic radiation emitted by the radar; of very short time duration. *See* pulse length.

**pulse length**—Pertaining to radar, the dimension of a radar pulse; may be expressed as the time duration or the length in linear units. Linear dimension is equal to time duration multiplied by the speed of propagation (approximately the speed of light).

## Q

**quasi-stationary front (commonly called stationary front)**—A *front* which is stationary or nearly so; conventionally, a front which is moving at a speed of less than 5 knots is generally considered to be quasi-stationary.

## R

**RADAR (contraction for radio detection and ranging)**—An electronic instrument used for the detection and ranging of distant objects of such composition that they scatter or reflect radio energy. Since *hydrometeors* can scatter radio energy, *weather radars*, operating on certain frequency bands, can detect the presence of precipitation, clouds, or both.

**radar altitude**—*See* altitude.

**radar beam**—The focused energy radiated by radar similar to a flashlight or searchlight beam.

**radar echo**—*See* echo.

**radarsonde observation**—A *rawinsonde observation* in which winds are determined by radar tracking a balloon-borne target.

**radiation**—The emission of energy by a medium and transferred, either through free space or another medium, in the form of electromagnetic waves.

**radiation fog**—*Fog* characteristically resulting when radiational cooling of the earth's surface lowers the air temperature near the ground to or below its initial dew point on calm, clear nights.

**radiosonde**—A balloon-borne instrument for measuring pressure, temperature, and humidity aloft. Radiosonde observation—a *sounding* made by the instrument.

**rain**—A form of *precipitation;* drops are larger than *drizzle* and fall in relatively straight, although not necessarily vertical, paths as compared to drizzle which falls in irregular paths.

**rain shower**—*See* shower.

**range attenuation**—*See* attenuation.

**range-height indicator (RHI) scope**—A radar indicator scope displaying a vertical cross section of *targets* along a selected azimuth.

**range resolution**—*See* resolution.

**RAOB**—A *radiosonde* observation.

**rawin**—A *rawinsonde* observation.

**rawinsonde observation**—A combined winds aloft and radiosonde observation. Winds are determined by tracking the *radiosonde* by radio direction finder or radar.

**refraction**—In radar, bending of the *radar beam* by variations in atmospheric density, water vapor content, and temperature.

(1) **normal refraction**—Refraction of the radar beam under normal atmospheric conditions; normal radius of curvature of the beam is about 4 times the radius of curvature of the Earth.

(2) **superrefraction**—More than normal bending of the radar beam resulting from abnormal vertical gradients of temperature and/or water vapor.

(3) **subrefraction**—Less than normal bending of the radar beam resulting from abnormal vertical gradients of temperature and/or water vapor.

**relative humidity**—The ratio of the existing amount of water vapor in the air at a given temperature to the maximum amount that could exist at that temperature; usually expressed in percent.

**relative vorticity**—*See* vorticity.

**remote scope**—In radar meteorology a "slave" scope remoted from weather *radar*.

**resolution**—Pertaining to radar, the ability of radar to show discrete *targets* separately, i.e., the better the resolution, the closer two targets can be to each other, and still be detected as separate targets.

(1) **beam resolution**—The ability of radar to distinguish between targets at approximately the same range but at different azimuths.

(2) **range resolution**—The ability of radar to distinguish between targets on the same azimuth but at different ranges.

**ridge (also called ridge line)**—In meteorology, an elongated area of relatively high atmospheric pressure; usually associated with and most clearly identified as an area of maximum anticyclonic curvature of the wind flow (*isobars*, *contours*, or *streamlines*).

**rime icing (or rime ice)**—The formation of a white or milky and opaque granular deposit of ice formed by the rapid freezing of supercooled water droplets as they impinge upon an exposed aircraft.

**rocketsonde**—A type of *radiosonde* launched by a rocket and making its measurements during a parachute descent; capable of obtaining *soundings* to a much greater height than possible by balloon or aircraft.

**roll cloud (sometimes improperly called rotor cloud)**—A dense and horizontal roll-shaped accessory cloud located on the lower leading edge of a *cumulonimbus* or less often, a rapidly developing *cumulus*; indicative of turbulence.

**rotor cloud (sometimes improperly called *roll cloud*)**—A turbulent cloud formation found in the lee of some large mountain barriers, the air in the cloud rotates around an axis parallel to the range; indicative of possible violent turbulence.

**runway temperature**—The temperature of the air just above a runway, ideally at engine and/or wing height, used in the determination of density *altitude*; useful at airports when critical values of density altitude prevail.

**runway visibility**—The *meteorological visibility* along an identified runway determined from a specified point on the runway; may be determined by a *transmissometer* or by an observer.

**runway visual range**—An instrumentally derived horizontal distance a pilot should see down the runway from the approach end; based on either the sighting of high intensity runway lights or on the visual contrast of other objects, whichever yields the greatest visual range.

## S

**St. Elmo's Fire (also called corposant)**—A luminous brush discharge of electricity from protruding objects, such as masts and yardarms of ships, aircraft, lightning rods, steeples, etc., occurring in stormy weather.

**Santa Ana**—A hot, dry, *foehn* wind, generally from the northeast or east, occurring west of the Sierra Nevada Mountains especially in the pass and river valley near Santa Ana, California.

**saturated adiabatic lapse rate**—The rate of decrease of temperature with height as saturated air is lifted with no gain or loss of heat from outside sources; varies with temperature, being greatest at low temperatures. *See* adiabatic process and dry-adiabatic lapse rate.

**saturation**—The condition of the atmosphere when actual *water vapor* present is the maximum possible at existing temperature.

**scud**—Small detached masses of stratus *fractus* clouds below a layer of higher clouds, usually *nimbostratus*.

**sea breeze**—A coastal breeze blowing from sea to land, caused by the temperature difference when the land surface is warmer than the sea surface. Compare *land breeze*.

**sea fog**—A type of *advection fog* formed when air that has been lying over a warm surface is transported over a colder water surface.

**sea level pressure**—The *atmospheric pressure* at *mean sea level*, either directly measured by stations at sea level or empirically determined from the *station pressure* and temperature by stations not at sea level; used as a common reference for analyses of surface pressure patterns.

**sea smoke**—Same as *steam fog*.

**sector visibility**—*Meteorological visibility* within a specified sector of the horizon circle.

**sensitivity time control**—A radar circuit designed to correct for range *attenuation* so that echo intensity on the scope is proportional to reflectivity of the *target* regardless of range.

**shear**—*See* wind shear.

**shower**—*Precipitation* from a *cumuliform* cloud; characterized by the suddenness of beginning and ending, by the rapid change of intensity, and usually by rapid change in the appearance of the sky; showery precipitation may be in the form of rain, ice pellets, or snow.

**slant visibility**—For an airborne observer, the distance at which he can see and distinguish objects on the ground.

**sleet**—*See* ice pellets.

**smog**—A mixture of *smoke* and *fog*.

**smoke**—A restriction to visibility resulting from combustion.

**snow**—Precipitation composed of white or translucent ice crystals, chiefly in complex branched hexagonal form.

**snow flurry**—Popular term for snow *shower*, particularly of a very light and brief nature.

**snow grains**—*Precipitation* of very small, white opaque grains of ice, similar in structure to *snow* crystals. The grains are fairly flat or elongated, with diameters generally less than 0.04 inch (1 mm.).

**snow pellets**—*Precipitation* consisting of white, opaque approximately round (sometimes conical) ice particles having a snow-like structure, and about 0.08 to 0.2 inch in diameter; crisp and easily crushed, differing in this respect from *snow grains;* rebound from a hard surface and often break up.

**snow shower**—*See* shower.

**solar radiation**—The total electromagnetic *radiation* emitted by the sun. *See* insolation.

**sounding**—In meteorology, an upper-air observation; a *radiosonde* observation.

**source region**—An extensive area of the earth's surface characterized by relatively uniform surface conditions where large masses of air remain long enough to take on characteristic temperature and moisture properties imparted by that surface.

**specific humidity**—The ratio by weight of *water vapor* in a sample of air to the combined weight of water vapor and dry air. Compare *mixing ratio.*

**squall**—A sudden increase in wind speed by at least 15 knots to a peak of 20 knots or more and lasting for at least one minute. Essential difference between a *gust* and a squall is the duration of the peak speed.

**squall line**—Any nonfrontal line or narrow band of active *thunderstorms* (with or without *squalls*).

**stability**—A state of the atmosphere in which the vertical distribution of temperature is such that a *parcel* will resist displacement from its initial level. (*See also* instability.)

**standard atmosphere**—A hypothetical atmosphere based on climatological averages comprised of numerous physical constants of which the most important are:

(1) A surface *temperature* of 59° F (15° C) and a surface pressure of 29.92 inches of mercury (1013.2 millibars) at sea level;

(2) A *lapse rate* in the troposphere of 6.5° C per kilometer (approximately 2° C per 1,000 feet);

(3) A *tropopause* of 11 kilometers (approximately 36,000 feet) with a temperature of −56.5° C; and

(4) An *isothermal* lapse rate in the stratosphere to an altitude of 24 kilometers (approximately 80,000 feet).

**standing cloud (standing lenticular altocumulus)**—*See* lenticular cloud.

**standing wave**—A wave that remains stationary in a moving fluid. In aviation operations it is used most commonly to refer to a *lee wave* or *mountain wave.*

**stationary front**—Same as *quasi-stationary front.*

**station pressure**—The actual *atmospheric pressure* at the observing station.

**steam fog**—Fog formed when cold air moves over relatively warm water or wet ground.

**storm detection radar**—A weather radar designed to detect *hydrometeors* of precipitation size; used primarily to detect storms with large drops or hailstones as opposed to clouds and light precipitation of small drop size.

**stratiform**—Descriptive of clouds of extensive horizontal development, as contrasted to vertically developed *cumuliform* clouds; characteristic of stable air and, therefore, composed of small water droplets.

**stratocumulus**—A low cloud, predominantly *stratiform* in gray and/or whitish patches or layers, may or may not merge; elements are tessellated, rounded, or roll-shaped with relatively flat tops.

**stratosphere**—The atmospheric layer above the tropopause, average altitude of base and top, 7 and 22 miles respectively; characterized by a slight average increase of temperature from base to top and is very stable; also characterized by low moisture content and absence of clouds.

**stratus**—A low, gray cloud layer or sheet with a fairly uniform base; sometimes appears in ragged patches; seldom produces precipitation but may produce *drizzle* or *snow grains.* A *stratiform* cloud.

**stratus fractus**—*See* fractus.

**streamline**—In meteorology, a line whose tangent is the wind direction at any point along the line. A flowline.

**sublimation**—*See* change of state.

**subrefraction**—*See* refraction.

**subsidence**—A descending motion of air in the atmosphere over a rather broad area; usually associated with *divergence.*

**summation principle**—The principle states that the cover assigned to a layer is equal to the summation of the sky cover of the lowest layer plus the additional coverage at all successively higher layers up to and including the layer in question. Thus, no layer can be assigned a sky cover less than a lower layer, and no sky cover can be greater than 1.0 (10/10).

**superadiabatic lapse rate**—A *lapse rate* greater than the *dry-adiabatic lapse rate.* See absolute instability.

**supercooled water**—Liquid water at temperatures colder than freezing.

**superrefraction**—*See* refraction.

**surface inversion**—An *inversion* with its base at the surface, often caused by cooling of the air near the surface as a result of *terrestrial radiation*, especially at night.

**surface visibility**—Visibility observed from eye-level above the ground.

**synoptic chart**—A chart, such as the familiar weather map, which depicts the distribution of meteorological conditions over an area at a given time.

# T

**target**—In radar, any of the many types of objects detected by radar.

**temperature**—In general, the degree of hotness or coldness as measured on some definite temperature scale by means of any of various types of thermometers.

**temperature inversion**—*See* inversion.

**terrestrial radiation**—The total infrared *radiation* emitted by the Earth and its atmosphere.

**thermograph**—A continuous-recording *thermometer*.

**thermometer**—An instrument for measuring *temperature*.

**theodolite**—An optical instrument which, in meteorology, is used principally to observe the motion of a *pilot balloon*.

**thunderstorm**—In general, a local storm invariably produced by a *cumulonimbus* cloud, and always accompanied by lightning and thunder.

**tornado (sometimes called cyclone, twister)**—A violently rotating column of air, pendant from a cumulonimbus cloud, and nearly always observable as "funnel-shaped." It is the most destructive of all small-scale atmospheric phenomena.

**towering cumulus**—A rapidly growing *cumulus* in which height exceeds width.

**tower visibility**—*Prevailing visibility* determined from the control tower.

**trade winds**—Prevailing, almost continuous winds blowing with an easterly component from the subtropical high pressure belts toward the *intertropical convergence zone;* northeast in the Northern Hemisphere, southeast in the Southern Hemisphere.

**transmissometer**—An instrument system which shows the transmissivity of light through the atmosphere. Transmissivity may be translated either automatically or manually into *visibility* and/or *runway visual range*.

**tropical air**—An air mass with characteristics developed over low latitudes. Maritime tropical air (mT), the principal type, is produced over the tropical and subtropical seas; very warm and humid. Continental tropical (cT) is produced over subtropical arid regions and is hot and very dry. Compare *polar air*.

**tropical cyclone**—A general term for a *cyclone* that originates over tropical oceans. By international agreement, tropical cyclones have been classified according to their intensity, as follows:

(1) **tropical depression**—winds up to 34 knots (64 km/h);

(2) **tropical storm**—winds of 35 to 64 knots (65 to 119 km/h);

(3) **hurricane or typhoon**—winds of 65 knots or higher (120 km/h).

**tropical depression**—*See* tropical cyclone.

**tropical storm**—*See* tropical cyclone.

**tropopause**—The transition zone between the *troposphere* and *stratosphere*, usually characterized by an abrupt change of *lapse rate*.

**troposphere**—That portion of the *atmosphere* from the earth's surface to the *tropopause;* that is, the lowest 10 to 20 kilometers of the atmosphere. The troposphere is characterized by decreasing temperature with height, and by appreciable water vapor.

**trough (also called trough line)**—In meteorology, an elongated area of relatively low atmospheric pressure; usually associated with and most clearly identified as an area of maximum cyclonic curvature of the wind flow (*isobars, contours,* or *streamlines*); compare with *ridge*.

**true altitude**—*See* altitude.

**true wind direction**—The direction, with respect to true north, from which the wind is blowing.

**turbulence**—In meteorology, any irregular or disturbed flow in the atmosphere.

**twilight**—The intervals of incomplete darkness following sunset and preceding sunrise. The time at which evening twilight ends or morning twilight begins is determined by arbitrary convention, and several kinds of twilight have been defined and used; most commonly civil, nautical, and astronomical twilight.

(1) **Civil Twilight**—The period of time before sunrise and after sunset when the sun is not more than 6° below the horizon.

(2) **Nautical Twilight**—The period of time before sunrise and after sunset when the sun is not more than 12° below the horizon.

(3) **Astronomical Twilight**—The period of time before sunrise and after sunset when the sun is not more than 18° below the horizon.

**twister**—In the United States, a colloquial term for *tornado*.

**typhoon**—A *tropical cyclone* in the Eastern Hemisphere with winds in excess of 65 knots (120 km/h).

# U

**undercast**—A cloud *layer* of ten-tenths (1.0) coverage (to the nearest tenth) as viewed from an observation point above the layer.

**unlimited ceiling**—A clear sky or a sky cover that does not meet the criteria for a *ceiling*.

**unstable**—*See* instability.

**updraft**—A localized upward current of air.

**upper front**—A *front* aloft not extending to the earth's surface.

**upslope fog**—Fog formed when air flows upward over rising terrain and is, consequently, adiabatically cooled to or below its initial *dew point*.

# V

**vapor pressure**—In meteorology, the pressure of water vapor in the atmosphere. Vapor pressure is that part of the total atmospheric pressure due to water vapor and is independent of the other atmospheric gases or vapors.

**vapor trail**—Same as *condensation trail.*

**veering**—Shifting of the wind in a clockwise direction with respect to either space or time; opposite of backing. Commonly used by meteorologists to refer to an anticyclonic shift (clockwise in the Northern Hemisphere and counterclockwise in the Southern Hemisphere).

**vertical visibility**—The distance one can see upward into a surface based *obscuration;* or the maximum height from which a pilot in flight can recognize the ground through a surface based obscuration.

**virga**—Water or ice particles falling from a cloud, usually in wisps or streaks, and evaporating before reaching the ground.

**visibility**—The greatest distance one can see and identify prominent objects.

**visual range**—*See* runway visual range.

**vortex**—In meteorology, any rotary flow in the atmosphere.

**vorticity**—Turning of the atmosphere. Vorticity may be imbedded in the total flow and not readily identified by a flow pattern.

(a) **absolute vorticity**—the rotation of the Earth imparts vorticity to the atmosphere; absolute vorticity is the combined vorticity due to this rotation and vorticity due to circulation relative to the Earth (relative vorticity).

(b) **negative vorticity**—vorticity caused by anticyclonic turning; it is associated with downward motion of the air.

(c) **positive vorticity**—vorticity caused by cyclonic turning; it is associated with upward motion of the air.

(d) **relative vorticity**—vorticity of the air relative to the Earth, disregarding the component of vorticity resulting from Earth's rotation.

# W

**wake turbulence**—*Turbulence* found to the rear of a solid body in motion relative to a fluid. In aviation terminology, the turbulence caused by a moving aircraft.

**wall cloud**—The well-defined bank of vertically developed clouds having a wall-like appearance which form the outer boundary of the *eye* of a well-developed *tropical cyclone.*

**warm front**—Any non-occluded *front* which moves in such a way that warmer air replaces colder air.

**warm sector**—The area covered by warm air at the surface and bounded by the *warm front* and *cold front* of a *wave cyclone.*

**water equivalent**—The depth of water that would result from the melting of snow or ice.

**waterspout**—*See* tornado.

**water vapor**—Water in the invisible gaseous form.

**wave cyclone**—A *cyclone* which forms and moves along a front. The circulation about the cyclone center tends to produce a wavelike deformation of the front.

**weather**—The state of the *atmosphere*, mainly with respect to its effects on life and human activities; refers to instantaneous conditions or short term changes as opposed to *climate.*

**weather radar**—Radar specifically designed for observing weather. *See* cloud detection radar and storm detection radar.

**weather vane**—A *wind vane.*

**wedge**—Same as *ridge.*

**wet bulb**—Contraction of either *wet-bulb temperature* or *wet-bulb thermometer.*

**wet-bulb temperature**—The lowest *temperature* that can be obtained on a *wet-bulb thermometer* in any given sample of air, by evaporation of water (or ice) from the muslin wick; used in computing *dew point* and *relative humidity.*

**wet-bulb thermometer**—A thermometer with a muslin-covered bulb used to measure wet-bulb temperature.

**whirlwind**—A small, rotating column of air; may be visible as a dust devil.

**willy-willy**—A *tropical cyclone* of hurricane strength near Australia.

**wind**—Air in motion relative to the surface of the earth; generally used to denote horizontal movement.

**wind direction**—The direction from which wind is blowing.

**wind speed**—Rate of wind movement in distance per unit time.

**wind vane**—An instrument to indicate wind direction.

**wind velocity**—A vector term to include both *wind direction* and *wind speed.*

**wind shear**—The rate of change of *wind velocity* (direction and/or speed) per unit distance; conventionally expressed as vertical or horizontal wind shear.

# X–Y–Z

**zonal wind**—A west wind; the westerly component of a wind. Conventionally used to describe large-scale flow that is neither cyclonic nor anticyclonic.

# END OF APPENDIX

# APPENDIX B
# THE WEATHER CHANNEL

> Please take a few minutes to study each of the concepts listed above and anticipate/imagine what they are and how they relate to the other listed concepts.

## A. The Weather Channel

1. The Weather Channel (TWC) is the only all-weather network on cable television that broadcasts local, regional, national, and international (mostly limited to Europe) weather reports and forecasts 24 hr. a day, 7 days a week.

   a. TWC began broadcasting in May 1982, and today it is available to over 63 million cable subscribers.

2. TWC's diversified programming is designed to meet everyone's weather needs.

3. TWC employs more than 67 staff meteorologists who analyze information from the NWS and other sources to produce local reports and forecasts for the entire nation.

   a. These forecasts are those made by the staff at TWC, not by the NWS.

4. TWC produces over 2,000 computer-generated graphics each day to display current and forecast weather conditions. These graphics help make the presentation more viewer-friendly.

   a. Local radar reports and surface observations are updated hourly, and sometimes more often.

   b. With this information, current surface analysis charts are also reviewed and updated hourly.

5. Regularly scheduled programs of interest to pilots are shown in the table below.

    a. TWC also provides a variety of specialty and seasonal programming.

| TIME (IN MINUTES PAST THE HOUR) | PROGRAM TITLE | DESCRIPTION |
|---|---|---|
| **EARLY MORNING  5-1pmET, 2-10amPT** | | |
| :00 | WeatherScope | A look at the day's national weather |
| :10/:40 (only at 7:10, 7:40 M-F; 9-10, 9:40 S-S) | Weather coverage | National and regional roundup for the day |
| :14 (during hours of 5,6, and 7 o'clock ET) | Aviation Weather | Report of weather information of interest to pilots |
| :20 | 5-day Planner | Tells impact of upcoming weather on travelers |
| :30 | WeatherScope | A look at the day's national weather |
| :42 (except 7:42 M-F & 9:42 S-S) | International Weather | Conditions across Europe |
| :50 | This Morning's Weather | Day's weather with seasonal weather update plus outlook |
| **MID-DAY  1 pm-5pmET, 10 am-2pmPT** | | |
| :00 (all other times) | WeatherScope | Forecast of national weather |
| :20 | 5-day Planner | Tells impact of upcoming weather on travelers |
| :30 | WeatherScope | Forecast of national weather |
| :42 | International Weather | Conditions across Europe and at 3:42 pm of Australia |
| :50 | This Afternoon's Weather | Day's weather with seasonal weather update plus outlook |
| **LATE DAY/EVENING  5pm-1amET, 2pm-10pmPT** | | |
| :10/:40 (at 9:10, 9:40 only) | Weather coverage | National weather report |
| :20 | 5-day Planner | Tells impact of upcoming weather on travel |
| :42 | International Weather | Conditions across Europe |
| :47 | This Evening's Weather | Day's weather with seasonal weather update plus outlook |
| **NIGHT    1am-5am ET, 10pm-2amPT** | | |
| :00 & :30 (at 1 amET/10 pmPT only) | Pacific/International | West Coast weather, international cities |
| :01 (all hours starting & after 2amET/ 11pmPT) | Good Morning Forecast | Weather forecast for day ahead with traveler's update |
| :12 (all hours after 2amET/11pmPT) | Pacific Regional Forecast | Provides Pacific coastal states with forecast |
| :20 | 5-day Planner | Tells impact of upcoming weather on travel |
| :15 (all hours after 2amET/11pmPT) | Satellite Review | National and regional satellite views with animation |
| :25 (at 1:25amET/11:25pmPT only) | Alaska/Hawaii Update | Provides Pacific weather with forecast for Alaska and Hawaii |
| :31 (all hours after 2amET/11pmPT) | Good Morning Forecast | Weather forecast for day ahead with traveler's update |
| :42 | International Weather | Conditions across Europe and in Australia |
| :44 (all hours after 2amET/11pmPT) | Pacific Regional Forecast | Provides Pacific coastal states with forecast |
| :50 | This Morning's Weather | Seasonal weather/storm update along with forecast for the day |

B.  **Local Weather**

    1.    TWC has developed a computerized system called the **Weather Star**, which allows TWC to broadcast different local reports simultaneously to 1,200 weather zones around the country.

        a.    The Weather Star system allows each cable system to receive local surface observations automatically from the nearest observation site.

            1)    This information normally appears at the bottom of the TV screen during other presentations.

        b.    The Weather Star system also provides the current and forecast conditions for each weather zone.

            1)    These forecasts are shown six times per hour (or every 10 min.) at 8, 18, 28, 38, 48, and 58 min. past the hour.

        c.    The part that is of interest to most pilots is the local radar image, which is also transmitted on the Weather Star system.

    2.    The radar presentations are enhanced versions of the raw radar images provided by NWS.

        a.    TWC uses a process that eliminates ground clutter and other false echoes.

            1)    This cleanup allows accurate radar imaging and contouring.

        b.    These enhancements make TWC's radar image very useful in preflight planning.

            1)    Radar shows the movement of the echoes over the last 90 min. to provide you with more information.

C.  **Using the Weather Channel**

    1.    TWC presents "Aviation Weather" in the morning at 5:14, 6:14, and 7:14 ET daily, and constantly updates the surface charts and radar imagery.

    2.    TWC is an excellent source to complement your weather briefing from an FSS specialist and to provide you with a visual presentation of the weather, which is currently impossible by telephone.

D.  **Internet Site**

    1.    TWC has a site on the Internet that contains, among a vast amount of weather information, various aviation-related charts. These aviation charts are the

        a.    Jet stream

            1)    Current-day location of the jet stream(s) over North America
            2)    Forecast location of the jet stream(s) over North America

        b.    Winds aloft observations (based on radiosonde data)

            1)    5,000 ft.
            2)    10,000 ft.
            3)    34,000 ft.

        c.    National airport overview

            1)    This chart depicts airports reporting either IFR or MVFR and also shows areas of precipitation. Check the time on the chart.

    2.    In addition to these aviation-related charts, you can select specific cities for forecasts. These forecasts also include access to local and regional radar images and satellite images.

3.   The site address is http://www.weather.com.

4.   Remember, this information will give you a background to use when you receive your official weather briefing from an FSS or through DUATS.

   a.   The more information you have, the more informed you are when making a "go/no go" decision.

# END OF APPENDIX

# APPENDIX C
# AUTOMATED FLIGHT SERVICE STATIONS

Please take a few minutes to study each of the concepts listed above and anticipate/imagine what they are and how they relate to the other listed concepts.

A. **Flight Service Stations (FSSs)**. Flight Service Stations (FSSs) are air traffic facilities that provide a variety of services to pilots.

1. Automated FSSs (AFSSs) have replaced most of the older FSSs, usually with one AFSS per state.

   a. Nonautomated FSSs were at a large number of airports around the country.

   b. The same services are available at AFSSs that were available at the FSSs, but with a much faster and more complete dissemination of information nationwide.

2. AFSS personnel are responsible for providing emergency, in-flight, and preflight services.

   a. Emergency situations are those in which life or property is in danger and include

      1) VFR search and rescue services
      2) Assistance to lost aircraft and other aircraft in emergency situations

   b. In-flight services are those provided to or affecting aircraft in flight or operating on the airport surface.

      1) NAVAID monitoring and restoration
      2) Local Airport Advisories (LAA)
      3) Delivery of ATC clearances, advisories, or requests
      4) Issuance of military flight advisory messages
      5) En Route Flight Advisory Service (EFAS) or Flight Watch
      6) Issuance of NOTAMs
      7) Transcribed or live weather broadcasts
      8) Weather observations
      9) PIREPs
      10) Radio pilot briefings

   c. Preflight services are those provided prior to actual departure, usually by telephone.

      1) Pilot briefings
      2) Flight plan filing and processing
      3) Aircraft operational reservations

## B. Transcribed or Live Broadcasts

1. A variety of reports and forecasts are broadcast over selected VOR and NDB frequencies.

2. Unscheduled broadcasts are made upon receipt of special weather reports, PIREPs, NOTAMs, weather advisories, radar reports, military training route data, alert notices, and other information considered necessary to enhance safety and efficiency of flight.

3. Transcribed weather broadcasts (TWEBs) are broadcast continuously over selected NDB and VOR frequencies.

   a. A TWEB is weather information presented in a route-of-flight format.

4. The Pilot's Automatic Telephone Weather Answering Service (PATWAS) is provided by nonautomated FSSs and is a recorded telephone briefing service with the forecast for the local area, usually within a 50-NM radius of the station.

   a. A few selected stations also include route forecasts similar to the TWEB.

5. The Telephone Information Briefing Service (TIBS) is provided by AFSSs and provides weather and/or aeronautical information for a variety of routes and/or areas.

   a. TIBS may provide any of the following:

      1) Area and/or route briefings

         a) METARs
         b) TAFs
         c) Winds aloft

      2) Airspace procedures
      3) Special announcements

   b. A pilot who calls an AFSS by touch-tone telephone can select from different TIBS briefing areas and/or routes.

      1) The information contained in a TIBS is the same as in a standard briefing.

6. Scheduled weather broadcasts are made at H+15 in Alaska only and include surface reports from a number of airports in the vicinity of the station.

7. The Hazardous In-flight Weather Advisory Service (HIWAS) is a continuous broadcast service of in-flight weather advisories, i.e., SIGMETs, convective SIGMETs, AIRMETs, CWAs, and AWWs over selected VORs.

## C. Pilot Briefing

1. AFSSs are equipped with computer terminals and graphic displays that can generate or display any desired report, forecast, chart, and/or satellite imagery.

2. A preflight briefing may be any one of the following types:

   a. A standard briefing is a summary and interpretation of all available data concerning an intended flight.

   b. An abbreviated briefing is intended to supplement mass disseminated data (e.g., TWEB, TIBS, etc.), update a previous briefing, or provide the pilot with only specifically requested information.

   c. An outlook briefing is provided when the proposed departure is 6 hr. or more from the time of the briefing.

## D. In-Flight Services

1. AFSSs provide services to aircraft on a "first come, first served" basis, except for the following, which receive priority:

   a. Aircraft in distress
   b. Lifeguard (air ambulance) aircraft
   c. Search and rescue aircraft

2.   The following types of radio communications are generally made by AFSSs:

    a.   Authorized transmission of messages necessary for ATC or safety

    b.   Routine radio contacts, e.g., in-flight briefings

    c.   ATC clearances, advisories, and requests

    d.   Flight progress reports upon receipt of pilot position reports

3.   AFSSs regularly monitor NAVAIDs within their area for proper operation.

    a.   When a malfunction is indicated or reported by several aircraft, the AFSS attempts to restore the NAVAID or discontinue its use and issue an appropriate NOTAM.

4.   A Local Airport Advisory (LAA) is a terminal service provided by an FSS physically located on an airport without an operating control tower.

    a.   LAA provides information to arriving and departing aircraft concerning wind direction and speed, favored runway, altimeter setting, pertinent known traffic, and pertinent known field conditions.

        1)   This information is advisory in nature and is not an ATC clearance.

    b.   LAA may also provide control of airport lighting, e.g., rotating beacon, runway lights, etc., after dark and at times of reduced visibility.

    c.   If Class D or E airspace exists at the surface of an airport served by an LAA, the FSS there may assist a pilot in obtaining a special VFR clearance, when appropriate.

5.   The En Route Flight Advisory Service (EFAS), or Flight Watch, provides en route aircraft with timely and pertinent weather data tailored to a specific altitude and route using the most current available information.

    a.   Briefings are intended to apply to the en route phase of flight (i.e., between climbout and descent to land).

    b.   When conditions dictate, information is provided on weather for alternate routes and/or altitudes.

    c.   EFAS may not be used for in-flight services, i.e., flight plan filing, position reporting, or full route (preflight) briefing.

    d.   PIREPs are solicited from pilots who contact Flight Watch.

## E.   Emergency Services

1.   An emergency can be either a distress or an urgent situation.

    a.   Distress is a condition of being threatened by serious and/or imminent danger and of requiring immediate assistance.

    b.   Urgency is a condition of being concerned about safety and of requiring timely but not immediate assistance, i.e., a potential distress condition.

2.   FSSs may provide assistance to aircraft in emergency situations by

    a.   Assisting in orienting a lost pilot

    b.   Coordinating communications between the aircraft, ATC, and search and rescue crews, as appropriate

    c.   Monitoring frequencies 121.5 and 243.0 for emergency radio calls and ELT signals

3.   The Direction Finder Service (DF Steer) allows an appropriately equipped FSS to determine an aircraft's bearing from the station using the two-way VHF radio communication transmitter on board the aircraft.

    a.   Under emergency conditions when a standard instrument approach cannot be executed, an FSS can provide DF guidance and instrument approach service.

4.   Using VOR or ADF facilities as appropriate to the aircraft's equipment, an FSS can assist a pilot in determining his/her position and provide guidance to the airport if desired.

F. **Flight Data**

1. FSSs accept domestic and international IFR, VFR, and DVFR flight plans and forward the flight plan data to the appropriate agencies.

2. Data from IFR flight plans are transmitted to the ARTCC as part of the IFR fight plan proposal.

   a. Search and rescue information from the flight plan is retained in the FSS and is available upon request.

3. VFR flight plans are forwarded to the FSSs serving the departure and the destination points.

   a. When an aircraft on a VFR flight plan changes destination, the FSS will forward a notification to the original and the new destination stations.

   b. When an aircraft changes its ETA, the information is passed by FSS to the destination station.

4. Military flight plans are handled by FSSs in much the same way as civilian flight plans.

5. For security control of air traffic, flight data and position reports from DVFR and IFR aircraft operating within an ADIZ are forwarded by FSS to ARTCC.

6. FSSs disseminate law enforcement alert messages, stolen aircraft summaries, and aircraft lookout alerts within FSS facilities and offices but not to the general public.

7. FSSs coordinate all pertinent information received from pilots prior to and during parachute jumping activity with other affected ATC facilities.

G. **International Operations**

1. AFSSs record and relay flight plans to and from Canada and Mexico to the appropriate stations.

   a. When the pilot requests Customs flight notification service (by including ADCUS in the remarks), the FSS will notify the appropriate Customs office.

      1) This service is available at airports so indicated in the *A/FD*.

   b. Round-robin flight plans to Mexico cannot be accepted.

2. AFSSs also relay International Civil Aeronautics Organization (ICAO) messages in four categories.

   a. Distress or urgency messages, or other messages concerning known or suspected emergencies, such as radio communications failures

   b. Movement and control

      1) Flight plans, amendments, and cancellations
      2) Clearances and flow control
      3) Requests and position reports

   c. Flight information

      1) Traffic information
      2) Weather information
      3) NOTAMs

   d. Technical messages

H. **Search and Rescue (SAR) Operations**

1. An aircraft on a VFR or DVFR flight plan is considered to be overdue when it fails to arrive 30 min. after its ETA and communications or location cannot be established.

   a. An aircraft not on a flight plan is considered to be overdue when it is reported to be at least 1 hr. late at its destination.

   b. In either case, the actions taken by an FSS are the same.

2. As soon as a VFR aircraft becomes overdue, the destination FSS will attempt to locate the aircraft by checking all adjacent flight plan area airports.

   a. Appropriate approach control and ARTCC facilities are also checked.

   b. If these measures do not locate the aircraft, a request is made for SAR information from the departure FSS.

   c. The departure FSS then checks locally for any information about the aircraft.

      1) If the aircraft is located, a notification is sent to the destination FSS.

      2) If not, all SAR information from the flight plan is sent, including any remarks that may be pertinent to the search.

3. If the aircraft has not been located within 30 min. after it becomes overdue, an Information Request (INREQ) is sent to all FSSs, Flight Watch stations, and ARTCCs along the route, as well as the Rescue Coordination Center (RCC).

   a. Upon receipt of an INREQ, a facility will check its records and all area airports along the proposed route of flight.

4. If the aircraft is not located within 1 hr. after transmission of the INREQ, an Alert Notice (ALNOT) is transmitted to all facilities within the search area.

   a. The search area is normally 50 mi. on either side of the proposed route of flight from the last reported position to the destination.

   b. Upon receipt of an ALNOT, each station within the search area conducts a search of airports within the area that were not checked previously.

   c. ALNOTs are also broadcast over VOR and NDB voice-capable frequencies.

5. When lake, island, mountain, or swamp reporting service programs have been established and a pilot requests the service, the FSS will establish contact every 10 min. with the aircraft while it is crossing the hazardous area.

   a. If contact with the aircraft is lost for more than 15 min., Search and Rescue is notified.
   b. Hazardous Area Reporting Service and chart depictions are published in the *AIM*.

I. **Aviation Weather Services**

1. Aviation routine weather reports (METAR, SPECI) are filed at scheduled and unscheduled intervals with stations having sending capability for dissemination.

2. FSSs actively solicit PIREPs when one or more of the following conditions are reported or forecast:

   a. Ceilings at or below 5,000 ft.
   b. Visibilities 5 SM or less
   c. Thunderstorms and related phenomena
   d. Turbulence of moderate degree or greater
   e. Icing of light degree or greater
   f. Wind shear

3. Radar weather reports (SD) are collected by the National Center Operations (NCO) in Washington, D.C. and transmitted to FSSs.

4. Winds and temperature aloft forecasts (FD) are computer-prepared and issued by the NCO to FSSs.

5. Terminal aerodrome forecasts (TAF) for selected U.S. airports are prepared by National Weather Forecast Offices and forwarded for distribution.

6. Area forecasts (FA) are issued by the Aviation Weather Center in Kansas City, MO and disseminated to all FSSs.

7. Severe weather forecasts are filed by the NWS and distributed to all FSSs.

8. SIGMETs and AIRMETs are issued by the Aviation Weather Center in Kansas City, MO to provide notice of potentially hazardous weather conditions.

9. The TWEB and synopsis for selected routes are prepared by National Weather Forecast Offices and distributed to the appropriate FSSs.

10. A Meteorological Impact Statement (MIS) is an unscheduled planning forecast intended for ARTCC specialists responsible for making flow control-related decisions.

    a. It enables these specialists to include the impact of expected weather conditions in making such decisions.

11. A Center Weather Advisory (CWA) is issued by the Central Weather Service Unit to reflect adverse weather conditions in existence at the time of issuance or conditions beginning within the next 2 hr.

J. **Using the AFSS**

1. A universal toll-free number has been established for AFSSs throughout the country.

    a. When you dial (800) WX-BRIEF [(800) 992-7433], you are switched automatically to the AFSS that serves your area.

        1) You are answered by a recording giving instructions for both touch-tone and rotary dial telephones.

        2) Touch-tone users can elect to talk to a briefer or select any of the direct-access services, e.g., TIBS or "fast-file" flight plan filing.

        3) If you are using a rotary or pulse dial telephone, you will be switched automatically to a briefer.

    b. So that your preflight briefing can be tailored to your needs, give the briefer the following information:

        1) Your qualifications, e.g., student, private, commercial, and whether instrument rated

        2) The type of flight, either VFR or IFR

        3) The aircraft's registration number or your name if you do not know the registration number

        4) The aircraft type

        5) Your departure point

        6) Your proposed route of flight

        7) Your destination

        8) Your proposed flight altitude(s)

        9) Your estimated time of departure (ETD)

        10) Your estimated time en route (ETE)

    c. At a minimum, a **standard briefing** should include the following information in sequence:

        1) **Adverse conditions** -- significant meteorological and aeronautical information that might influence you to alter your proposed route of flight or even cancel your flight entirely. Expect the briefer to emphasize conditions that are particularly significant, e.g., low-level wind shear, embedded thunderstorms, reported icing, or frontal zones.

2) **VFR flight not recommended** -- a statement issued when, in the briefer's judgment, a proposed VFR flight is jeopardized by conditions present or forecast, surface or aloft. The briefer will describe the conditions and affected locations, and announce, "VFR flight is not recommended."

    a) This is advisory in nature.

    b) You are responsible for making a final decision as to whether the flight can be conducted safely.

3) **Synopsis** -- a brief statement describing the type, location, and movement of weather systems and/or air masses that may affect the proposed flight

4) **Current conditions** -- a summary from all available sources of reported weather conditions applicable to the flight

    a) This summary is omitted if the proposed time of departure is over 2 hr. later, unless you request it.

5) **En route forecast** -- a summary of forecast conditions for the proposed route presented in logical order, i.e., departure/climbout, en route, and descent

6) **Destination forecast** -- a forecast of any significant changes within 1 hr. before and after the ETA at the planned destination

7) **Winds aloft** -- a summary of forecast winds aloft for the proposed route and altitude

8) **NOTAMs** -- information from any NOTAM (D) or NOTAM (L) pertinent to the proposed flight, and pertinent FDC NOTAMs within approximately 400 mi. of the FSS providing the briefing

    a) NOTAM (D) and FDC NOTAMs that have been published in the *Notices to Airmen Publication* are not included, unless requested by you.

9) **ATC delays** -- any known ATC delays and flow control advisories that might affect the proposed flight

10) The following may be obtained on your request:

    a) Information on military training routes (MTR) and military operations area (MOA) activity within the flight plan area and a 100-NM extension around the flight plan area

    b) Approximate density altitude data

    c) Information regarding such items as air traffic services and rules, customs/immigration procedures, ADIZ rules, etc.

    d) LORAN-C NOTAMs

    e) GPS NOTAMs

    f) Other assistance as required

d. Request an **abbreviated briefing** when you need to supplement mass disseminated data (e.g., weather channel or TIBS), update a previous briefing, or ask for only one or two specific items.

1) Provide the briefer with appropriate background information, the time you received the previous information, and/or the specific items needed.

2) The briefer can then limit the briefing to the information that you have not received and/or to changes in conditions since your previous briefing.

    a) To the extent possible, the briefing will be given in the same sequence as a standard briefing.

3) If you request only one or two specific items, the briefer will advise you if adverse conditions are present or forecast.

 a) Details on these conditions will be provided at your request.

e. You should request an **outlook briefing** whenever your proposed time of departure is 6 hr. or more from the time of the briefing.

1) The briefer will provide available forecast data applicable to the proposed flight for planning purposes only.

2) You should obtain a standard briefing prior to departure in order to obtain such items as current conditions, updated forecasts, winds aloft, and NOTAMs.

2. To provide in-flight services such as weather updates, filing a flight plan, etc., a number of sources exist.

a. Transcribed weather broadcasts such as TWEB or HIWAS are available on certain VOR and NDB frequencies, as indicated in the identifier boxes on your navigational charts.

1) To hear such a broadcast, listen on the appropriate frequency through your VOR or ADF receiver.

b. To obtain current weather along your route of flight or to file a PIREP, contact Flight Watch (i.e., EFAS) on 122.0 MHz below FL 180 and as published at and above FL 180.

c. For weather outlooks and to file, activate, or extend a flight plan, contact the nearest FSS.

1) Use the frequency shown on your navigational chart atop the identifier box of the VOR nearest you.

3. Following any briefing, feel free to ask for any information that you or the briefer may have missed.

a. It helps to save your questions until the briefing has been completed.

b. This enables the briefer to present the information in a logical sequence and reduces the chance of important items being overlooked.

K. **Visiting the AFSS**

1. You should visit an AFSS and tour the facility. A visit is an excellent way for you to learn what services are available and how you can best use them.

a. AFSSs are open 24 hr., but usually the best time to visit is during normal business hours.

b. You do not need to call ahead, but you may want to so you do not visit during a normal peak time or when staffing is reduced.

2. Your tour will be given by a specialist and usually begins at a telephone briefing station.

a. The specialist is able to call up the various weather reports and forecasts for the departure, en route, and destination phases of a flight.

b. Various charts (e.g., surface analysis chart) and a sequence of satellite images (either NOAA or GOES) can be displayed on another monitor.

c. Here, also, the specialist will input the data for a flight plan. The specialist can also view information on aircraft that have already filed, either departing from or arriving in the AFSS's area.

3.  You will also see the radio communication station used to contact aircraft.

    a.  Here you will learn why it is very important that you identify the frequency on which you are transmitting and your aircraft's location.

        1)  Remember that most states have only one AFSS, so it is important for the specialist to know where you are in order to select the proper remote transmitter to talk to you.

    b.  At these positions, the specialist can provide the same services as those positions that answer incoming telephone calls, in addition to accepting PIREPs.

    c.  At a separate station will be the specialist who provides the En Route Flight Advisory Service (EFAS) for aircraft en route.

4.  As you tour the facility, you may also observe the other services provided at the AFSS. These may include the VHF direction finding (VHF/DF) equipment to help lost pilots and information on how weather observations are taken.

5.  By touring the AFSS and talking with specialists, you will gain a better understanding of how the AFSS supports pilot operations.

    a.  By understanding the strengths and weaknesses of the AFSS, you can maximize the services provided by the AFSS, and you can provide assistance (i.e., by making a PIREP).

L.  **Kavouras Weather Graphics**

1.  A company in Minneapolis called Kavouras is the FAA's contractor to provide weather graphics to AFSSs.  It provides both hardware and software.

2.  The capability of each computer console includes four graphics menus:

    a.  Current menu -- analysis charts
    b.  Forecast menu -- forecast charts
    c.  Satellite menu -- satellite imagery
    d.  Radar menu -- radar data

3.  A final menu provides maps, charts, etc., that are produced at the specific request of a particular AFSS, e.g., Bahamas maps for Florida AFSSs.

4.  A paragraph on each chart follows.

    a.  Current menu (analysis charts)

        1)  Weather depiction

            a)  The Weather Depiction Analysis is a contoured and shaded depiction of MVFR and IFR areas.  This chart gives the user a general overview of the country in terms of ceiling and visibility.  Major synoptic features such as highs, lows, and fronts are included to aid in interpreting the chart. Individual station data and models are not shown on this chart.  The Weather Depiction Analysis is quite valuable in assessing the large-scale picture in the vicinity of a route of flight; however, the latest surface observations at the destination airport must also be checked.

        2)  North American surface

            a)  The North American Surface Analysis depicts isobars, high and low pressure centers, and fronts.  Individual station data and station models are not shown.  The analysis depicts synoptic features -- those of fairly large scale.  The placement of frontal features is determined by the Kavouras meteorologist using computer-generated surface plots which are hand analyzed every 2 to 3 hr.  Before any features are placed, they are compared to the previous 3-hourly position and also to recent trends noticed over the last 12 hr.

3) National radar summary

    a) The National Radar Summary Analysis is a composite of 211 National Weather Service, military, and ARTCC radars. The chart depicts precipitation areas using the standard VIP scale of six intensity levels. Echo top data, movement, and "Out for Maintenance" and "Not Available" sites are also included.

| VIP Level | Contour Color | Intensity Level |
|---|---|---|
| 1 | Light green | Light |
| 2 | Dark green | Moderate |
| 3 | Light yellow | Heavy |
| 4 | Dark yellow | Very heavy |
| 5 | Light red | Intense |
| 6 | Dark red | Extreme |

4) Upper air (850, 700, 500 mb/hPa)

    a) Upper-air charts at 850, 700, and 500 mb/hPa display height contours in decameters, temperature in degrees Celsius, and relative moisture content. The following table gives an approximate relationship between millibar/hectoPascal level and altitude:

| Pressure (mb/hPa) | Altitude (ft.) |
|---|---|
| 500 | 18,000 |
| 700 | 10,000 |
| 850 | 5,000 |

The actual altitude of these levels varies significantly with season and latitude. All levels are lower in winter and in northern latitudes since the atmosphere is colder and more compact (denser).

5) Upper air (300, 200 mb/hPa)

    a) Upper-air charts at 300 mb/hPa and 200 mb display height contours in decameters and wind speeds in knots. The 300-mb/hPa level is at approximately 30,000 ft., and 200 mb/hPa is at approximately 39,000 ft. The exact altitude of these levels varies significantly with season and latitude.

6) Freezing level

    a) The freezing level chart displays the height of the lowest freezing level in thousands of feet above the surface. Data are derived from NWS balloon soundings, taken twice daily at 0000Z and 1200Z. The map is computer-generated, with a contour color of red and an interval of 4,000 ft.

7) Lifted Index/K Index

    a) The Lifted Index/K Index chart is a measure of atmospheric stability. Data are derived from National Weather Service radiosondes launched at 0000Z and 1200Z. Two values are displayed for every radiosonde site. The top value is the Lifted Index, with negative values indicating an unstable atmosphere and positive values indicating a more stable atmosphere. The lower value is the K Index. The larger this number, the greater the likelihood of precipitation.

8) Precipitable water

    a) Precipitable water is a measure of the amount of liquid water in a vertical column of air. This chart displays a contoured analysis of precipitable water, which can be correlated with precipitation total. Higher values indicate the atmosphere is holding a greater amount of moisture and imply that more significant precipitation is possible from the air mass, given that other conditions are favorable.

9) Average relative humidity (SFC -- 500 mb/hPa)

    a) The Surface to 500 mb/hPa Relative Humidity Chart gives the average humidity in the lower 18,000 ft. of the atmosphere. Relative humidity values are contoured every 10% in red.

10) Winds aloft (FL 040, FL 140, FL 240, FL 340)

    a) Winds aloft charts are an analysis of wind flow at various levels in the atmosphere. Data are displayed using the conventional wind barb format. The data are derived from NWS radiosondes, with the analysis computer-generated. Barbs are yellow and temperatures are black for FL 240 and FL 340 (temperatures are not reported for FL 040 and FL 140).

b. Forecast menu (forecast charts)

1) North American surface

    a) The North American Surface Forecast depicts high- and low-pressure centers, fronts, and precipitation. The chart depicts synoptic features, those of fairly large scale. Therefore, weak pressure centers and features produced by terrain are generally ignored.

2) Low-level significant weather

    a) Low-level significant weather forecasts display important weather features from the surface to 24,000 ft. Freezing levels, shaded regions of MVFR and IFR, and turbulence are provided.

3) Winds/temperatures aloft (FD winds)

    a) Forecast winds aloft are generated twice daily by the National Weather Service in Suitland, Maryland. Forecasts are available for 12-, 24-, 36-, and 48-hr. time periods. At this time, only the 12-hr. chart (actually the 9- to 18-hr. forecast) is displayed since it is the most pertinent and often the most accurate. These data are reproduced without any changes by Kavouras, Inc. Wind data are displayed using the conventional wind barb format in yellow. Forecast temperatures are shown in degrees Celsius and colored black.

| | |
|---|---|
| FL 390 .. | approximately 200 mb/hPa |
| FL 340 .. | approximately 250 mb/hPa |
| FL 300 .. | approximately 300 mb/hPa |
| FL 240 .. | approximately 400 mb/hPa |
| FL 180 .. | approximately 500 mb/hPa |
| FL 120 .. | approximately 650 mb/hPa |
| FL 090 .. | approximately 700 mb/hPa |
| FL 060 .. | approximately 800 mb/hPa |

4)  U.S. high-level significant weather

    a)  High-level significant weather forecasts cover events occurring at levels above 24,000 ft. Forecasts are available four times daily, covering a 6-hr. time period (3 hr. either side of 0000Z and 1200Z). Each chart displays jet stream axes with altitude and wind maximums, tropopause heights, areas of broken thunderstorm coverage, areas of moderate or greater turbulence, and surface fronts.

5)  36-hour thickness/sea-level pressure

    a)  This forecast gives a depiction of how the atmosphere will look 36 hr. in the future. Sea-level pressure is depicted, along with frontal features. The thickness of the lower half (approximately) of the atmosphere is also displayed.

c.  GOES menu (GOES satellite imagery)

1)  GOES (Geostationary Operational Environmental Satellite) imagery is available to AFSSs every half hour. Kavouras receives the data directly from the satellite via its communications facility in Minneapolis. The transmission schedule is dictated partly by NOAA and partly by Kavouras.

d.  Radar menu (NWS radar data)

1)  National radar composite

    a)  The national radar composite consisting of the National Weather Service WSR-88D radars is available at 5-min. intervals. This composite is assembled at Kavouras's central processing facility in Minneapolis. Option 1 displays the animation of the most recent images. Option 6 displays the latest image received by the system. The ANIMATE UP and DOWN keys can be used to change the speed of the animation loop when option 1 is chosen or to single step through the images when option 6 is chosen.

2)  Regional radar composite

    a)  Twenty regional radar composites are assembled and transmitted every 5 min. Each AFSS facility will receive one region, likely one which correlates with the most air traffic handled by the facility. Each region is about the size of five or six states and gives the advantage of seeing high resolution radar data on a larger basemap.

3)  Single-site radar imagery

    a)  Data from two single-site radars can be combined and either animated or viewed singly. The options are either to have one radar site with two ranges or to have two individual radar sites with one range from each. It is up to the AFSS facility to choose whether two radar sites or two ranges from a single site are needed.

# END OF APPENDIX

# APPENDIX D
# DIRECT USER ACCESS TERMINAL SYSTEM (DUATS)

> Please take a few minutes to study each of the concepts listed above and anticipate/imagine what they are and how they relate to the other listed concepts.

A.  **DUATS Introduction**.  The FAA's Direct User Access Terminal System (DUATS) is a computerized weather briefing and flight planning system that provides pilots with the most up-to-date and reliable briefing information possible.

1.  DUATS is accessed over toll-free 800-number telephone lines, and there are no usage fees for the basic service.

2.  The service is provided under contract from the FAA by two companies, Data Transformation Corporation (DTC) and GTE.

3.  You should try both DTC and GTE DUATS before deciding which one you will use regularly.

    a.  This is a personal decision, i.e., choosing the service that you feel is the most comfortable.

    b.  You, of course, may also use both systems.

4.  Both systems are being upgraded continuously.

B.  **Accessing DUATS**.  To access DUATS, you need an IBM-compatible or Apple (e.g., MacIntosh) personal computer, a modem, a telephone line, and one of the many telecommunications programs on the market.

1.  You may need assistance setting up your computer modem.  Consult a friend or computer store personnel.

2.  Set your telecommunications program to use 8 bits, one stop bit, no parity, full duplex, and echo off to access either service.

    a.  DTC's modem line number is (800) 245-3828.
    b.  GTE's modem line number is (800) 767-9989.

3.  Both services also offer customer assistance lines that are open 24 hr. a day.

    a.  DTC's help line is (800) 243-3828.
    b.  GTE's help line is (800) 345-3828.

4.  In the following discussion, you will see 14 DUATS screens, seven from DTC and seven from GTE.  We recommend both vendors to you equally.

C. **Signing Up**. Once you establish a connection, both systems walk you through a sign-up procedure that confirms your authorization to use the system by asking you to enter your access code and password, as shown below.

**DTC's Sign-Up Procedure**        **GTE's Sign-Up Procedure**

```
Welcome to Data Transformation's

DIRECT USER ACCESS TERMINAL SYSTEM

***********************************
* Message Space for News and Info  *
*  on new DUAT System Features     *
***********************************

Transaction 2433828 6/14/96 1245 (UTC)

If you do not have an Access Code
Press <ENTER>:

Enter Access Code. . . . . . . . . ? __
Enter Password  . . . . . . . . . . ? __

    For HELP enter a ? at any prompt.
```

```
        GTE Contel DUAT System

Session number:  26155

Enter DUAT access code  -or-  last name:

Enter your password:

Transaction number:  006994
Fri June 14 19:23:48 1996 (UTC)
```

1. Only medically current pilots, student pilots, balloonists, ultralight and glider pilots, and flight dispatchers are authorized by the FAA to access the free DUATS functions.

2. If you do not have an access code and a password and you are an authorized user, follow the instructions on the screen and DUATS will assign you an access code (which you can change if you wish) and ask you to choose a password.

     a. These must then be used in all subsequent DUATS briefings.

D. **Storing Your Aircraft Profile**. After signing onto DUATS, or anytime during a briefing, you can store a profile of your aircraft that is retained permanently in the DUATS.

1. This aircraft profile consists of search and rescue information such as aircraft registration number, type of aircraft, aircraft color, etc.

2. When you use DUATS to file a flight plan, this information is entered automatically in the appropriate spaces unless you wish to change it.

E. **DUATS Main Menu**. The DUATS main menu allows you to access a weather briefing, plan a flight, and use various other options, as shown below.

**DTC's Main Menu**        **GTE's Main Menu**

```
          DUAT MAIN MENU

Weather Briefing  . . . . . . . . . . . . . . . . . . . . 1
Flight Planning . . . . . . . . . . . . . . . . . . . . . 2
Encode Function . . . . . . . . . . . . . . . . . . . . 3
Decode Function . . . . . . . . . . . . . . . . . . . 4

Bulletin Board  . . . . . . . . . . . . . . . . . . . . B
System Information . . . . . . . . . . . . . . . . . S
Expanded Help  . . . . . . . . . . . . . . . . . . . . H
Download Library . . . . . . . . . . . . . . . . . . . L

Exit DUAT system  . . . . . . . . . . . . . . . . . . X
Selection  . . . . . . . . . . . . . . . . . . . . . . . . ?
```

```
          DUAT MAIN MENU

Weather Briefing                        1
Flight Plan and Planner                 2
Encode                                  3
Decode                                  4
Modify Personal Data Profile            5
Service Information                     6
Extended Decode                         7
FAA/NWS Contractions                    8

Select function (or 'Q' to quit):
```

1.   You are allowed 15 min. on the FAA functions, after which the system terminates your briefing automatically.

   a.   If you have not completed your briefing by that time, you must sign back on and begin again.

F.   **Weather Briefing**.  If you select a weather briefing from the main menu, the screen displays the Weather Briefing Menu, as shown below.

**GTE's Weather Briefing Menu**

| | |
|---|---|
| Standard Weather | 1 |
| Outlook Weather | 2 |
| Abbreviated Weather | 3 |
| Plain Language Weather | 4 |

**DTC's Weather Briefing Menu**

```
Standard/Outlook Briefing
    Route Briefing  . . . . . . . . . . . . . . . . . . 1
    Local Area Briefing  . . . . . . . . . . . . . . 2

Abbreviated Briefing
    State Collectives  . . . . . . . . . . . . . . . . 3
    Specific Locations  . . . . . . . . . . . . . . . 4
    Select Route  . . . . . . . . . . . . . . . . . . 5
    Select Local Weather  . . . . . . . . . . . . 6

Graphic Products  . . . . . . . . . . . . . . . . . . 7
Briefing Parameters  . . . . . . . . . . . . . . . . B
Return to Main Menu  . . . . . . . . . . . . . . . M
```

1.   Both DUATS services offer similar options here.

   a.   The route options are used to retrieve weather products along a proposed route of flight.

      1)   You are required to enter your departure point, time of departure, altitude, destination, route of flight, and estimated time en route.

      2)   If a preferred IFR route is available, you have a choice of selecting it or entering your own route.

   b.   The local options are used to retrieve weather products available at or near a specific location.

      1)   You are prompted for the location identifier and your time of departure.

   c.   Abbreviated Weather (DTC and GTE) allows you to request specific weather products from specified weather reporting locations.

      1)   Regional and state collectives (part of Abbreviated Briefing) allow you to request weather products by state or geographical regions.

2.   As with an FSS briefer, DUATS allows you the option of a standard, an abbreviated, or an outlook briefing.

   a.   Selecting a route or local briefing provides you with a standard briefing, including all the reports and forecasts commonly provided by a briefer.

   b.   Abbreviated weather briefings provide the report types you request for selected locations.

**GTE's Weather Data Types**

o Enter a sequence of weather data types separated by spaces (e.g., METAR TAF FD)

Valid types are:

| | |
|---|---|
| AIRMETS | WA |
| Amended Severe Weather Forecasts | WW-A |
| Area Forecast | FA |
| Center Weather Advisory | CWA |
| Convective SIGMETs | WST |
| Flight Data Center NOTAMs | FDC |
| Flow Control Advisories | ATC |
| Hurricane and Tropical Depression | WH |
| NOTAMs (Notice to Airmen) | NO |
| NOTAM Summary (Notice to Airmen) | NS |
| Pilot Report | UA |
| Radar Summaries | SD |
| Surface Obs Weather Trends (3 hours) | TW |
| Severe Weather Forecast Alerts | AWW |
| Severe Weather Outlook | AC |
| Severe Weather Warnings | WW |
| SIGMETs | WS |
| Surface Observation | METAR |
| Terminal Forecasts | TAF |
| Winds Aloft Forecast | FD |

   c. Requesting weather information more than 6 hr. in advance of your proposed time of departure provides you with an outlook briefing.

3. The weather data presented are normally displayed in the standard abbreviated format unless you request the expanded (plain English) version.

   a. EXAMPLES:

      1) Standard format:

         METAR KLAL 011454Z 19004 8SM SCT040 OVC120 21/16 A3014

      2) Expanded format (GTE DUATS):

         LAKELAND FL (LAKELAND LINDER REGIONAL) [LAL] ROUTINE OBSERVATIONS AT 9:54 A.M. EST: WIND 190 AT 4, VISIBILITY 8 MILES, 4,000 SCATTERED, CEILING 12,000 OVERCAST, TEMPERATURE 21, DEWPOINT 16, ALTIMETER 30.14

G. **Flight Plan Filing**. The Flight Plan Menu is used to file, amend, or cancel flight plans.

**GTE's Flight Plan Menu**

| | |
|---|---|
| File Flight Plan | 1 |
| Amend Flight Plan | 2 |
| Cancel Flight Plan | 3 |
| View Flight Plan | 4 |
| Close VFR Flight Plan | 5 |
| Flight Planner | 6 |
| Modify Flight Planner Profile | 7 |
| Flight Planner Users Guide | 8 |

1.   Note that, if you have not previously entered your access code and password, the system will not allow you to file or amend any flight plan information.

2.   When filing a flight plan, you may use any previously stored data on your aircraft or enter all the values at this time.

    a.   After you answer all the flight data prompts, your entire flight plan is displayed, and you are prompted for any changes you wish to make.

    b.   When all the fields are correct, enter an "F" to file your flight plan.

        1)   DUATS displays the ARTCC or FSS to which your flight plan will be sent and the time it will be transmitted.

        2)   Up to the time it is sent, you can amend or cancel with DUATS service.

        3)   After it is sent, you must contact an FSS.

3.   To amend or cancel a flight plan you filed with DUATS, you must enter the aircraft ID number used in the flight plan.

    a.   If you attempt to amend or cancel your flight plan after it is forwarded, you are advised to contact the appropriate FSS.

4.   DTC's Flow Control Messages provide current advisories issued by the Air Traffic Control Systems Command Center, as shown below.

**DTC's Flow Control Message Menu**

```
Advisories  . . . . . . . . . . . . . 1
Weather Outlook  . . . . . . . . . 2
North Atlantic Tracks   . . . . . 3
Miscellaneous Messages  . . 4

Return to Main Menu   . . . . . M
```

    a.   Selection 1 retrieves departure, en route, and arrival delay messages.

    b.   Selection 2 retrieves the ATC System Outlook which includes a weather outlook and potential or current problem areas for delays.

    c.   Selection 3 retrieves the daily routes for the North Atlantic tracks.

    d.   Selection 4 retrieves miscellaneous messages that may concern NOTAMs, missile firing, etc.

5.   DTC's Data File and GTE's Modify Flight Planner Profile allow you to store or modify any data on your aircraft profile.

    a.   This data may be used when requesting a weather briefing, filing a flight plan, or creating a flight log.

6.   DTC's Flight Log and GTE's Flight Planner both create a navigation log of a proposed flight based on your estimated departure time, airspeed, and the forecast winds aloft at your proposed altitude.

    a.   To plan a flight, you must supply DUATS with certain information.

        1)   Departure point
        2)   Destination
        3)   Departure time
        4)   Route selection
        5)   Aircraft performance information
        6)   Cruise altitude

b. You may input your own route or allow DUATS to generate one of its own.

    1) DTC will automatically generate the preferred IFR route for you, if one exists.

    2) GTE will generate one of the following routes at your request.

        a) Low-Altitude Airway Auto-Routing selects the shortest path from your origin to the destination using Victor airways.

            i) This option may not be possible for certain airports that are very remote from any navigational facility.

        b) Jet Route Auto-Routing selects the shortest path from your origin to the destination using jet routes.

            i) You will be prompted for departure and arrival routings (e.g., SIDs or STARs).

        c) VOR-Direct Auto-Routing is similar to Low-Altitude Airway Auto-Routing.

        d) Direct Routing for LORAN

        e) Direct Routing for RNAV

        f) User Selected Routing

c. Once you select one of the routing options, DUATS computes and displays the optimal route, along with the distance along the route.

    1) Note that DUATS does NOT take into account obstacles, terrain, controlled airspace, and special-use airspace.

    2) You must verify the suggested route against your charts to see that it can be flown safely.

d. If you approve the suggested route, DTC's DUATS generates a flight log as shown below, complete with forecast winds aloft and heading and groundspeed computations.

**DTC's Flight Log**

---

The IFR Low Altitude Preferred Route(s) between GNV and MIA:

1. V157 LBV V529 V035 CURVE
   100 AND BLO : 1300-0300 :
2. Return to Previous Menu

Selection....? 1

                    ===> Data Transformation's Flight Log <===
          Altitude--7000 Ft   Air Speed--150 Knots   Departure Time--2200Z

| LEG | MAG CRS | MAG HDG | GND SPD | DIST (NM) | ETE (MIN) | WIND | ATE |
|-----|---------|---------|---------|-----------|-----------|------|-----|
| GNV ARPT | 217 | 225 | 138 | 9 | 0+04 | 276/023 | _____ |
| GNV VORTAC | 162 | 170 | 157 | 25 | 0+09 | 276/023 | _____ |
| OCF VORTAC | 171 | 179 | 160 | 72 | 0+27 | 290/024 | _____ |
| LAL VORTAC | 153 | 160 | 166 | 77 | 0+28 | 290/024 | _____ |
| LBV VORTAC | 157 | 163 | 160 | 61 | 0+23 | 284/018 | _____ |
| XING V035 | 109 | 110 | 167 | 19 | 0+07 | 284/018 | _____ |
| CURVE | 087 | 085 | 167 | 19 | 0+07 | 284/018 | _____ |
| MIA ARPT | TOTAL | | | 282 | 1+45 | | _____ |

      YOUR ETA = 2345Z

e.    GTE will prompt you for your aircraft's climb, cruise, and descent airspeeds, climb and descent rates, and fuel consumption rates throughout and then generate the following flight log.  Other flight log formats are available in Modify Flight Planner Profile menu.

### GTE's Flight Log

From:    KGNV -- Gainesville FL
To:        KMIA -- Miami FL
Time:    Fri Feb 26 22:00 (UTC)

Routing options selected:  Automatic low altitude airway.
Flight plan route:
    KGNV OCF V295 BAIRN V267 GREMM KMIA
Flight totals:  fuel: 30 gallons, time: 1:37, distance 265.7 nm.

| Ident  Type/Morse Code<br>Name or Fix/radial/dist<br>Latitude  Longitude  Alt. | Route<br>Winds<br>Temp | Mag<br>Crs<br>Hdg | KTS<br>TAS<br>GS | Fuel<br>Time<br>Dist | Fuel<br>Time<br>Dist |
|---|---|---|---|---|---|
| 1.  KGNV      Apt.<br>Gainesville FL<br>29:41:24  82:16:18      2 | Direct<br>291/15<br>+16C | 179<br>184 | 130<br>136 | 4.8<br>0:14<br>31 | 0.0<br>0:00<br>266 |
| 2.  OCF        --- -.-. ..-.<br>d113.7 Ocala<br>29:10:39  82:13:35      70 | V295<br>284/24<br>+10C | 132<br>136 | 155<br>176 | 6.1<br>0:20<br>60 | 4.8<br>0:14<br>235 |
| 3.  ORL        --- .-. .-..<br>d112.2 Orlando<br>28:32:34  81:20:06      70 | V295<br>284/24<br>+10C | 165<br>172 | 155<br>166 | 4.1<br>0:14<br>38 | 10.9<br>0:34<br>175 |
| 4.  BAIRN Int.<br>ORLr162/37  VRBr300<br>27:56:53  81:06:54      70 | V267<br>284/24<br>+10C | 165<br>173 | 155<br>166 | 8.0<br>0:26<br>73 | 15.0<br>0:48<br>137 |
| 5.  PHK        .--. .... -.-<br>d115.4 Pahokee<br>26:46:58  80:41:29      70 | V267<br>277/17<br>+13C | 158<br>164 | 161<br>169 | 4.3<br>0:15<br>41 | 23.0<br>1:14<br>64 |
| 6.  GREMM Int.<br>PHKr155/41  MIAr020<br>26:09:57  80:22:36      40 | Direct<br>263/17<br>+17C | 171<br>177 | 170<br>171 | 2.4<br>0:08<br>23 | 27.3<br>1:29<br>23 |
| 7.  KMIA  Apt.<br>Miami FL<br>25:47:35  80:17:25      0 | | | | | 29.7<br>1:37<br>0 |

NOTE:  Fuel calculations do not include required reserves.
Flight totals:  fuel: 30 gallons, time: 1:37, distance 265.7 nm.
Average groundspeed  165 knots.
Great circle distance is 256.4 nm -- this route is 4% longer.

7. DTC's Special Use Airspace selection provides you with information on Alert Areas, MOAs, Prohibited Areas, Restricted Areas, and Warning Areas within 5 NM of your route of flight.

8. GTE's Flight Planner Users Guide gives complete user-friendly directions on how to use the Flight Planner option.

H. **Encode and Decode Functions**. The Encode and Decode Functions are used to find the identifier for a known location and to receive information on NAVAIDs and airports for any valid location identifier, respectively.

1. In the Encode Function, all locations that use the name you enter are shown.

**DTC's Encode Function**

```
ENTER LOCATION, STATE...?  PHILI, PA

IDENT CITY, STATE                 AIRPORT/NAVAID/WX RPT
1N3   PHILIPSBURG, PA             ALBERT//
PSB   PHILIPSBURG, PA             MID-STATE/VORTAC/WX


ENTER LOCATION, STATE...?  FERGUSON

IDENT CITY, STATE                 AIRPORT/NAVAID/WX RPT
82J   PENSACOLA, FL              FERGUSON//
12M0  WINDSOR, MO                FERGUSON FARMS//
TN09  PHILADELPHIA, TN           FERGUSONS FLYING CIRCUS//
```

2. In the Decode Function, information concerning a given identifier or contraction is provided.

**GTE's Decode Function**

```
ENTER IDENT(S):  GNV

GNV  airport GAINESVILLE REGIONAL   GAINESVILLE, FL
        fss: GNV(GAINESVILLE)  artcc: ZJX(JACKSONVILLE)
GNV  navaid    GAINESVILLE, FL                        EFAS
        fss: GNV(GAINESVILLE)  artcc:
GNV  navaid GAINESVILLE   GAINESVILLE, FL             VORTAC
        fss: GNV(GAINESVILLE)  artcc:  JACKSONVILLE
GNV  weather  GAINESVILLE, FL
```

3.  At any prompt that requires entry of a location identifier, both DTC and GTE allow you to retrieve Encode/Decode information.

    a.  Whenever you are prompted for an aircraft type, the escape to Encode/Decode Function is changed to search for aircraft type designators and model numbers, rather than for location identifiers.

    b.  EXAMPLE:

**DTC's Aircraft Decode/Encode Function**

Aircraft Type/
Special Equipment . . . ? ? **PA24**

| DESI | MODEL | MANUFACTURER |
|------|-------|--------------|
| PA24 | COMANCHE | PIPER AIRCRAFT CORP. |

Aircraft Type/
Special Equipment . . . ? ? **DAUP**

| DESI | MODEL | MANUFACTURER |
|------|-------|--------------|
| HH65 | DAUPHINE 2 | AEROSPATIALE |
| HR3C | DAUPHINE 2 | AEROSPATIALE |

# END OF APPENDIX

# APPENDIX E
# AVIATION WEATHER RESOURCES ON THE INTERNET

A. In recent years, an explosion of data has been available on the Internet, and much of it is of interest to pilots, especially aviation weather. By careful use of this seemingly endless, constantly changing resource, any pilot with access to a personal computer equipped with a modem, communications software, and an Internet Service Provider (ISP) account can obtain current official and unofficial aviation weather products.

1. These products include both text and color graphics in addition to the black and white National Weather Service charts covered in Part III of this book.

2. The typical cost for an ISP account with a major nationwide company such as America Online (AOL), Netcom, or AT&T Worldnet, which allows the user unlimited access to the Internet, is about $20 per month.

3. Use caution. So many excellent sources of weather information are on the Internet that it is very easy to lose track of time exploring them.

    a. Some sites provide aviation weather data free of charge to the user. These sites are normally sponsored by commercial vendors, and their billboard commercials are placed around the site. If you are in the market for any of the advertised products, your support of these companies, with a mention that you saw their ads on the Internet, will help continue this free service to all users.

    b. Other sites require one to sign up for membership, generally at no cost, but user demographic data are often gathered for marketing purposes.

    c. Other sites provide aviation weather data to subscribers for a monthly subscription fee, typically about $10.

4. Once your favorite list of book marks (short names of Internet sites that can be accessed with a simple mouse click) is established, you can get the weather information you need quickly.

5. By having as many different sites as you choose, you can have backups for those times when technical problems with a particular site may make it unavailable.

B.  **Glossary of Terms**. Before you begin to navigate around the Internet, you should become familiar with the terms you will be using.

1.  **Internet** -- a catchall word to describe the massive global network of computers; a network of networks. The Internet is the physical side of the global network. Gleim Publications' information network of electronic documents may be accessed by anyone, worldwide, with a minimum of equipment and software.

2.  **World-Wide Web (WWW)** -- a wide-area (global) information retrieval system giving everyone access to a large number of electronic documents. Gleim Publications has a site, or Home Page, on the World-Wide Web.

3.  **Universal resource locator (URL)** -- a web site address or location description that must be used precisely to navigate to the specific web site desired. Gleim Publications' URL is http://www.gleim.com.

4.  **Browser** -- a software program such as Netscape or Microsoft Internet Explorer that allows one to access the Internet

5.  **Search engine** -- a software program such as Yahoo or Excite that allows one to search for a group of web sites with common characteristics. By entering key words, such as aviation weather, one can generate a list of URLs to be explored.

6.  **Hypertext/links** -- special underlined and highlighted text in a document that, when clicked on with a mouse, sends the user to the site described by the hypertext. At the Gleim Publications Home Page, a link reads: "Links to other Aviation Sites." If this hypertext is clicked on, the list of aviation links will appear. One link is "National Weather Service." If this hypertext is clicked on, the user will leave the Gleim Home Page and go to the National Weather Service's Home Page, which has a URL of http://www.nws.noaa.gov.

    a.  By navigating through this electronic document, one can find, download, save, and print all of the black and white weather charts covered in the Gleim series of books and tested on the FAA pilot knowledge tests.

    b.  Many other products are available at this site.

        1)  If the user adds this URL to the bookmarks list, it can be accessed directly in the future by simply selecting it from that list.

        2)  The URL may also be typed in at the ISP Home Page.

C.  **Sites**. The following is a list of eight selected aviation-weather-related Internet sites and their URLs that will serve to get you started in using the Internet as a quick, practical, inexpensive source of aviation weather data. These sites represent just the tip of the iceberg and are representative of the hundreds of sites available. Also, you will want to try typing weather or aviation weather in your search engine to see how many sites are listed for you to explore.

1.   **Gleim Publications**. http://www.gleim.com/aviation.html -- contains several links to other aviation and aviation weather sites on the Internet, plus other information about Gleim aviation products

Pilot Training | Books, Software, Tapes, Supplies | FIRC| Non-Airplane Questions and Answers
Book and Software Updates | Register Your Software | CFI Directory | Flight School Directory
FBO/Bookstore Directory| Aviation Medical Examiner Directory| Testing Sites
Aviation Links and College Aviation Programs| Unsolicted Comments About Gleim's Aviation Products

*Gleim Publications, Inc.*
*P.O. Box 12848 University Station*
*Gainesville, Florida 32604*
***Telephone:*** *(800) 87-GLEIM or (352) 375-0772*
***FAX:*** *(888) 375-6940*
*All information Copyright © 1995-2000*
*Gleim Publications, Inc. and Gleim Internet, Inc.*

2.   **Landings**.  http://www.landings.com -- a very large site with many different aviation links. The Aviation Weather Information sources section has 43 different links, including information (e.g., key to ASOS Weather Observations) and current data (e.g., convective SIGMETs from NWS/NOAA).  This Internet site is very popular with pilots.

## Landings: Aviation Weather information

**Index:**

Aviation Weather

Satellite Images and
Weather Maps:
World Wide
Australia/New
Zealand
Antarctica
North America
South America
Caribbean
Europe
Middle East
Asia/Pacific Rim
Africa

Weather by Region:
North America
South America
Europe
Middle East
Caribbean
Australia
New Zealand
Antarctica
Asia
Africa

Weather Servers:
Telnet
FTP

Weather Related
Other services

Aviation's Directory

### Aviation weather information sources

- GTE DUAT Aviation Weather: This is the GTE DUATS website for certificated pilots. GTE's Direct User Access Terminal Service (GTE DUATS) provides current FAA weather and flight plan filing services to all certified civil pilots. The service is available 24 hours a day, 7 days a week, at no charge to the user -- fees to operate the BASIC GTE DUAT Service (weather briefing, flight plan filing, encode/decode) are paid by FAA. You need to be a certificated pilot to use this service.
- GTE DUAT Aviation Weather: This is a telnet connection to GTE DUATS for certificated pilots. GTE's Direct User Access Terminal Service (GTE DUATS) provides current FAA weather and flight plan filing services to all certified civil pilots. The service is available 24 hours a day, 7 days a week, at no charge to the user -- fees to operate the BASIC GTE DUAT Service (weather briefing, flight plan filing, encode/decode) are paid by FAA. You need to be a certificated pilot to use this service.
- GTE DUATS Web Site User's Guide: This is the users guide for the GTE DUATS web site.
- GTE DUATS User's Guide: This is the users guide for the GTE DUATS.
- GTE DUATS Flight Planning User's Guide: This is the users guide for the GTE DUATS Flight Planning System.
- Aviation weather forecasts: Massachussetts Institute of Technology: Aviation forecast maps at MIT.
- Aviation Weather Center and Storm Prediction Center: Aviation Weather information.
- NOAA Experimental forecast products: Icing, mountain wave, convective forecasts (from NOAA/AWC).
- Radar Observations: Radar observations at the National Weather Service

3.  **AVweb**. http://www.avweb.com -- another very large, popular site with many aviation links, including aviation weather sources worldwide. The AVweb site requires membership but no fee. Membership includes a weekly aviation news summary (very interesting and informative) e-mailed to the user every Monday morning.

 **Weather on the Web**

*Real-time weather reports, all manner of weather images, weather news from CNN and USA Today, GTE DUATS, and other cool WX-related stuff.*

## First, The Basic Info We Need

### METAR/TAF Information
Lots of cool tips, tricks, and help with METAR/TAF weather. If you ever wanted one source for learning your way around, this is it.

## Weather Sites

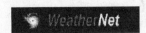

**University of Michigan WeatherNet**
One of the best lists of U.S. weather sites on the Web, including a huge number of local and regional NOAA offices, TV stations, university sites, and weather-cams.

**U.S. National Weather Service - Aviation Weather Center**
Textual aviation weather products from NWS, including area forecasts, airmets, convective SIGMETs, non-convective SIGMETs, oceanic SIGMETs, and neural-net icing forecasts. There's even a page devoted to METAR/TAF info!

**Unisys Weather**
One of the most comprehensive sources of world-wide weather images.

**USA Today Weather**
Weather maps, regional forecasts, and weather-related news stories from the foremost U.S. national daily newspaper. This site includes interesting text-and-graphics tutorials on weather for pilots and how weather works.

**CNN Interactive Weather**
World-wide weather maps, forecasts, and weather news stories from the WWW service of Cable News Network.

4.  **Aircraft Owners and Pilots Association (AOPA)**. http://www.aopa.org -- a superb aviation site with many aviation databases and an excellent source of aviation weather data, including text products (e.g., METARs and TAFs) and charts. You must be an AOPA member to access the databases and weather links since they are in the members-only section.

5.  **National Weather Service**.  http://www.nws.noaa.gov -- official aviation weather products, including text products and charts that are available in several different formats.  It is a great site with many products that are not normally seen.

National Oceanic and Atmospheric Administration

[Organization & Services|Mission Statement|Strategic Plan]

**Direct Access to U.S.**
**OFFICIAL WEATHER FORECAST PRODUCTS &**
**OBSERVATIONS**

6.  **Aviation Weather Center products**.
    http://www.awc-kc.noaa.gov/awc/aviation_weather_center.html -- official products themselves, such as AIRMETo, TAFs, winds aloft forecasts, etc., and excellent descriptions of the products.  New products at the site include satellite pictures from GOES-8 and 9 with aviation flight conditions (IFR, MVFR, and VFR) and sky cover overlaid, and National Radar with tops and convective SIGMETs overlaid, updated twice hourly.

7.  **Intellicast**.  http://intellicast.com -- superb WSR-88D NexRad (Doppler radar) color radar summary graphics of the entire U.S. and individual parts of the country (e.g., Jacksonville Radar Summary).  The graphics show precipitation returns, tops, movement, etc.

8.  **Tropical Prediction Center**.  http://www.nhc.noaa.gov -- the best color tropical cyclone graphics available, including histories and current storm movement predictions

# END OF APPENDIX

## FOR CHOOSING GLEIM

We dedicate ourselves to providing pilots with knowledge transfer systems, enabling them to pass the FAA knowledge (written) tests and FAA practical (flight) tests. We solicit your feedback. Use the last page in this book to make notes as you use *Aviation Weather and Weather Services* and other Gleim products. Tear out the page and mail it to us when convenient. Alternatively, e-mail (irvin@gleim.com) or FAX (352-375-6940) your feedback to us.

# GLEIM'S E-MAIL UPDATE SERVICE

## update@gleim.com

Your message to Gleim must include (in the subject or body) the acronym for your book or software, followed by the edition-printing for books and version for software. The edition-printing is indicated on the book's spine and at the bottom right corner of the cover. The software version is indicated on the CD-ROM label.

|  | Written Exam | | Flight Maneuvers |
| --- | --- | --- | --- |
|  | Book | Software | Book |
| **Private Pilot** | PPWE | FAATP PPWE | PPFM |
| **Instrument Pilot** | IPWE | FAATP IPWE | IPFM |
| **Commercial Pilot** | CPWE | FAATP CPWE | CPFM |
| **Flight/Ground Instructor** | FIGI | FAATP FIGI | FIFM |
| **Fundamentals of Instructing** | FOI | FAATP FOI | |
| **Airline Transport Pilot** | ATP | FAATP ATP | |
| **Flight Engineer** | FEWE | FAATP FEWE | |
|  | | Reference Book | |
| **Pilot Handbook** | | PH | |
| **Aviation Weather and Weather Services** | | AWWS | |
| **Private Pilot Syllabus and Logbook** | | SYLLOG | |
| **FAR/AIM** | | FARAIM | |

**E
X
A
M
P
L
E
S**

For *Aviation Weather and Weather Services,* third edition-first printing:

> To: update@gleim.com
> From: your e-mail address
> Subject: AWWS 3-1

For *FAA Test Prep* software, Private Pilot, version 3.4:

> To: update@gleim.com
> From: your e-mail address
> Subject: FAATP PPWE 3-4

**IT
ONLY
TAKES
A
MINUTE**

If you do not have e-mail, have a friend send e-mail to us and print our response for you.

# USE GLEIM'S *FAA TEST PREP* -- A POWERFUL TOOL IN THE GLEIM KNOWLEDGE TRANSFER SYSTEM

Give yourself the competitive edge! Because all of the FAA's "written" tests have been converted to computer testing, Gleim has developed software specifically designed to prepare you for the computerized pilot knowledge test.

➡ *FAATP* emulates the computer testing vendor of your choice -- CATS, LaserGrade, Sylvan, or AvTEST. You will be completely familiar with the computer testing system you will be using.

➡ *FAATP* has two interactive modes: "Study" and "Test." Study mode permits you to select questions from specific sources, e.g., Gleim modules, questions that you missed from the last session, etc. You can also determine the order of the questions (Gleim or random), and you can randomize the order of the answer choices for each question.

➡ *FAATP* precludes you from looking at the answers before you commit to an answer and provides the actual testing environment. This is a major difference from the book.

➡ *FAATP* contains the well-known Gleim answer explanations which are intuitively appealing and easy to understand.

➡ *FAATP* maintains a history of your proficiency in each topic. This enables you to focus your study only on topics that need additional study.

➡ *FAATP* is the most versatile and complete software available.

**AN OVERVIEW OF GLEIM'S *FAA TEST PREP* SOFTWARE FOR WINDOWS**

Gleim's *FAA Test Prep for Windows*™ contains many of the same features found in earlier versions. However, we have simplified the study process by incorporating the outlines and figures from our books into the new software. Everything you need to study for any of the FAA knowledge tests will be contained in one unique, easy-to-use program. Below are some of the enhancements you will find with our new study software.

# AUTHOR'S RECOMMENDATIONS

**The Experimental Aircraft Association, Inc.** is a very successful and effective nonprofit organization that represents and serves those of us interested in flying, in general, and in sport aviation, in particular.  I personally invite you to enjoy becoming a member:

$35 for a 1-year membership
$20 per year for individuals under 19 years old
Family membership available for $45 per year

| |
|---|
| Membership includes the monthly magazine *Sport Aviation*. |

*Write to:*   Experimental Aircraft Association, Inc.
P.O. Box 3086
Oshkosh, Wisconsin  54903-3086

*Or call:*   (414) 426-4800
(800) 564-6322
*Web site:*   www.eaa.org

**The annual EAA Oshkosh AirVenture** is an unbelievable aviation spectacular with over 12,000 airplanes at one airport!  Virtually everything aviation-oriented you can imagine!  Plan to spend at least 1 day (not everything can be seen in a day) in Oshkosh (100 miles northwest of Milwaukee).

*Convention dates:*   2000 -- July 26 through August 1
2001 -- July 25 through July 31

**The annual Sun 'n Fun EAA Fly-In** is also highly recommended.  It is held at the Lakeland, FL (KLAL) airport (between Orlando and Tampa).  Visit the Sun 'n Fun web site at http://www.sun-n-fun.org.

*Convention dates:*   2001 -- April 8 through 14
2002 -- April 7 through 13

## BE-A-PILOT:  INTRODUCTORY FLIGHT

Be-A-Pilot  is an industry-sponsored marketing program designed to inspire people to "Stop dreaming, start flying."  Be-A-Pilot has sought flight schools to participate in the program and offers a $35 introductory flight certificate that can be redeemed at a participating flight school.

The goal of this program is to encourage people to experience their dreams of flying through an introductory flight and to begin taking flying lessons.

For more information, you can visit the Be-A-Pilot home page at http://www.beapilot.com or call 1-888-BE-A-PILOT.

## CIVIL AIR PATROL:  CADET ORIENTATION FLIGHT PROGRAM

The Civil Air Patrol (CAP) Cadet Orientation Flight Program is designed to introduce CAP cadets to general aviation operations.  The program is voluntary and primarily motivational, and it is designed to stimulate the cadet's interest in and knowledge of aviation.

Each orientation flight includes at least 30 min. of actual flight time, usually in the local area of the airport.  Except for takeoff, landing, and a few other portions of the flight, cadets are encouraged to handle the controls.  The Cadet Orientation Flight Program is designed to allow five front-seat and four back-seat flights.  But you may be able to fly more.

For more information about the CAP cadet program nearest you, visit the CAP home page at http://www.capnhq.gov or call 1-800-FLY-2338.

# PRIVATE PILOT FAA WRITTEN EXAM

Gleim's *Private Pilot FAA Written Exam* book is designed to help you prepare for and successfully take the FAA pilot knowledge test for the private or recreational pilot certificate.

**A**ll of the 711 questions in the private pilot test bank that are applicable to airplanes have been grouped into the following 11 chapters:

Chapter 1:  Airplanes and Aerodynamics
Chapter 2:  Airplane Instruments, Engines, and Systems
Chapter 3:  Airports, Air Traffic Control, and Airspace
Chapter 4:  Federal Aviation Regulations
Chapter 5:  Airplane Performance and Weight and Balance
Chapter 6:  Aeromedical Factors
Chapter 7:  Aviation Weather
Chapter 8:  Aviation Weather Services
Chapter 9:  Navigation:  Charts, Publications, Flight Computers
Chapter 10:  Navigation Systems
Chapter 11:  Cross-Country Flight Planning

Add Gleim's *FAA Test Prep* software and *FAA Audio Review* to your test preparation and change the FAA pilot knowledge test process from an answer memorization marathon to a rewarding learning experience. Obtain higher test scores with less study time.

**T**he FAA's question-numbering scheme does **not** group questions together by topic. We unscramble them for you in *Private Pilot FAA Written Exam*.

**W**ithin each of the chapters listed, questions relating to the same subtopic (e.g., duration of medical certificates, stalls, carburetor heat, etc.) are grouped together to facilitate your study program.

**S**even major differences between Gleim *FAA Written Exam* products and other available products:

❍ Only airplane questions are included in our books.

❍ Comprehensive explanations of each question are provided, including explanations of why each incorrect answer is incorrect.

❍ Answer explanations are **NEXT** to each question, **NOT** at the back of the book.

❍ FAA test questions are reorganized into logical topics.

❍ Easy-to-study outlines explain the concepts tested at the beginning of each chapter.

❍ A detailed index helps you find topics and questions you wish to study.

❍ We provide a practice test using the FAA's mix of question topics.

Other books useful to your training are *Private Pilot Flight Maneuvers and Practical Test Prep*, *Pilot Handbook*, and *Aviation Weather and Weather Services*. Thus, for less than $100, you have all the written material and software you should need to earn your Private Pilot Certificate.

# DO IT TODAY!

# NEW!!  GLEIM AUDIO LECTURES FOR THE PRIVATE PILOT KNOWLEDGE TEST!!

Many of our customers have requested audio and/or video lectures to accompany our **FAA Written Exam** book and **FAA Test Prep** software.  Videos already abound and require video equipment and your full attention.  Audio lectures are NOT widely available and, most importantly, do NOT require the same "full attention" as do videos.

We believe audios will work better for many people because

1.  Audio lectures can be studied while you are in your car, mowing your lawn, exercising, etc.  Thus, audio lectures do NOT require much, if any, extra time to use.

2.  Audio lectures allow you to concentrate on the lectures so you can visualize the concepts; i.e., you are forced to learn as you listen and visualize.  Conversely, videos allow you to passively view without storing the concept(s) in long-term memory.

Our **Private Pilot FAA Written Exam** book and **FAA Test Prep** software are organized into 11 chapters which are covered in eleven 30-45 minute audio lectures.  We believe most people can carefully listen to an audio lecture and then correctly answer virtually all of the FAA questions.  Accordingly, you can prepare for and take your FAA pilot knowledge test in less than 30 hours total time.  Each chapter should take about 2 hours:  30-45 minutes of audio lecture and 1-2 hours studying and answering questions in our Gleim **FAA Test Prep** software or Gleim book.

Order your **Private Pilot Audio Review** today, or call for a free demonstration tape.  Alternatively, ask your CFI or FBO to order a set for all students to share.

# PILOT KNOWLEDGE (WRITTEN EXAM) BOOKS AND SOFTWARE

Before pilots take their FAA pilot knowledge tests, they want to understand the answer to every FAA test question. Gleim's pilot knowledge test books are widely used because they help pilots learn and understand exactly what they need to know to pass. Each chapter opens with an outline of exactly what you need to know to pass the test. Additional information can be found in our reference books and flight maneuver/practical test prep books.

Use **FAA Test Prep** software with the appropriate Gleim book to prepare for success on your FAA pilot knowledge test.

## PRIVATE PILOT AND RECREATIONAL PILOT FAA WRITTEN EXAM ($13.95)

The test for the private pilot certificate consists of 60 questions out of the 736 questions in our book. Also, the FAA's pilot knowledge test for the recreational pilot certificate consists of 50 questions from this book.

## INSTRUMENT PILOT FAA WRITTEN EXAM ($18.95)

The test consists of 60 questions out of the 899 questions in our book. Also, become an instrument-rated flight instructor (CFII) or an instrument ground instructor (IGI) by taking the FAA's pilot knowledge test of 50 questions from this book.

## COMMERCIAL PILOT FAA WRITTEN EXAM ($14.95)

The test consists of 100 questions out of the 595 questions in our book.

## FUNDAMENTALS OF INSTRUCTING FAA WRITTEN EXAM ($9.95)

The test consists of 50 questions out of the 160 questions in our book. This test is required for any person to become a flight instructor or ground instructor. The test needs to be taken only once. For example, if someone is already a flight instructor and wants to become a ground instructor, taking the FOI test a second time is not required.

## FLIGHT/GROUND INSTRUCTOR FAA WRITTEN EXAM ($14.95)

The test consists of 100 questions out of the 833 questions in our book. This book is to be used for the Flight Instructor--Airplane (FIA), Basic Ground Instructor (BGI), and the Advanced Ground Instructor (AGI) knowledge tests.

## AIRLINE TRANSPORT PILOT FAA WRITTEN EXAM ($26.95)

The test consists of 80 questions each for the ATP Part 121, ATP Part 135, and the flight dispatcher certificate. Studying for the ATP will now be a learning and understanding experience rather than a memorization marathon -- at a lower cost and with higher test scores and less frustration!!

## FLIGHT ENGINEER FAA WRITTEN EXAM ($26.95)

The FAA's flight engineer turbojet and basic knowledge test consists of 80 questions out of the 688 questions in our book. This book is to be used for the turbojet and basic (FEX) and the turbojet-added rating (FEJ) knowledge tests.

# REFERENCE AND FLIGHT MANEUVERS/PRACTICAL TEST PREP BOOKS

Our Flight Maneuvers and Practical Test Prep books are designed to simplify and facilitate your flight training and will help prepare pilots for FAA practical tests as much as the Gleim written exam books help prepare pilots for FAA pilot knowledge tests. Each task, objective, concept, requirement, etc., in the FAA's practical test standards is explained, analyzed, illustrated, and interpreted so pilots will gain practical test proficiency as quickly as possible.

| | | |
|---|---|---|
| Private Pilot Flight Maneuvers and Practical Test Prep | 368 pages | ($16.95) |
| Instrument Pilot Flight Maneuvers and Practical Test Prep | 432 pages | ($18.95) |
| Commercial Pilot Flight Maneuvers and Practical Test Prep | 336 pages | ($14.95) |
| Flight Instructor Flight Maneuvers and Practical Test Prep | 544 pages | ($17.95) |

## PILOT HANDBOOK ($13.95)

A complete pilot ground school text in outline format with many diagrams for ease in understanding. This book is used in preparation for private, commercial, and flight instructor certificates and the instrument rating. A complete, detailed index makes it more useful and saves time. It contains a special section on biennial flight reviews.

## AVIATION WEATHER AND WEATHER SERVICES ($22.95)

A complete rewrite of the FAA's Aviation Weather 00-6A and Aviation Weather Services 00-45E into a single easy-to-understand book complete with maps, diagrams, charts, and pictures. Learn and understand the subject matter much more easily and effectively with this book.

## FAR/AIM ($14.95)

The purpose of this book is to consolidate the common Federal Aviation Regulations (FAR) parts and the Aeronautical Information Manual into one easy-to-use reference book. The Gleim book is better because of bigger type, better presentation, improved indexes, and full-color figures. FAR Parts 1, 43, 61, 67, 71, 73, 91, 97, 103, 105, 119, Appendices I and J of 121, 135, 137, 141, and 142 are included.

# GLEIM'S PRIVATE PILOT KIT

Gleim's Private Pilot FAA Written Exam book, Private Pilot Flight Maneuvers and Practical Test Prep, Pilot Handbook, FAR/AIM, a combined syllabus/logbook, a flight computer, a navigational plotter, and a durable, all-purpose flight bag. Our introductory price (substantial savings over purchasing items separately) is far lower than similarly equipped kits found elsewhere. The Gleim Kit retails for **$99.95**. Gleim's FAR/AIM, Private Pilot Syllabus/Logbook, flight computer, navigational plotter, and flight bag are also available for individual sale. See our order form for details.

| **Gleim Publications, Inc.**<br>P.O. Box 12848<br>University Station<br>Gainesville, FL 32604 | TOLL FREE: (800) 87-GLEIM<br> or (800) 874-5346<br>LOCAL: (352) 375-0772<br>FAX: (352) 375-6940<br>INTERNET: www.gleim.com<br>E-MAIL: sales@gleim.com | Customer service is available:<br>8:00 a.m. - 7:00 p.m., Mon. - Fri.<br>9:00 a.m. - 2:00 p.m., Saturday<br>Please have your credit card ready<br>or save time by ordering online. |
| --- | --- | --- |

## GLEIM'S PRIVATE PILOT KIT

Includes everything you need to pass the FAA pilot knowledge (written) test and FAA practical (flight) test: Gleim's *Private Pilot FAA Written Exam* book; *Private Pilot Flight Maneuvers and Practical Test Prep*; *Pilot Handbook*; *FAR/AIM*; a combined syllabus/logbook; flight computer; navigational plotter; and an attractive, durable, all-purpose flight bag. Our price is far lower than similarly equipped kits found elsewhere. ORDER TODAY!

**BIG savings➡** Original kit (with demo software) ........................... $ 99.95 _____
With CD-ROM (For Windows 95, Windows 98, or NT 4.0) .............. $119.95 _____

### *FAA Test Prep* Software

| WRITTEN TEST BOOKS AND SOFTWARE | | Books | CD-ROM‡ | ☐ free demo | Audio | ☐ free demo | |
| --- | --- | --- | --- | --- | --- | --- | --- |
| *Private/Recreational Pilot* .... | Eighth Edition | ☐ @ $13.95 | ☐ @ $49.95 | | ☐ @ $60.00 | | _____ |
| *Instrument Pilot* .......... | Seventh Edition | ☐ @ 18.95 | ☐ @ 59.95 | | | | _____ |
| *Commercial Pilot* ......... | Seventh Edition | ☐ @ 14.95 | ☐ @ 59.95 | | | | _____ |
| *Fundamentals of Instructing* .. | Sixth Edition | ☐ @ 9.95 | ☐ @ ] $59.95 for both | | | | _____ |
| *Flight/Ground Instructor* ..... | Sixth Edition | ☐ @ 14.95 | | | | | _____ |
| *Airline Transport Pilot* ....... | Third Edition | ☐ @ 26.95 | ☐ @ 59.95 | | | | _____ |
| *Flight Engineer* ........... | First Edition | ☐ @ 26.95 | ☐ @ 59.95 | | | | _____ |

‡CD-ROM version includes all questions, figures, charts, and outlines for each of the pilot knowledge tests.
Must have Windows 95, Windows 98, or NT 4.0 or higher and a CD-ROM drive to use.

## FLIGHT MANEUVERS/PRACTICAL TEST PREP BOOKS

*Private Pilot Flight Maneuvers and Practical Test Prep* .......... (Third Edition) ......... $16.95 _____
*Instrument Pilot Flight Maneuvers and Practical Test Prep* ....... (Third Edition) ......... 18.95 _____
*Commercial Pilot Flight Maneuvers and Practical Test Prep* ...... (Third Edition) ......... 14.95 _____
*Flight Instructor Flight Maneuvers and Practical Test Prep* ....... (Second Edition) .............. 17.95 _____

## REFERENCE BOOKS AND SYLLABUS

*FAR/AIM* ........................................................ (Most Recent Edition) ......... $14.95 _____
*Aviation Weather and Weather Services* ...................... (Third Edition) ......... 22.95 _____
*Pilot Handbook* .................................... (Sixth Edition) ......... 13.95 _____
*Private Pilot Syllabus and Logbook* ............................ 9.95 _____

## OTHER

*Flight Computer* .............................................. $ 9.95 _____
*Navigational Plotter* ......................................... 5.95 _____
*Flight Bag* ................................................... 29.95 _____

**Shipping and Handling (nonrefundable): 1 item = $5; each additional item = $1** _____

Add applicable sales tax for shipments within the State of Florida. Sales Tax _____

*Please FAX or write for additional charges for outside the 48 contiguous United States.*

**Printed 07/00. Prices subject to change without notice.** TOTAL $_____

| | |
| --- | --- |
| *1. We process and ship orders (via UPS in the 48 contiguous states) within 1 business day over 98.8% of the time. Call by noon for same-day service!*<br><br>*2. Please copy this order form for friends and others. Orders from individuals must be prepaid. No CODs.*<br><br>*3. Gleim Publications, Inc. guarantees the immediate refund of all resalable texts and unopened software and audiotapes returned in 30 days. Applies only to items purchased direct from Gleim. No refunds on shipping and handling.*<br><br>*4. If your local FBO or aviation bookstore does not stock the books you are ordering from us directly, please provide us with a name and address, so we can invite them to stock Gleim.* | Name _____<br> (please print)<br><br>Shipping Address _____<br> (street address required for UPS) Apt. #<br><br>_____<br><br>City _____ State _____ Zip _____<br><br>☐ VISA/MC/DISC ☐ Check/M.O. Expiration Date (month/year) _____/_____<br><br>Credit Card No. _____-_____-_____-_____<br><br>Daytime Phone (____) _____ Evening Phone (____) _____<br><br>Signature _____ E-mail _____ |

020

## GLEIM'S PRIVATE PILOT KIT

Includes everything you need to pass the FAA pilot knowledge (written) test and FAA practical (flight) test: Gleim's *Private Pilot FAA Written Exam* book; *Private Pilot Flight Maneuvers and Practical Test Prep*; *Pilot Handbook*; *FAR/AIM*; a combined syllabus/logbook; flight computer; navigational plotter; and an attractive, durable, all-purpose flight bag. Our price is far lower than similarly equipped kits found elsewhere. ORDER TODAY!

**BIG savings** ➤ Original kit (with demo software) . . . . . . . . . . . . . . . . . . . . . . . . . . . . . . . . . $ 99.95 _____
With CD-ROM (For Windows 95, Windows 98, or NT 4.0) . . . . . . . . . . . . . $119.95 _____

*FAA Test Prep* Software

### WRITTEN TEST BOOKS AND SOFTWARE

| | | Books | CD-ROM‡ | ☐ free demo | Audio | ☐ free demo |
|---|---|---|---|---|---|---|
| *Private/Recreational Pilot* . . . . | Eighth Edition | ☐ @ $13.95 | ☐ @ $49.95 | | ☐ @ $60.00 | |
| *Instrument Pilot* . . . . . . . . . . | Seventh Edition | ☐ @ 18.95 | ☐ @ 59.95 | | | |
| *Commercial Pilot* . . . . . . . . . | Seventh Edition | ☐ @ 14.95 | ☐ @ 59.95 | | | |
| *Fundamentals of Instructing* . . | Sixth Edition | ☐ @ 9.95 | ☐ @ ] $59.95 for both | | | |
| *Flight/Ground Instructor* . . . . . | Sixth Edition | ☐ @ 14.95 | | | | |
| *Airline Transport Pilot* . . . . . . . | Third Edition | ☐ @ 26.95 | ☐ @ 59.95 | | | |
| *Flight Engineer* . . . . . . . . . . . | First Edition | ☐ @ 26.95 | ☐ @ 59.95 | | | |

‡CD-ROM version includes all questions, figures, charts, and outlines for each of the pilot knowledge tests.
Must have Windows 95, Windows 98, or NT 4.0 or higher and a CD-ROM drive to use.

### FLIGHT MANEUVERS/PRACTICAL TEST PREP BOOKS

*Private Pilot Flight Maneuvers and Practical Test Prep* . . . . . . . . . (Third Edition) . . . . . . . . . . . . . . . $16.95 _____
*Instrument Pilot Flight Maneuvers and Practical Test Prep* . . . . . . . (Third Edition) . . . . . . . . . . . . . . . 18.95 _____
*Commercial Pilot Flight Maneuvers and Practical Test Prep* . . . . . . (Third Edition) . . . . . . . . . . . . . . . 14.95 _____
*Flight Instructor Flight Maneuvers and Practical Test Prep* . . . . . . . (Second Edition) . . . . . . . . . . . . . 17.95 _____

### REFERENCE BOOKS AND SYLLABUS

*FAR/AIM* . . . . . . . . . . . . . . . . . . . . . . . . . . . . . . . . . . . . . . . . . . (Most Recent Edition) . . . . . . . . . $14.95 _____
*Aviation Weather and Weather Services* . . . . . . . . . . . . . . . . . . . (Third Edition) . . . . . . . . . . . . . . . 22.95 _____
*Pilot Handbook* . . . . . . . . . . . . . . . . . . . . . . . . . . . . . . . . . . . . (Sixth Edition) . . . . . . . . . . . . . . . 13.95 _____
*Private Pilot Syllabus and Logbook* . . . . . . . . . . . . . . . . . . . . . . . . . . . . . . . . . . . . . . . . . . . . . . . . . 9.95 _____

### OTHER

*Flight Computer* . . . . . . . . . . . . . . . . . . . . . . . . . . . . . . . . . . . . . . . . . . . . . . . . . . . . . . . . . . . . . $ 9.95 _____
*Navigational Plotter* . . . . . . . . . . . . . . . . . . . . . . . . . . . . . . . . . . . . . . . . . . . . . . . . . . . . . . . . . . . 5.95 _____
*Flight Bag* . . . . . . . . . . . . . . . . . . . . . . . . . . . . . . . . . . . . . . . . . . . . . . . . . . . . . . . . . . . . . . . . . . 29.95 _____

**Shipping and Handling (nonrefundable): 1 item = $5; each additional item = $1** _____

Add applicable sales tax for shipments within the State of Florida.  Sales Tax _____

*Please FAX or write for additional charges for outside the 48 contiguous United States.*

**Printed 07/00. Prices subject to change without notice.** TOTAL $_____

1. *We process and ship orders (via UPS in the 48 contiguous states) within 1 business day over 98.8% of the time. Call by noon for same-day service!*

2. *Please copy this order form for friends and others. Orders from individuals must be prepaid. No CODs.*

3. *Gleim Publications, Inc. guarantees the immediate refund of all resalable texts and unopened software and audiotapes returned in 30 days. Applies only to items purchased direct from Gleim. No refunds on shipping and handling.*

4. *If your local FBO or aviation bookstore does not stock the books you are ordering from us directly, please provide us with a name and address, so we can invite them to stock Gleim.*

Name _____
(please print)

Shipping Address _____
(street address required for UPS)      Apt. #

_____

City _____ State _____ Zip _____

☐ VISA/MC/DISC ☐ Check/M.O.    Expiration Date (month/year) _____ / _____

Credit Card No. _____ - _____ - _____ - _____

Daytime Phone (_____) _____    Evening Phone (_____) _____

Signature _____ E-mail _____

# INDEX

Please forward your suggestions, corrections, and comments to **Irvin N. Gleim • c/o Gleim Publications, Inc. • P.O. Box 12848 • University Station • Gainesville, Florida • 32604** for inclusion in the next edition of *Aviation Weather and Weather Services*. Please include your name and address on the back of this page so we can properly thank you for your interest.

1. _____

2. _____

3. _____

4. _____

5. _____

6. _____

7. _____

8. _____

9. _____

10. _____

11. _____

12. _____

13. _____

14. _____

15. _____

Remember for superior service:   Mail, e-mail, or fax questions about our books or software. Telephone questions about orders, prices, shipments, or payments.

Name: _____

Company: _____

Address: _____

City/State/Zip: _____

Phone:   Home: _____   Work: _____   Fax: _____

E-mail: _____